Clusters ~~and~~ ~~Industrial~~
Specialisation
On Geography
Technology
and Networks

European research in regional science

1   **Infrastructure and regional development**
    edited by R W Vickerman

2   **Sustainable development and urban form**
    edited by M J Breheny

3   **Regional networks, border regions and European integration**
    edited by R Cappellin, P W J Batey

4   **Issues in environmental planning**
    edited by H Voogd

5   **Convergence and divergence among European regions**
    edited by R W Vickerman, H W Armstrong

6   **Microsimulation for urban and regional policy analysis**
    edited by G P Clarke

7   **Regional governance and economic development**
    edited by M Danson, assisted by M G Lloyd, S Hill

8   **Clusters and regional specialisation**
    edited by M Steiner

p Pion Limited, 207 Brondesbury Park, London NW2 5JN

European research in regional science 8

Clusters and Regional Specialisation
On Geography Technology and Networks

Editor **M Steiner**

Series editor J H L Dewhurst

European research in regional science

Series editor
J H L Dewhurst   University of Dundee

Editorial board
M J Breheny   University of Reading
J R Cuadrado-Roura   Universidad de Alcalá, Madrid
M Madden   University of Liverpool
R Maggi   University of Zurich
J B Parr   University of Glasgow

ISBN 0 85086 168 3
ISSN 0960-6130

© 1998 Pion Limited

All rights reserved. No part of this book may be reproduced in any form by photostat microfilm or any other means without written permission from the publishers.

British Library Cataloguing in Publication Data
A catalogue record for this book is available from the British Library.

published by Pion Limited, 207 Brondesbury Park, London NW2 5JN

p

Printed in Great Britain by Page Bros (Norwich) Limited

Contributors

E M Bergman   *Institute for Urban and Regional Studies*
University of Economics and Business Administration
Augasse 2 – 6, A-1090 Vienna, Austria

R Cappellin   *Department of Economics, University of Rome "Tor Vergata"*
Via di Tor Vergata, 00133 Rome, Italy

J Cullen   *Evaluation Development and Review Unit, The Tavistock Institute*
30 Tabernacle Street, London EC2A 4DD, England

M Danson   *Department of Accounting, Economics and Languages*
University of Paisley, High Street, Paisley PA1 2BE, Scotland

R A Dubin   *Weatherhead School of Management*
Department of Economics, Case Western Reserve University
10900 Euclid Avenue, Cleveland, OH44106-7206, USA

X Feng   *Faculty of Economics and Organisational Sciences*
Universität der Bundeswehr München
Werner-Heisenberg-Weg 39, D85577 Neubiberg, Germany

E J Feser   *Department of City and Regional Planning*
The University of North Carolina at Chapel Hill
New East Building, CB# 3140, Chapel Hill, NC 27599-3140, USA

P Friedrich   *Faculty of Economics and Organisational Sciences*
Universität der Bundeswehr München
Werner-Heisenberg-Weg 39, D85577 Neubiberg, Germany

O M Fritz   *Institute of Technology and Regional Policy*
Joanneum Research, Wiedner Haupstrasse 76
A-1040 Vienna, Austria

C Hartmann   *Department of Economics, University of Graz*
A-8101 Graz, Austria and
Institute of Technology and Regional Policy
Joanneum Research, Elisabethstrasse 20, A-8010 Graz, Austria

S Helper   *Weatherhead School of Management*
Department of Economics, Case Western Reserve University
10900 Euclid Avenue, Cleveland, OH 44106-7206, USA

G J D Hewings   *Regional Economics Applications Laboratory*
University of Illinois, 607 S Mathews 220
Urbana, IL 61801-3671, USA

P R Israilevich	*Regional Economics Applications Laboratory* *University of Illinois, 607 S Mathews 220* *Urbana, IL 61801-3671, USA* and *Federal Reserve Bank of Chicago, Chicago, IL, USA*
H Mahringer	*Institute of Technology and Regional Policy, Joanneum Research,* *Wiedner Haupstrasse 76, A-1040 Vienna, Austria*
S Mizrahi	*School of Management, The Cukier Goldstein-Goren Building* *Ben-Gurion University, POB 653, Beer-Sheva 84105, Israel*
A Rallet	*IRIS, University of Paris — Dauphine* *75775 Paris Cedex 16, France*
G R Schindler	*Regional Economics Applications Laboratory* *University of Illinois, 607 S Mathews 220* *Urbana, IL 61801-3671, USA*
R Shotton	*European Commission, Directorate-General XVI* *Regional Policy and Cohesion, Rue de la Loi 200* *Brussels, Belgium*
M Sonis	*Regional Economics Applications Laboratory* *University of Illinois, 607 S Mathews 220,* *Urbana, IL 61801-3671, USA* and *Department of Geography, Bar-Ilan University,* *52 100 Ramat-Gan, Israel*
M Steiner	*Department of Economics, University of Graz, A-8101 Graz, Austria* and *Institute of Technology and Regional Policy,* *Joanneum Research, Elisabethstrasse 20, A-8010 Graz, Austria*
G Tichy	*Institute of Technology Assessment, Austrian Academy of Sciences* *Postgasse 7/4/3, A-1010 Vienna, Austria*
A Torre	*National Institute of Agricultural Research* *147 rue de l'Université, 75007 Paris, France*
M T Valderrama	*Department of Finance, Institute for Advanced Studies* *Stumpergasse 56, A-1060 Vienna, Austria*
G Whittam	*Department of Accounting, Economics and Languages* *University of Paisley, High Street, Paisley PA1 2BE, Scotland*

Contents

The Discrete Charm of Clusters: An Introduction   1
M Steiner

Part 1 Theoretical Considerations

Old and New Theories of Industry Clusters   18
E J Feser

On Geography and Technology: Proximity Relations   41
in Localised Innovation Networks
A Rallet, A Torre

The Transformation of Local Production Systems: 57
International Networking and Territorial Competitiveness
R Cappellin

Regional Cooperation and Innovative Industries: 81
Game-theoretical Aspects and Policy Implications
S Mizrahi

Industrial Trade Clusters in Action: Seeing Regional Economies   92
Whole
E M Bergman

Part 2 Methodological Approaches and Case Studies

Agglomeration, Clustering, and Structural Change:
Interpreting Changes in the Chicago Regional Economy
G J D Hewings, P R Israilevich, G R Schindler, M Sonis *

In Search of Agglomeration Economies: The Adoption   111
of Technological Innovations in the Automobile Industry
R A Dubin, S Helper

Cluster Formation in the Framework of the Treuhand   129
Approach: From Socialist to Market-oriented Clusters
P Friedrich, X Feng

A Risk-oriented Analysis of Regional Clusters   181
O M Fritz, H Mahringer, M T Valderrama

Networks, Innovation, and Industrial Districts: 192
The Case of Scotland
M Danson, G Whittam

Learning with Clusters: A Case Study from Upper Styria 211
M Steiner, C Hartmann

Part 3 Policy Implications

Clusters: Less Dispensable and More Risky than Ever 226
G Tichy

Promoting Competitiveness for Small Business Clusters through 238
Collaborative Learning: Policy Consequences from a European
Perspective
J Cullen

Clusters: New Developments in Austria and their Relevance 254
in Economic Policy
W Clement

Clusters in the Context of the European Union's Cohesion 268
Policies
R Shotton

Index 275

* For technical reasons and as an editorial innovation
the paper by Hewings et al is available only on the Pion
website: http://www.pion.co.uk/ep/errs/inderrs.html

The Discreet Charm of Clusters: An Introduction

M Steiner
University of Graz, and Joanneum Research, Graz/Vienna

1 Introduction

Clusters have the discreet charm of obscure objects of desire. This charm rests on the assumption that regional specialisation on interlinked activities of complementary firms (in production and service sectors) and their cooperation with public, semipublic, and private research and development institutions creates synergies, increases productivity, and leads to economic advantages (however defined). Hence, regions should specialise and policy should create, develop, and support such clusters.

In recent times, therefore, clusters have become an object of desire for many regions. They may be based on quite different foundations: concentrated forms of economic activity with strong connections to the knowledge infrastructure (sometimes euphemistically called 'knowledge clusters'), vertical production chains of rather narrowly defined branches where subsequent stages of production form the core of clusters (for example, 'textile clusters' with the different stages of fabric manufacture, dying, tailoring, design, and couture), sectorial concentrations on different levels of aggregation (for example, 'automotive' or 'electronic clusters'), and sometimes collections of firms which share a basic technology ('biotechnology clusters') or a common demand or need ('eco-clusters').

The obscurity of clusters stems from this multidimensionality: clusters are based on different economic dimensions, take different forms, are measured and quantified with quite different methods and empirical approaches, and are legitimated by a range of theories and hypotheses. Not surprisingly they have also become a desired object of research: the still vague character of clusters poses problems of theoretically sound definition, of empirical measurement, of policy recommendation and evaluation.

The contributors to this volume want to lift the spell and to clear the obscurity from the object. The chapters are not primarily intended as a 'state of the art' description of the cluster debate (although some of them contain good surveys on different aspects of it). Starting from a common understanding and positive acceptance of a basic concept of clusters the authors concentrate on finding answers to open questions, raise some new ones, and cover new ground. Building on experiences in empirical analysis they provide diverse methodological approaches for the conceptualisation and identification of clusters. They also address the policy relevance of a cluster approach and openly discuss the risks of regional specialisation. Although most of the contributions combine all three elements they are—according to their particular emphasis—grouped into three parts: theoretical considerations, methodological approaches and case studies, and policy implications.

2 From agglomeration economies to innovative networks—a short history of the cluster idea

Clusters have been a matter of fact since the earliest history of economic activity and existed long before the industrial revolution. Many goods were produced cooperatively and in specialised regions, such as silk in China, trade services in the cities of the Hanse, and steel in differentiated value chains of production in Styria in the 17th century. During and after the industrial revolution clusters magnified and multiplied: steel and shipbuilding in Glasgow, cars in Detroit, watches in Switzerland, machinery in Southern Germany, to name but a few. Some of them declined and became old, others arose as new centres of economic activity in regions such as Silicon Valley, or industrial districts with different economic foundations in Italy or along the M4 in southern England. There are numerous examples today: many regions in Europe and in the USA, but also in other parts of the world, are either proud of having a cluster of their own or strive to concentrate and specialise on a more and more limited set of interlinked activities.

Theoretical considerations about the advantages of specialisation go back, of course, to Adam Smith: his famous dictum in the third chapter of the first volume of his *Wealth of Nations* that "the division of labour is limited by the extent of the market" is the first hint that specialisation depends on globalisation and that the enlargement of markets is also a precondition for regional specialisation leading to higher productivity and calling for cooperation. This classic statement was not only used to explain vertical disintegration as industry output grows (Stigler, 1951), but also to account for—among other aspects—differences in the degree of the division of labour across local markets (Baumgardner, 1988). Another early apologist for cluster formation was List: in his *Das nationale System der politischen Ökonomie* of 1841 he argued in support of networks to diffuse knowledge and to train workers to encourage the underdeveloped Germany industry. Marshall's contribution in 1890 is a cornerstone of cluster—and especially industrial district—theorising, emphasising the dynamics of external economies associated with learning, innovation, and increased specialisation, differing in this respect from industrial location theory (Weber, 1909; Hotelling, 1929; Hoover, 1937). Schumpeter (1934)—although never much concerned with the spatial dimensions of economic activity—nevertheless pointed to a clustering of innovations in time.

Florence (1944) started a discussion on the measurement of agglomerative tendencies and spatial complexes, taken up later by Streit (1969), Richter (1969), Czamanski and Czamanski (1977), Harrigan (1982), and Kubin and Steiner (1987), testing the assumption that industrial agglomerations are not the result of or not only the result of a common attraction to urban centres but are also the result of interaction among the various industries. Perroux's growth pole approach (1955) shows clear affinities to today's cluster debate; his ideas were partly taken up by the 'filière' approach in the 1970s in France (for a survey see Quelin, 1993). Chinitz's (1961) contribution paved the way for a new understanding of agglomeration economies and their contrasting manifestations, pointing to elements such as market structure, entrepreunerial behaviour, and industrial organisation. In the 1970s a series of studies using a production function approach tried to measure regional productivity differences via different indicators for agglomeration economies (Aberg, 1973; Sveikauskas, 1975; Segal, 1976; Carlino, 1978; 1979; 1980; Moomaw, 1981).

In the 1980s industrial districts (with an emphasis on the Italian example) were at the centre of attention as new manifestations of the advantages of interfirm cooperation at a concentrated geographical scale (for a survey see Pyke et al, 1990). A similar focus was taken by GREMI (Groupe de Recherche Européen sur les Milieux Innovateurs), analysing territorial innovative processes and the production – reproduction modalities of the competitive advantages of the complex socioeconomic fabric (Aydalot, 1986; for a recent assessment see Ratti et al, 1997).

Then came Porter (1990) reviving the cluster debate, although his approach is more a theory of competitiveness. Yet his contribution was taken as a starting point, offering a holistic framework to show how interdependence affects innovation and growth. This led to notions and concepts such as embeddedness (Grabher, 1993), social networks (Scott, 1991), untraded interdependencies (Storper, 1995), and led to a new emphasis on economies of scale in a geographical context (Krugman, 1991).

3 What we believe we know, what we don't know, and what we agree to disagree on

Definitions of clusters are abundant and various, yet they have several elements in common. The first element is specialisation, based on a sophisticated division of labour and leading to interlinked activities and a need for cooperation. These linkages can assume quite different forms: simple input – output relations between firms (not only in production, but also in services), exchange of knowledge between firms and with research institutions, and contacts with public and semipublic policy bodies and development agencies. These linkages may be based on formal contracts, on long-term relations of trust, or on implicit exchanges of tacit knowledge. They are therefore more than market-oriented forms of exchange, and require a subtle net of social, political, and cultural ties, a certain 'milieu'.

A second element is a precondition for the emergence of these kinds of linkages: proximity. Cooperation needs closeness: for just-in-time delivery, given that many steps of production have been outsourced, for communication, for the exchange of knowledge (especially in its tacit form), for the emergence of a social fabric of regional production. To a first approximation there are good reasons to interpret the necessary closeness of the participating actors in a geographical sense, implying a regional dimension to clusters.

From these two elements results a third: cooperation between specialised actors in close proximity leads to spillovers and synergies. These resulting synergetic effects are the very basis of the charm of clusters. They are a precondition for an improved competitiveness of regions leading to economic advantages benefits both for the participating firms and for the region as a whole, an increase in productivity and growth, stabilisation, and an increase in employment. Hence the demand for policy intervention: clusters are a good thing, policy should therefore create, develop, and support such clusters.

The assumptions underlying these constitutive elements represent quite a challenge to orthodox economic concepts, to traditional regional policy, but also to the understanding of the causes of regional and national competitiveness. It is presumed that:
(1) The existence of clusters is the decisive element for the competitiveness of regions and nations, not cheap land, labour, or energy, nor high subsidies and low social costs, nor even high technology and strong leading industries.
(2) There is a need for firms to adopt new strategies: instead of an individual 'search for excellence' they have to go for a cooperative approach because 'as a group we are stronger'. This partly reflects the new strategy of business behaviour; it also reflects new insights into innovative and organisational approaches.
(3) Feedback mechanisms replace linear causalities; the multidimensionality of factors influencing competitiveness escapes a monocausal relationship.
These presumptions are indeed irritating. Cost factors predominate in everyday talk about locational advantages. Cooperation is usually not a first-order strategy for firms. Entrepreneurs were trained in a philosophy of competition—they are frightened by the difficulties of trust and teamwork. And for theorists a world of monocausality and linearity is much nicer to model and understand than a world of complex interrelationships.

There exists a widespread consensus about these elements as constitutive parts of clusters. Yet there is also a widespread consensus that these elements are open to interpretation, that other elements are also important, and that they imply further questions. The most important ones are mentioned briefly.

What are the causes of cooperation and resulting forms of clusters?
Ardent Marshallians have different views on this from practising Porterians or French filière lovers. Different theoretical concepts accordingly lead to varying legitimisation of cluster concepts: as industrial districts being an expression of spatially concentrated forms of economic activity, as large aggregations of connected sectors, as vertical production chains (Jacobs and de Man, 1996). A related question touches on the economic dimension of cooperation in clusters: is it a point on the Coaseian line between market exchange and vertical integration or does it mainly have a character of its own (Chesnais, 1996)? Each aspect may lead to a different typology of clusters: working (overachieving), latent (underachieving), potential (wannabe) clusters (Rosenfeld, 1997, quoting Enright); a list open to extension (for example, see Cullen in this volume), it also leads to a distinction between clusters and networks. This still leaves open the question about the incentive structure for more or less cooperation and how firms in clusters actually cooperate.

What is the right level of cluster analysis and what kinds of techniques can be used for identifying clusters?
This is a question confronting any kind of empirical analysis: any new concept of analysis—be it a system of national accounts or of input–output analysis—arises from criticism of an existing statistical nomenclature and its inability to define the 'correct' units of economic activity. The challenge for an adequate cluster analysis is to find a better description of a regional production system. The answer also depends on the focus of analysis: input–output relations, trade patterns, spillovers of different kinds (technological, labour market), or patterns of communication and cooperation.

What constitutes proximity?
The spatial dimension has already been mentioned as a first approximation. How local or regional is this geographic dimension and can it be extended to national and international scales? Do modern communication technologies reduce the need for closeness and allow cooperation at longer distances? Yet the discussion can be extended: is proximity a purely geographic concept or can it be substituted by other conceptualisations of closeness?

Is policy intervention in favour of clusters legitimate, and if so, what kinds of instruments are available?
For someone with a strong belief in the allocative and dynamic power of markets this is of course an economically incorrect question. For someone who believes in the possibility of the economic incorrectness of markets there is ample legitimation of policy intervention. Tichy in this volume cites List (1841) as an early disbeliever in market power (especially for the rise of infant industries) and as a supporter for cluster strategies. Jacobs and de Man (1996) point to different phases of industrial policy with the cluster approach—after a first phase of 'backing losers' and a second of 'picking winners', there is a third stage with a more modest ambition combining the strong aspects of the previous phases ('backing winners'). They distinguish between two policy alternatives: knowledge intensification of existing clusters and the creation of new cooperative structures.

Are today's cluster-specialised regions the problem areas of tomorrow?
The nightmare of cluster adherents and politicians is of course the decline of formerly innovative and prosperous specialised regions. Their stories were told one or two decades ago and connected with a life cycle of regional development (Norton and Rees, 1979; Suarez-Villa, 1984; Markusen, 1985; Steiner, 1987; 1990a; Tichy, 1987; 1991). But are all clusters destined to become old or are there strategies to keep them young?

Can clusters learn?
Learning—often addressed as an important phenomenon in economics since Arrow (1962)—has only recently received attention in the interpretation and analysis of clusters. Its importance is nevertheless acknowledged, not least because of the acceptance of the relevance of modern growth theory to the

cluster concept. Yet the external effects of human capital (Lucas, 1988), the dimensions of the learning curve, and especially of organisational learning are only beginning to be integrated into the cluster concept (Florida, 1995; Bellini, 1996; Coombs et al, 1996; Hassink, 1997).

How successful are clusters?
This is a good question because everybody presumes there are synergies. Feser argues in this volume that there are few explicit clues in the recent literature to a link between clusters and regional growth, how it works and what it looks like. This essential question therefore remains unanswered. There is also hardly any quantitative evidence for a superior economic performance by clusters. There is some evidence that firms in specific sectors of industrial districts are better off than firms not present in such districts (Signorini, quoted by Pezzini, 1998). There are weak, and sometimes contradictory, associations between performance measures of firms with stronger or weaker cooperation (DELOS, 1998). There is a long list of studies showing higher productivity in agglomerations (sometimes tending to decline). But there is no real empirically significant test showing the success of clusters. There are several difficulties: clusters are open systems with a latent social ecology and are therefore hard to pin down with quantitative data; the dimensions of economic performance are multidimensional and not only restricted to growth and employment (Steiner, 1990b); and we still lack adequate methods and criteria for the evaluation of regional development in general, and of clusters in particular.

So the charm of clusters still remains discreet. The contributors to this volume do not want to remove the charm, but to make it a little less discreet. In the following section, therefore, the main ideas representing further developments of the concept of clusters will be presented and a synthesis for their policy relevance made. Partly for technical reasons (to give the authors a little more time, as they were approached late, and to allow the inclusion of coloured figures), but also as an editorial innovation to attract additional readers to the volume, one of the contributions (that by Hewings et al) is published not in the book but on Pion's website and can be accessed there under http://www.pion.co.uk/ep/errs/inderrs.html

4 Tentative answers, some new ground
4.1 Changing character of regional specialisation
Regional specialisation is a result of globalisation, of the extension of markets, of the contraction of time and space, leading to significant modifications in the way economic activity is organised. This process of internationalisation is more than an ever-increasing international flow of goods, services, labour, and capital, more than a uniform market mechanism determining prices on worldwide markets. It also implies a change in behaviour of individual economic actors not necessarily leading to an increase in homogeneity but rather to an increase in interdependence. Several contributors to this volume document the resulting change in regional economies.

Cappellin points to several aspects of the transformation of local production systems which no longer specialise in traditional production mainly oriented to the regional and national market. First, there is the rising importance of local production systems within the pattern of industrial development characterised by strong interfirm cooperation among small and medium-sized enterprises (SMEs). This system of production is more than just a territorial concentration of specific firms working in the same sector but involves complex organisations with tight transsectorial relationships, implying a change from 'industrial district' to 'network' forms of organisation at the interregional and international level.

An important characteristic of this change is an internationalisation of the constituent elements of the local production system itself, namely SMEs. This process can be interpreted as the extension to an international framework of the same model of specialisation and cooperation with other firms which has long existed within a regional framework. Instead of interpreting the globalisation process as an external constraint and a risk to their survival, the increasing internationalisation of local production systems, has to be described as the gradual extension to the international level of tight relationships of comanufacture which have traditionally existed at the local and interregional level.

Yet a precondition for the development of the ability to compete at the international scale is the competitiveness of the overall regional economy in which the firms are located—the organisation and quality of the territory become key factors which also call for new relationships between the urban centres and their rural hinterlands. This leads, in contrast to general perceptions, to a shift in the industrial system of the advanced economies from an 'international' structure to an 'interregional' structure, as the economies of the various countries become integrated in a similar way to regional economies within the same country.

Some of these tendencies are corroborated by the findings of Hewings et al. Their conceptual framework for exploring changes in the spatial dependence observed in the Chicago region leads to an emphasis on market orientation as an even more dominant force in locational decisionmaking, resulting from a reduction in transport and communication costs and from the integration of the region into the global economy. The spatial scale of agglomeration economies has shifted from the urban or metropolitan to the regional; returns to scale are now complemented by returns to scope. This might lead to a change in the composition of interregional trade, with interindustry trade replaced by intraindustry interregional trade. Spatial clustering of activities will focus on different attributes of the regional.

Using a sophisticated regional econometric input – output model for the Chicago region and the Midwest US economy, Hewings et al can trace important changes in regional specialisation and agglomeration over time. Their most important finding suggests that the Chicago region seems to be hollowing out, becoming more dependent on the rest of the country (or even the rest of the world) as a source of inputs and as a market for outputs. Extending their analysis to the spatial exchange between the Midwestern states provides evidence that trade is more heavily concentrated in intraindustry flows and that the detected differences in structure of the economic landscape may reflect the exploitation of economies of scope and the formation of niche production centres with major sectors in each state. Yet, although these regions are clearly diversifying away from prior domination by one or two sectors, the region as a whole is still very much dependent upon itself.

The development of the theoretical concepts and the trajectory of the recent literature also emphasise the changing character: regional specialisation is less and less founded on agglomeration economies but on active forms of cooperation in innovative networks. Agglomeration economies are external, available to anyone located in the region. Clusters certainly do have elements of agglomeration economies, but the more they assume a network-like character the more the gains are internalised: they are like clubs where you have to pay a membership fee and/or have to take an active part in the activities to remain a member and enjoy the benefits. As Feser notes, there is no cluster theory per se, rather a broad range of theories and ideas that constitute the logic of clusters. This resulting logic is increasingly focused on the social or cultural dimensions of the interdependence, rather than on more traditional economic or technical relations. This favours a systems-oriented view of economic processes rather than an atomistic perspective of interdependence among economic actors.

4.2 Different dimensions of proximity

Closely related to the above tendency is the discussion about the role played by geographical proximity as a constituting element of clusters and their regional dimension, especially in the process of technological transfer. The necessity of closeness is the conventional reason for the high degree of concentration of innovative activities.

This theoretical assumption that geographical proximity is a necessary condition for efficient sharing of knowledge, especially for tacit knowledge-intensive activities such as innovation, is put under close examination by Rallet and Torre. They ask if advances in information and communication technologies alter the need for spatial proximity between knowledge users. They first put four arguments in favour of the need for face-to-face relations, leading one to expect an extension to, but no basic modification of, the geographical scale of the coordination process in research and innovative activities.

Yet two arguments moderate this thesis: mobility of human resources and another kind of proximity—organisational proximity—as an alternative basis for knowledge exchange that no longer requires permanent colocation. Rallet and Torre advance this thesis by examining the role of geographical proximity in networks of innovation: lessons drawn from case studies on localised networks of innovation supported by public institutions in three French regions reveal that organisational proximity appears to be a stronger support to technology transfer and innovation diffusion than geographical proximity.

Dubin and Helper argue alone the same lines: in this age of telecommunications, geographic distance may not be the main barrier to information flow. Social distance may be far more important: people are more likely to communicate with those who have close ties to them than with those who are geographically close. In their study of the adoption of technological innovations in the US automobile industry they use two measures for proximity: geographical closeness and similarity. Their results lead them to conclude that distance in similarity space is more important than geographical space: firms which are similar in size and industry appear to be more inclined to share information about technology than those which are geographically close.

This then leads one to acknowledge that the cluster concept itself has two core dimensions: an economic and a geographic one. Feser proposes a typology of clusters according to these two dimensions: some end-market sectors may be both economically and geographically clustered, others may be more or less clustered on either of the two dimensions. To complicate matters, he points to the relativity of the degree of economic and geographic clustering according to space, time, and scale. An end-market sector clustered geographically from a national perspective may not be spatially clustered from a regional or local one (and vice versa); it may become more economically clustered as vertical, horizontal, and lateral linkages expand and deepen; and there is no reason, a priori, to assume that clustering along either dimension need increase over time.

4.3 How do clusters really work, how does cluster formation really happen?
We do not really know yet, but this volume suggests some ideas for analysis and an agenda for future research. There is a widespread consensus that the (changing) behaviour of businesses and of the entrepreneur as the central figure in SME network formation should be the focus of analysis. For Hewings et al the most important element for the identification of clusters and changes in them are changes in business practices. Tichy mentions that a fourth factor (besides Marshall's three causes) for the importance of clusters today is the new organisational principles of firms—today's clusters are a substitute for yesterday's hierarchical firms. Cappellin emphasises the changing behaviour of the individual economic actors as the most important aspect on the supply side of regional and national economies. Cullen points to the entrepreneur in (especially) SMEs as the pivotal figure in decisionmaking for networks. Danson and Whittam see networking as something to be encouraged at the level of the individual entrepreneur on an atomistic basis and strongly dependent of his or her commitment.

Approaching the question of why and to what extent there is cooperation Mizrahi uses public choice and game theory. He models the various conflicts related to regional cooperation as a collective action problem and analyses possible strategies to solve it. He shows that most of these conflicts can be described by the prisoner's dilemma. Discussing possible solutions for that dilemma he argues that strategies which create proper structural conditions for players to learn for themselves the advantages of cooperation are more efficient than strategies which use subsidies and/or sanctions to motivate people to cooperate. There is, however, a size limit which specifies in individual cases the number of required participants for optimal regional cooperation.

Addressing the question of whether clusters and cluster policy also work in the process of transformation Friedrich and Feng formulate a model in which sales contracts and property transfers are analysed on the basis of different cluster-formation policies. Five general scenarios of clustering with three main types of policies are explored. Taking account of different incentive structures and contract-determining factors they show that reclustering is necessary in the transformation process, that it is necessary to build up public and private institutions when establishing a cluster, that there are high costs within a privatisation of clustering, yet nevertheless—despite extreme difficulties in reshaping traditional clusters or forming new ones—successful cluster transformation is possible.

In another attempt to look inside the black box of cluster formation and cluster activities Cullen, and Steiner and Hartmann apply concepts of organisational learning. In both contributions learning is regarded as a communicative process rather than as a cognitive performance, requiring new thinking about the nature and forms of the transmission and dissemination of knowledge within a social and organisational context, such as the firm or a cluster. Cullen suggests a typology of organisational learning within SMEs that reflects interactions between structural and sociocultural characteristics. He identifies five main types of firms and reflects on how organisational learning within SMEs is shaped according to the interrelationship between information-gathering, knowledge-acquisition, and competence-development behaviour, and the relationship between the SME and its industrial milieu, and other structural characteristics, such as size, length of time established, and decision-making style. Steiner and Hartmann specifically show the correspondence between the intensity of network relations and higher levels of organisational learning—clusters may act more as learning organisations the more they have a real network structure.

4.4 Analytical approaches and quantitative methods

There is certainly no universally accepted and defined methodology for the analysis of clusters. Bergman speaks of a 'Rashomon effect' referring to Kurosawa's famous film where the same scene is seen completely differently by the persons involved. The perspective and the reason for looking at clusters therefore influence the tools applied to cluster analysis.

One of the fundamental interests is the identification of clusters. Even here there are a multitude of approaches depending on a range of possible levels of analysis. Feser and Bergman differentiate between macro, meso, and micro levels. At a national (macro) level, clusters are conceived as broad industry groups linked within the overall macroeconomy—transaction networks of universally documented flows are usually analysed here. Clusters at the industry (meso) level constitute the extended value chains of given end-market products involving best-practice benchmarking and studies of cluster-specific technology adoption and innovation processes. At the firm (micro) level, clusters are conceived as one enterprise or a few linked enterprises often restricted to a single visible collection of similar-sector firms, overlooking linkages that some of the members may have with regionally colocated firms from very different sectors—not surprisingly microstudies often document only one cluster per region. As Bergman shows, different perspectives applied to the same region may produce quite different results leading him to a regional mesolevel resolution to find some common ground between the approaches at the extremes.

Clement applies different scales to describe successful industrial areas in Austria that cannot be illustrated with the traditional official statistics. In his attempt to identify regionally based clusters he uses as a starting point, measures for the international competitiveness of export products and their regional distribution. This material is enriched and completed by a broad questionnaire accompanied by numerous personal interviews to trace actual interdependencies, thus combining a top-down approach with bottom-up methods. In a similar way Steiner and Hartmann want to discover slowly emerging forms of cooperation in a former 'old industrial area': are there any signs of links between newly founded SMEs? They proceed in three steps: identification of cluster firms through the database of the regional technology transfer centre; qualitative interviews with selected firms; and a baseline survey for quantitative evaluation.

Hewings et al also use different levels to identify clusters and changes in them. Applying a regional econometric input–output model they move the spatial perspective from a level analysing the competition among macroregions of the USA to the Midwest economy and finally to the Chicago region. To them this movement across different spatial levels is necessary to provide a more complete interpretation of the changes at the microlevel—the impact of globalisation forces will be manifest at the macrolevel, gradually filtering down to the microregional level.

Yet methods vary according to the questions asked. Friedrich and Feng use for their sales model a long list of cluster factors—ranging from production techniques to fees for municipal services, tax rates, capital and labour prices, and transportation costs—to show their influence on the formulation of the contract and on the bargaining solution. Dubin and Helper use different indicators for closeness in their statistical model. Fritz et al apply regional portfolio analysis as a framework to analyse the potential trade-offs between the positive returns for the region from cluster formation and the risk that may be involved. Cullen combines the results of a baseline survey of decisionmakers in over 320 SMEs in six different European countries to present a typology of clusters associated with different types of organisational learning.

5 Seeing cluster policy as a whole

Should regions specialise and should regional policy thus support cluster and network formation? The first answer is straightforward: yes, clusters are less dispensable than ever (Tichy). This straightforward response is of course based on the common explicit and implicit understanding of the competitive advantages of regions. In terms of recent developments in economic theory, locational advantages have turned from comparative advantage (being relatively cheaper) to competitive advantage relying on qualitative elements. This locational specification is founded on a special profile—what a region is able to do along specific lines of production. Without a coherent set of special attributes and abilities (an alternative definition of a cluster) it will not be able to attract partners for cooperation in the long term. In the short term it has to compete constantly with other similar (most cheaper) suppliers anywhere in the world: without recognisable competences based on highly qualified lines of cooperation (a further alternative cluster definition) it gets caught in the so-called 'globalisation trap'.

If a region's special image is linked to a set of specific competences, if its innovative ability and, consequently, the competitiveness of its firms depends on being embedded in interlinked activities with other firms and institutions (that is, depends on being part of a cluster), there is no longer room for traditional instruments of regional policy. Cluster orientation and specialisation then have to be understood as the active response to the potential dangers of globalisation. Or to put it differently, Adam Smith's argument of the division of labour has to be seen from the other side: if the extent of the market is (nearly) unlimited, you have to specialise.

The directness of the answer is of course subject to revision and qualification. The first counterargument—presented by Tichy and Fritz et al—points to the dangers of specialisation: specialisation increases efficiency but is also associated with risk and makes the specialised region more vulnerable. Regional policymakers are therefore confronted with a risk–return trade-off.

According to Fritz et al, there are two types of risk associated with regional clusters: in analogy to the regional portfolio concept a 'structural' and a 'cyclical' risk can be differentiated. The first kind of risk is manifest in the well-known case of 'old industrial areas' where a permanent decline in specific sectors leads to the decline of whole regions which specialised in these industries. The broad literature describing these cases provides ample evidence of the imminent dangers of this risk. Regions become closed systems, the petrification of the cluster structures impedes innovation and subsequently leads to an inability to adjust to new situations. According to regional theories of the product cycle, different stages of petrification can be differentiated and—according to the stage—strategies and instruments to avoid the ageing of clusters can also be conceived. Tichy argues here not for more, but for better intervention, with a feeling for long-term consequences.

The second type of risk refers to economic stability: unnecessary fluctuations cause inefficiency and welfare losses. Because business cycles are unevenly distributed among industries, the region's industrial structure and its optimal mix is a goal per se, yet one that is in competition with the gains of regional specialisation—the more a regional economy is dominated by a cluster the more risk is associated with the regional portfolio.

These risks are partly dependent on the kind and, of course, number of clusters within a region; their existence should nevertheless raise awareness of the long-term structural consequences as well as the short-term instability implications of a cluster strategy. Tichy therefore points to a two-sided cluster trap between full specialisation (with deep knowledge, yet few new ideas) on the one side, and nonspecialisation (with a consequent skill deficit) on the other.

Another point is the spatial scale and the geographical dimension of cluster orientation: geographical proximity is but one aspect of cluster formation and—as Rallet and Torre, and Dubin and Helper argue—too much emphasis has been placed on it. Of equal—if not higher—importance is organisational proximity: firms of equal size and knowledge orientation (for example, geographically dispersed plants belonging to the same legal firm) show stronger forms of cooperation than do accidentally spatially close enterprises. Consequently, local and regional development policies should diversify and gradually move from the exclusive search for local synergies to more open strategies of development.

This gradual extension from the intraregional to the interregional level of close relationships of comanufacture leads Cappellin to call for new tasks of policy support. Investments in the development of modern logistics and distribution services become a critical factor for wider interregionalisation and internationalisation of the firms involved. There is a need for major decisions to undertake large projects on the infrastructure network and on other modern collective services. Simply because of an enlargement of the spatial dimension of cooperation, the organisation and quality of the territory will become key factors in international competitiveness.

This qualitative organisation of the territory implies a new relationship between the urban centres and their rural hinterlands, leading to the question of to what extent a cluster orientation disregards the periphery: does it reduce geographic disparities or does it mean the acceptance of a potential worsening in regional inequality?

The optimistic variant—as argued by Cappellin—emphasises the interdependence between the larger territory and the urban centres which have to be considered as a pool of infrastructures or as a centre of 'soft' location. This task extends not only to large metropoles as nodes in the transport and communication network or as gateways between regions and the outside world, but also to intermediate cities. Therefore a diffused urban system made up of various intermediate and small cities may be more efficient than a centralised one. According to this perspective the traditional conflict between rural areas and cities is being transformed into a more synergetic relationship.

The more critical (not to say pessimistic) position points to the basic fact that a cluster policy is a 'backing the winner' strategy (Jacobs and de Man,

1996) based on already existing strengths, a policy therefore which is only relevant to those who are already successful. Defenders point to positive results of rural clustering (Rosenfeld, 1997) and—as does Shotton—to the need for patience: such regions have to build their capabilities in successive phases where a classic starting point is to aim for excellence in research and education through public funding in targeted areas. Nevertheless, there remains competition between the many prospective starters amongst the underprivileged regions. The final answer will probably remain an empirical matter.

One of the threads running through most of the contributions is the emphasis on changed business behaviour and the importance of learning and knowledge exchange at the level of the firm as well as between firms. Yet as Cullen, and Steiner and Hartmann show, learning within clusters (that is, organisational learning) is not a universal phenomenon amongst the different forms of clusters. Training and logistical support policies and initiatives, especially for SMEs, need to be carefully targeted rather than generic. Policy instruments have to reflect the different configurations of clusters and their organisational learning behaviour. From this experience a number of policy implications for the support of training and joint learning in favour of SMEs are made by Cullen.

(1) In relation to training and support policies three main constituent components of 'organisational learning' need to be targeted: information gathering, knowledge acquisition, competence consolidation and development, implying different training and logistical support capabilities. They should incorporate provision of 'formal' services, together with actions designed to enhance informal networking arrangements.

(2) SMEs are relatively active in knowledge-acquisition activities, but not in lower level market-intelligence gathering or higher level competence development. This underlines the need for awareness-raising campaigns—SMEs need to be made aware of the need to balance these three different components in their human resource development planning and management.

(3) The entrepreneur turned out to be the pivotal figure in decisionmaking within SMEs, yet the evidence suggested that a large proportion of SMEs are 'crisis-management' rather than proactive learning organisations. There is therefore a need to encourage SMEs to adopt a more participative style of collective learning.

(4) Microenterprises and new start-ups are particularly prone to crisis management and often lack coherent organisational learning strategies. Because this situation is almost certainly associated with lack of resources, it would suggest the need for support services that can provide pooled resources for SMEs.

Another thread: clusters presuppose a mentality of partnership, a willingness to cooperate, an insight into the necessity for clusters, an incentive structure, and, in particular, the existence of trust. This suggests above all that no kind of policy can substitute for the dynamism and social organisational skills that must exist on a local and regional level for cluster-building policies to succeed: "If the spark is not there, the lift-off cannot be artificially generated", argues Shotton. This also implies that, without decentralisation, without greater political autonomy of local and regional institutions, without federalist reforms, network creation becomes difficult.

Yet, as Mizrahi shows with his game-theoretical approach to regional cooperation, there are conflicts embodied in the dynamics of decentralisation and regionalisation of innovative industries in two aspects: free-rider problems may arise and certain players may prefer noncooperation rather than cooperation with decentralisation policies. At the core of a long-term strategy of decentralisation stands the transformation of games of conflict, to produce less conflict, creating the conditions for players to learn for themselves the advantages of cooperation.

In a similar vein Danson and Whittam argue that, to ensure the success of potential networks, regional policy should move from trust based on contracts to trust based on goodwill within networking arrangements. With the background of the experience of Scottish Enterprise over the last decade they point to the necessity of involving all potential firms in the decisionmaking process to ensure that the needs of the firms are being catered for within the networks. This necessity stems from the fact that the services being delivered within the network of firms will consist of elements of public goods such as information.

After all these revisionist and limiting qualifications, is there any legitimation left for straightforward support for a strategy of regional specialisation without again referring to its inevitability? Maybe we could revert to a metaphor used in theological circles where a second level of naivity—being aware of all doubts—allows for a reflected kind of belief. Such a belief in the usefulness of a cluster orientation is then founded on the necessity of "seeing regional economies whole" (Bergman).

This enlightened view would suggest different complementary levels of cluster analysis and different strategies and instruments for different segments of the resulting clusters with specific needs. Their combined needs require a portfolio of suitable policies. These policies would then have to adopt strategies and instruments suited to a continually changing mix of industries.

This view also requires us to see cluster policy—as do Clement and Feser—as a whole. One of the hallmarks of a cluster-oriented regional policy would be the implementation of holistic development strategies, a comprehensive attempt to nurture a given value chain through a range of carefully crafted demand-side and supply-side policy interventions. A cluster approach may serve as a structuring principle of different domains of industrial policy and lead to an institutional renewal of economic policy in general. In this sense clusters may also serve as a measure to develop creativity and as a disciplining instrument for fragmented, not only regional, policy responsibilities. In this respect they may contribute to a learning process of modern economic policy.

References

Aberg Y, 1973, "Regional productivity differences in Swedish manufacturing" *Regional and Urban Economies* **3** 131–156

Arrow K, 1962, "The economic implications of learning by doing", *Review of Economic Studies* **29** 155–173

Aydalot P, Ed., 1986 *Milieux innovateurs en Europe—Innovative Environments in Europe* (GREMI, Paris)

Baumgardner J R, 1988, "The division of labour, local markets and worker organization" *Journal of Political Economy* **96** 509–527

Bellini N, 1996, "Italian industrial districts: evolution and change" *European Planning Studies* **4** 3–4

Carlino G, 1978 *Economies of Scale in Manufacturing Location: Theory and Measurement* (Martin Nijhoff, Leiden)

Carlino G, 1979, "Increasing return to scale in metropolitan manufacturing" *Journal of Regional Science* **19** 363–374

Carlino G, 1980, "Contrasts in agglomeration: New York and Pittsburgh reconsidered" *Urban Studies* **17** 343–351

Chesnais F, 1996, "Technological agreements, networks and selected issues in economic theory", in *Technological Cooperation: The Dynamics of Cooperation in Industrial Innovation* Eds R Coombs, A Richards, P P Saviotti, V Walsh (Edward Elgar, Cheltenham, Glos) pp 18–33

Chinitz B, 1961, "Contrasts in agglomeration: New York and Pittsburgh" *American Economic Review* **51** 363–374

Coombs R, Richards A, Saviotti P P, Walsh V (Eds) 1996 *Technological Cooperation: The Dynamics of Cooperation in Industrial Innovation* (Edward Elgar, Cheltenham, Glos)

Czamanski D, Czamanski S, 1977, "Industrial complexes: their typology, structure and relation to economic development" *Papers of the Regional Science Association* **38** 93–111

DELOS, 1998, "Project report", DELOS (DEveloping Learning Organisation Models in SME-Clusters), TSER project within the 4th Framework of the EC; copy available from Istituto G Tagliacame, Via Appia Pignatelli 62, 1-00178, Rome

Florence S, 1944, "The selection of industries suitable for dispersion into rural areas" *Journal of the Royal Statistical Society* **107** 93–116

Florida R, 1995, "Toward the learning region" *Futures* **27** 527–536

Grabher G (Ed.), 1993 *The Embedded Firm: On the Socio-economics of Industrial Networks* (Routledge, London)

Harrigan F, 1982, "The relationship between industrial and geographical linkages: a case study of the United Kingdom *Journal of Regional Science* **22** 19–31

Hassink R, 1997, "Localized industrial learning and innovation policies" *European Planning Studies* **5** 279–282

Hoover E, 1937 *Location Theory and the Shoe and Leather Industries* (Harvard University Press, Cambridge, MA)

Hotelling H, 1929, "Stability in competition" *Economic Journal* **39** 41–57

Jacobs D, de Man A-P, 1996, "Clusters, industrial policy and firm strategy: a menu approach" *Technology Analysis and Strategic Management* **8** 425–437

Krugman P, 1991 *Geography and Trade* (MIT Press, Cambridge, MA)

Kubin I, Steiner M, 1987, "Funktionale Verflechtung und räumliche Nähe—einige weitere Differenzierungen", mimeo, Joanneum Research, Graz

List F, 1841 *Das nationale System der politischen Ökonomie* published in English in 1885 as *The Naitonal System of Political Economy* (Longman, London)

Lucas R E, 1988, "On the mechanics of economic development" *Journal of Monetary Economics* **22** 3–42

Markusen A, 1985 *Profit Cycles, Oligopoly and Regional Development* (MIT Press, Cambridge, MA)
Moomaw R, 1981, "Productive efficiency and region" *Southern Economic Journal* **48** 344–357
Norton R, Rees J, 1979, "The product cycle and the spatial decentralisation of American manufacturing" *Regional Studies* **13** 141–151
Perroux F, 1955 *Les Techniques Quantitatives de la Planification* (Presses Universitaires de France, Paris)
Pezzini M, 1998, "Building competitive regional economies", paper presented at the OECD conference, Modena; copy available from Territorial Development, OECD, Paris
Porter M E, 1990 *The Competitive Advantage of Nations* (Free Press, New York)
Pyke F, Becattini G, Sengenberger W (Eds) 1990 *Industrial Districts and Inter-firm Co-operation in Italy* (International Institute for Labour Studies, Geneva)
Quelin B, 1993, "Les analyses de la filiére: bilan et perspectives", WP 5, Industriewissenschaftliches Institut, Vienna
Ratti R, Bramanti A, Gordon R (Eds), 1997 *The Dynamics of Innovative Regions: The GREMI Approach* (Ashgate, Aldershot, Hants)
Richter C, 1969, "The impact of industrial linkages on geographic association" *Journal of Regional Science* **9** 19–27
Rosenfeld S A, 1997, "Bringing business clusters into the mainstream of economic development" *European Planning Studies* **5** (1) 3–23
Schumpeter J, 1934 *The Theory of Economic Development* (Harvard University Press, Cambridge, MA)
Scott J, 1991 *Social Network Analysis: A Handbook* (Sage, London)
Segal D, 1976, "Are the returns to scale in city size?" *Review of Economics and Statistics* **53** 339–350
Steiner M, 1987, "Contrasts in regional potentials: some aspects of regional economic development" *Papers of the Regional Science Association* **61** 79–92
Steiner M, 1990a *Regionale Ungleichheit: Studien zu Politik und Verwaltung* **32** (Böhlau, Vienna)
Steiner M, 1990b, " 'Good' and 'bad' regions? Criteria to evaluate regional performance in face of an enforced internationalization of the European economy" *Built Environment* **16** 52–68
Stigler G J, 1951, "The division of labour is limited by the extent of the market" *Journal of Political Economy* **59** 185–193
Storper M, 1995, "The resurgence of regional economies ten years later: the region as a nexus of untraded interdependencies" *European Urban and Regional Studies* **2** 191–221
Streit M, 1969, "Spatial associations and economic linkages between industries" *Journal of Regional Science* **9** 177–188
Suarez-Villa L, 1984, "Industrial export enclaves and manufacturing change" *Papers of the Regional Science Association* **54** 89–112
Sveikauskas L, 1975, "The productivity of cities" *Quarterly Journal of Economics* **89** 392–413
Tichy G, 1987, "A sketch of a probabilistic modification of the product-cycle hypothesis to explain the problems of old industrial areas", in *International Economic Restructuring and the Regional Community* (Avebury, Aldershot, Hants) pp 64–78
Tichy G, 1991, "The product-cycle revisited: some extensions and clarifications" *Zeitschrift für Wirtschafts- und Sozialwissenschaften* **111** (1) 27–54
Weber A, 1909 *Über den Standort der Industrien* (Mohr, Tübingen)

Old and New Theories of Industry Clusters

E J Feser
University of North Carolina at Chapel Hill

1 Introduction
Despite the intense interest in industry clusters in economic development policymaking in Europe and North America, there is presently little consensus about the precise meaning of an industry cluster, the dynamics that underlie cluster growth and development, and the policy initiatives that would best build and strengthen clusters. To be sure, there has been movement toward consensus as the literature deepens, with most progress probably occurring in the area of methodologies for identifying and documenting clusters. But the great variety of development strategies that are either nominally or fully based on some notion of industry clusters testifies to a high degree of malleability in the cluster concept (Marceau, 1997; van der Laan, 1997). In Europe, Boekholt (1997, page 1) writes that the "multitude of cluster initiatives has led to a widespread confusion of what clusters really are, and in what way they differ from related phenomena, such as industrial districts, technopoles, networks, and industry-research collaborations". And in a critique of cluster policy in the USA, Held (1996, page 249) notes: "Sadly, in the rush by various governments to employ clusters, some fundamental issues have been slighted, including appropriate research methods and even the definition of the cluster itself."

As Jacobs and de Man (1996) assert, a degree of flexibility in the cluster concept and a variety in policy approaches can be useful. However, Jacobs and de Man also go on to provide a careful outline of various dimensions of an industrial cluster in an effort to clarify terms, definitions, and policy implications. Clearly, the concern that "if clusters are everything, maybe they are nothing" is one worth addressing.

The experience of two related concepts that are similarly broad in scope is instructive. Harrison (1992) questions whether the theory of new industrial districts is merely "old wine in new bottles", noting its many similarities to more traditional, and well-worn, regional theories. In a paper published two decades earlier, Gilmour (1974, page 336) has this to say about the concept of external scale economies: "...this explanation of agglomeration is so intellectually appealing as to cause us to assume it must be right (it probably is).... It has never been demonstrated that it is completely invalid, but neither has the converse been demonstrated. More than anything else this noteworthy state of affairs reflects the theory's difficulty of verification". Of course, Harrison goes on to conclude that new industrial district theory is, in fact, more than just old wine, and Gilmour does attempt to verify the presence of external scale economies. So all is probably not lost with industry clusters either.

In this chapter I provide a brief review of the broad range of theories and ideas that constitute, often implicitly, the logic behind strategic cluster policies. The title of the chapter notwithstanding, there is no theory of industry clusters, per se. Even Porter's (1990) seminal contribution is more a theory of firm competitiveness than of clusters (Kaufman et al, 1994). There is, instead, a variety of older and newer theories of (1) the interrelationships between economic actors that clusters describe, and (2) the implications of such interrelationships for economic growth and development. Industry clusters have proven a useful way of characterising webs of relationships between and among firms and other institutions. Policymakers designing cluster strategies attempt to leverage such relationships in the interest of growth and development objectives. For their part, regional analysts must strive to specify and test clearer hypotheses about the workings and impacts of such relationships in order to verify the efficacy of cluster policies.

In section 2 I describe several ways of classifying the wide range of development initiatives related to industry clusters. Policies differ based on varying definitions of clusters, possible levels of analysis, and the degree to which clustering constitutes the central focus. Industry cluster principles are often used to improve the implementation of traditional development schemes. In general, cluster-related development policies represent attempts to leverage synergies arising from economic and spatial interdependence between economic actors. Section 3 contains a summary of major theories of such interdependence in economic and geographic space, as well as models of the link between interdependence and regional growth. Section 4 summarises a set of unresolved theoretical questions and their implications for cluster policy.

2 Clusters in development policy

A number of authors have attempted to clarify the range and appropriateness of different cluster policies (Boekholt, 1997; Jacobs and de Man, 1996; Roelandt et al, 1997; Rosenfeld, 1995; 1997). Naturally, much of the focus is on how clusters are defined. Boekholt (1997) demonstrates how different definitions of clusters imply different development strategies. She develops a typology of cluster initiatives based on how such policies define: (a) the types of collaborative links among cluster firms (for example, simple buyer-supplier versus knowledge or technology transfer); (b) the types of constituent firms and actors included in the cluster (firms only or firms and supporting institutions); (c) the appropriate level of aggregation (micro versus macro); (d) the position of firms in the value chain (that is, horizontal, vertical, or lateral); (e) the spatial level of intervention (local, regional, national, international); and (f) the specific policy mechanisms employed (general business assistance, network brokering, technology transfer, information provision, and so on). The typology is helpful for classifying the fast-growing number of policy applications, from the development of joint facilities for related warehousing and distribution industries in the Hudson Valley, New York (Held, 1996), to the

Table 1. Value-chain approach at different levels of analysis (source: Roelandt et al, 1997).

Level of analysis	Cluster concept	Focus of analysis
National (macro)	Industry groups' linkages in overall economic structure.	Patterns of specialisation in national or regional economy. Innovation and technology upgrade needs in megaclusters.
Branch or industry (meso)	Interindustry and intraindustry linkages in different stages of production chain of a single end product.	Industry benchmarking and SWOT (strengths, weaknesses, opportunities, threats) analysis. Innovation needs.
Firm (micro)	Specialised suppliers around one or a few core enterprises (interfirm linkages).	Strategic business development needs. Value-chain analysis and chain management. Need for collaborative innovation projects.

reform of telecommunications industry regulations in Austria (Peneder and Warta, 1997). The typology also helps clarify the difference between related concepts, such as clusters and networks (see also Rosenfeld, 1997).

Roelandt et al (1997) suggest another typology based on a range of possible levels of analysis (table 1). At the national level, clusters are conceived as broad industry groups linked within the overall macroeconomy. Relevant types of cluster analysis at this level include the study of patterns of industrial specialisation and the examination of general innovation processes as well as the adoption and use of more generic production and management technologies (those applicable to a wide range of industries). Clusters at the industry (or meso) level constitute the extended value chains of given end-market products (as revealed through patterns of interindustry and intraindustry linkage). Cluster analysis at the mesolevel involves best-practice benchmarking and studies of cluster-specific technology adoption and innovation processes. At the firm (or micro) level, clusters are conceived as one or a few linked or related enterprises along with their important specialised suppliers. Microlevel cluster analysis includes needs assessments with regard to networking and other types of interfirm collaboration, 'chain analysis and chain management' (for example, examination of buyer–supplier contracting practices), and design of business development programmes (perhaps involving marketing, recruitment, and entrepreneurship strategies). Clearly, the type of analysis implies different types of policy intervention, from the establishment of framework conditions and general technology policies at the national level to the implementation of networking initiatives and business development at the microlevel.

Another way to distinguish policy applications of the industrial cluster concept is in terms of those that focus specifically on identified clusters (*cluster-specific strategies*) and those that use cluster-related principles and techniques to *inform* traditional development policies (*cluster-informed strategies*). The difference is subtle but important. Figure 1 (see over) is a summary of the first type of approach for a hypothetical environmental technologies cluster. Under the cluster-specific policy approach, the objective is to encourage the emergence or development of a distinct identified cluster. The first step is to map out the cluster itself which, in the case of environmental technologies, includes four major end-market industries (equipment and supplies, sustainable goods, environmental services, and environmental resources), their numerous specialised suppliers, providers of advanced services, related and supporting public and quasi-public institutions, industry trade groups, and so forth.

But the broad environmental technologies value chain constitutes only the supply-side element in a comprehensive cluster strategy. To influence the development of the cluster, policymakers must also consider the demand side, which is made up of the major users of environmental technologies goods (government, private industry, and consumers). Thus, a cluster strategy designed to build or extend the environmental technologies value chain must include supply-side elements (many of which are outlined in table 1, including encouragement of interindustry linkages, technology transfer, networking, improved innovation systems) as well as demand-side initiatives. In the case of environmental technologies, government has a strong role to play in ensuring adequate demand through environmental regulation, enforcement, and resource pricing, as well as purchasing (as it adopts monitoring, prevention, and cleanup systems and as it subsidises sustainable infrastructure). The provision of assistance with compliance as well as tax and spending incentives would encourage demand in private industry, and education, various incentives, and adequate recycling facilities would help boost demand for environmentally sustainable goods among consumers.

The primary characteristic of the cluster-specific approach is the comprehensive attempt to nurture a given value chain through a range of carefully crafted demand-side and supply-side policy interventions. It is not the individual policy initiatives that are unique, but rather the way in which they are formulated and targeted; that is, in a manner that establishes—and mutually reinforces—the conditions for the growth and development of key or promising end-market sectors and their supporting industries. Note also that cluster policies from this perspective may involve many economic interventions that are not development strategies per se, but are instead traditional functions of government (regulation, enforcement, pricing, education). Indeed, one of the hallmarks of the cluster approach is the implementation of holistic and comprehensive development strategies that account for the full range of factors influencing the success of a given sector or set of sectors. It stands to reason that many of these factors will fall outside the purview of the typical set of individual development initiatives and tools.

Policies
University–industry
linkages, technology
transfer, commercialisation

Incubators

Targeted training
and trade schools

Buyer–supplier
conferences, networking
trade shows

Technical assistance

Venture capital

Supplier network
- Equipment and supplies
  instruments, chemicals,
  transport
- Sustainable goods
  Transit, agriculture, construction,
  retail, ecotourism
- Services
  testing, treatment, management,
  remediation, consulting, education
- Resources
  water, recycling, energy

Value chain/supply

Policies
Regulation and
enforcement
Resource pricing
Purchasing

Assistance with
compliance

Tax credits and
incentives

Education, marketing,
and incentives

Recycling services

Government

Private industry

Consumers

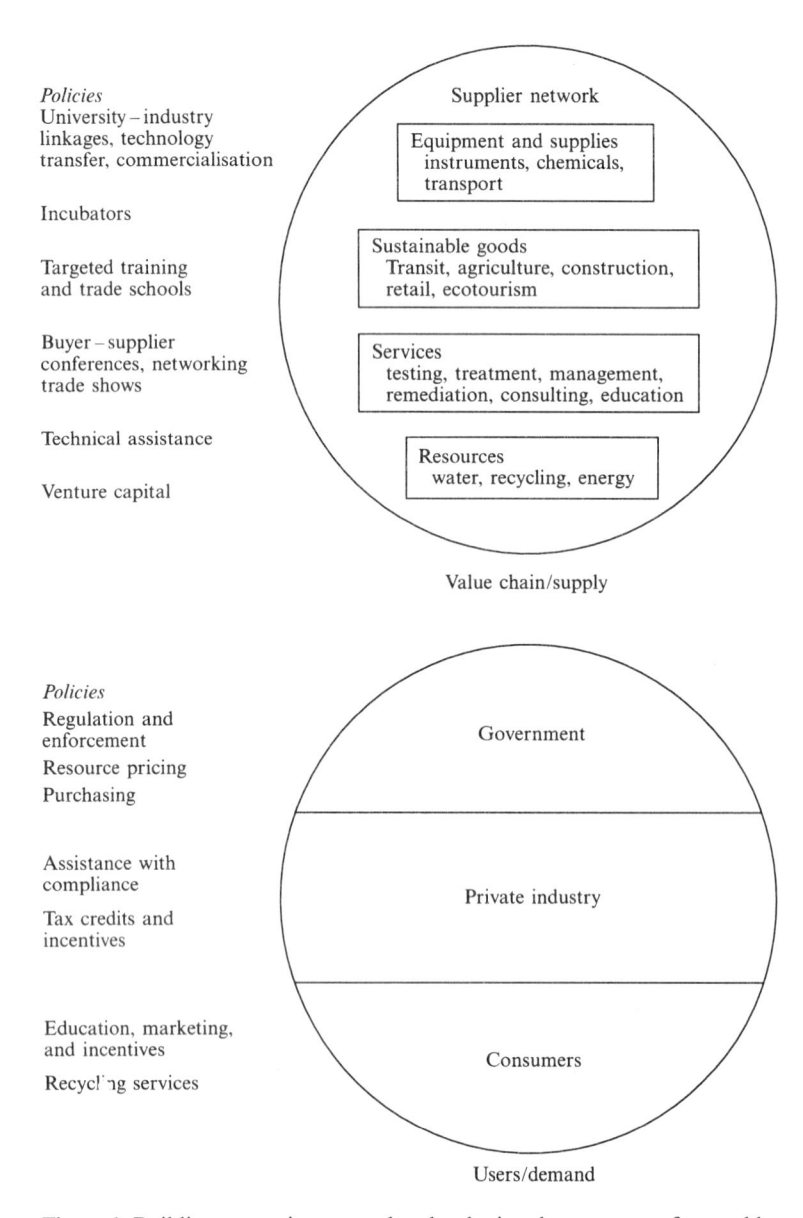

Users/demand

Figure 1. Building an environmental technologies cluster: create favourable conditions for emergence and development. Adapted from a diagram presented by C DeBresson at the OECD Workshop on Cluster Analysis and Cluster-Based Policies in Amsterdam, 10–11 October 1997. Information on the environmental technologies cluster is drawn from Kirkpatrick and Gavaghan (1996).

In contrast to what I have termed the cluster-specific approach, the principal policy objective from the cluster-informed perspective is the improved implementation of individual (or isolated) development initiatives. The logic is that particular development schemes or interventions will be more effective if they take account of economic and spatial interdependencies. Consider an example. In 1995 a state technology-planning agency commissioned a study of industry clusters in North Carolina to determine where best to focus modernisation programmes (Bergman et al, 1996). The agency was particularly interested in better diffusing advanced production technologies and practices through extended supplier chains by targeting programmes to specific sectors within given chains. If, for example, end-market sectors play an important role in encouraging investment in new technology among their core suppliers, the agency could *leverage* modernisation resources by ensuring best-practice techniques among end-market sectors, thus initiating a 'modernisation ripple effect' within the value chain. Alternatively, the agency could use the information to diagnose better the obstacles to technology adoption among given segments of supplier chains. Also, the organisation planned to identify gaps in particular chains that were preventing technology diffusion, limiting networking, or otherwise reducing the advantages of cooperative relationships between local producers. These gaps would be considered possible areas for targeted business development initiatives (entrepreneurship or recruitment). Thus, from the cluster-informed policy perspective, clusters are primarily an analytical device used to improve the efficacy of narrower policy tools or types of policy tools.

All cluster policies, however classified, have two important and related goals: resource *targeting* and resource *leveraging*. The analysis of industry clusters in national or regional economies helps policymakers identify appropriate targets for scarce development resources; rather than a scattershot approach to economic development, clusters provide a useful device for strategic planning and investment. This goal is most apparent in cluster-specific initiatives. But more importantly, clusters presumably hold promise for leveraging development resources through the encouragement of synergies, external economies, and increasing returns in the clusters themselves, a goal most apparent in cluster-informed types of strategies. To achieve this leveraging goal, however, we require a solid understanding of how clusters work, what the specific types of synergies are, and how (indeed, if) they can be cultivated. These questions are presumably the subject of cluster theory. Unfortunately, attention to theory in the industry-cluster debate has been relatively limited. If there is a (new) 'cluster theory', it is more a complex amalgamation of 'old' development theories, than a self-contained model of cluster growth and development.

3 Clusters in regional development theory

Although Porter's (1990) seminal study has been criticised as vague in application (Rouvinen and Ylä-Anttila, 1997), it nevertheless represents the most useful starting point for a discussion of how industrial clusters fit into the broader framework of regional development theory. Although one objective of this chapter is to demonstrate that the concept of industrial clusters has been around in various guises for some time, *The Competitive Advantage of Nations* has clearly galvanised policy interest at both the national and the regional level. More importantly, Porter provides a holistic framework for summarising what has become a general trend in regional analysis. This is the study of how interdependence between firms, industries, and public and quasi-public institutions affects innovation and growth in regional agglomerations. More specifically, the regional agglomeration literature is increasingly focused on the social or cultural dimensions of such interdependence, rather than on the more traditional economic or technical relations between firms (Hassink, 1997; Malmberg and Maskell, 1997).

Porter's ideas are also consistent with recent developments in growth and trade theory that highlight the role of increasing returns and the consequent likely spatial concentration of industrial activity. Although these basic ideas also have something of a history (as described below), it is notable that economists traditionally unconcerned with spatial questions have begun finding implications for geography in their models (Martin and Sunley, 1996). And though Porter is no regional economist either, his own studies of the determinants of economic competitiveness have led him to highlight the role of location.

Porter's model of national competitiveness—the 'diamond'—is by now well known and warrants only a brief summary here. Porter used industries' success in international markets as the primary barometer of the competitive strength of a nation. Accordingly, he traced the competitiveness of firms to four major national factors: (1) the nature of firm strategy, structure, and rivalry in the country, including attitudes toward competition, market institutions, the degree of local competition, and other cultural and historical factors affecting how firms do business with each other, their workers, and the government; (2) factor conditions, or the basic endowments or conditions on which local firms seek to compete (for example, cost-related basic factors such as ready supplies of natural resources, or inexpensive unskilled labour versus knowledge and/or technology-related advanced factors); (3) demand conditions or the nature of local demand (for example, the needs and wants of the consumer for foreign and domestic goods as well as the existence of local industrial demand for related intermediate goods); (4) the presence of related and supporting industries, including suppliers and successful competitors (both to stimulate cooperation, the latter also to stimulate rivalry).

Competitive companies must depend, to a degree, on the competitiveness of their intermediate input suppliers, who must depend on the capabilities of their suppliers, and so on, back through all links in the value chain. But such companies also depend on service providers (management, marketing, financing, legal, etc), sources of basic and applied R&D (for example, universities and/or contract research organisations), capital goods suppliers, wholesalers and distributors, and suppliers of trained workers (again, universities and colleges). Even competitors are important, including direct competitors to the company as well as competitors to the company's suppliers, because their presence maintains pressure continually to upgrade processes and techniques and to seek new opportunities. Competitors also provide opportunities for cooperation in solving joint problems or addressing industry-wide issues (see Best, 1990).

Thus, the success of an individual company may be partly traced to the size, depth, and nature of the *cluster* of related and supporting enterprises—both private and public—of which it is a part. Much of Porter's analysis, which turns on the findings from an impressive set of national case studies, focuses on outlining the basic conditions determining cluster competitiveness. His framework leads naturally to a focus on end-market sectors as the point of departure for studying clusters. But such end-market industries should not be studied in isolation; the critical function of interdependence in the process of economic growth and change—not just in terms of how it has traditionally been viewed, that is, in technical or input–output terms—is the guiding principle in his study.

The dynamics that presumably characterise industry clusters need not be localised in scope, though Porter argued that clusters tend to be geographically concentrated. Thus we have two core dimensions of the cluster concept: economic and geographic (see figure 2). Some end-market sectors (for example, automobiles) may be both economically and geographically clustered (quadrant 1 in figure 2). Others may be more or less clustered on either of the two dimensions. For example, we could imagine an extractive industry (for example, raw timber that is exported with little subsequent processing) as falling into quadrant 2: little economic or geographic clustering. Other examples are possible for quadrants 3 and 4, particularly in the services industries.

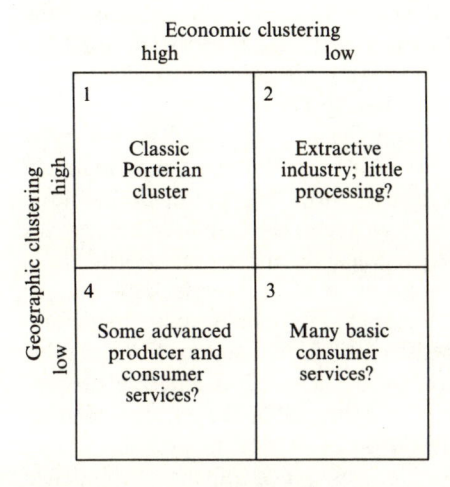

	Economic clustering high	Economic clustering low
Geographic clustering high	1 Classic Porterian cluster	2 Extractive industry; little processing?
Geographic clustering low	4 Some advanced producer and consumer services?	3 Many basic consumer services?

Figure 2. Two core cluster dimensions.

To complicate matters, the degree of economic and geographic clustering one observes for a particular end-market industry is relative to space, time, and scale. An end-market sector clustered geographically from a national perspective may not be spatially clustered from a regional or local one (or vice versa). Moreover, a given sector becomes more economically clustered as vertical, horizontal, and lateral linkages and relationships expand and deepen (with growth in related and supporting industries and/or the establishment of stronger ties or networking with existing enterprises). And there is no reason, a priori, to assume that clustering along either dimension need only increase over time, even with economic growth and nominal increases in various elements of the cluster. Changes in the social, cultural, or political environment could lead to altered relations between cluster firms such that the positive synergies described by Porter are reduced. Alternatively, improvements in the transportation or communication infrastructure may lead to some spatial dispersal of cluster firms and a reduction in geographic clustering. Last, there is the element of scale. True clusters are probably large, perhaps exceeding some threshold, but size alone does not guarantee clustering.

Even with this brief description, it should be apparent how complicated Porter's industrial cluster concept is likely to be in application. One problem is that there is a distinct normative element both to the geographic and to the economic dimensions that makes identifying clusters for policy purposes difficult. Economic clusters are not just related and supporting industries and institutions, but rather related and supporting institutions that are more competitive *by virtue* of their relationships (Rosenfeld, 1996; 1997). Enterprises are not clustered in space merely because they are located next door, but rather because they are located next door *and* because they enjoy the types of interdependence described by the economic clustering dimension. As there are few direct measures of interfirm relationships outside of input–output relations, which clearly do not capture all of the richness of the dynamics described by Porter, it is not surprising that clusters specified for policy attention often constitute little more than large industries and suppliers in which the nation or region has a particular specialisation.

Another problem that significantly complicates nationally or regionally administered development policies is that economic clustering may manifest itself over widely varying geographic scales. How, for example, should a regional agency foster (or leverage) positive synergies between local enterprises and cluster members located in other regions or across the world? It is possible that the competitiveness of a given local enterprise may depend more on a global cluster than on a regional one. Could some regionally focused cluster strategies hinder competitiveness in this context?

These problems, among others, have been discussed at length elsewhere (Jacobs and de Jong, 1992; Jacobs and de Man, 1996; Rouvinen and Ylä-Anttila, 1997). More important for the purposes of this chapter is the fact that the theoretical links between (a) various kinds of interdependence between economic actors; (b) the time-dependent and space-dependent economic and geographic clustering dimensions, and (c) regional growth and change, are only weakly specified in the Porter framework. On the whole, the model, descriptive and stylised as it is, is presented almost as a fait accompli. This means that regional scientists must determine what Porter's analysis adds to a deeper and richer (though perhaps less comprehensive) body of literature on interdependence and regional development in order to begin to understand how clusters work, how they change over time, and what the implications of such changes are for regional policy.

In this chapter I attempt to contribute to this effort by summarising the major regional theories and models relevant to the cluster concept. The discussion follows a three-part organisational scheme: (1) theories addressing interdependence and economic space; (2) theories of interdependence and the formation of agglomerations; and (3) theories of the relationship between interdependence or agglomeration and regional growth. There are undoubtedly other ways in which this somewhat disparate material might be effectively organised. The framework used here is based primarily on the two core cluster dimensions.

3.1 Interdependence in economic space

By economic space is meant the nonspatial sphere of relations between firms and other economic actors. Clearly one cannot discuss economic interdependence without reference to Perroux's (1950) theory of abstract economic space as a field of forces in which relations between firms and their buyers and suppliers take place. For Perroux there is no reason why physical space should necessarily bear any relationship to economic space; enterprise linkages will extend without spatial limit throughout the globe, at least where they are economically justified. Directing one's analysis to particular regions will provide only a distorted picture of the growth and development process [geographic space as 'banal', though see Perroux (1988)]. In Perroux's framework, to understand economic growth and change, analysts need to focus on the role of propulsive industries, those industries that dominate other sectors because of their large size, considerable market power, and role as lead innovators. Propulsive industries (or even individual firms) represent poles of growth which attract, focus, and direct other economic resources (Darwent, 1969).

The similarities between the cluster concept and Perroux's theory of growth poles should be readily apparent [see also Dahmén's (1984; 1988) notion of 'development blocks']. The cluster focus on how end-market industries drive the deep and broad value chains of which they are a leading part is consistent with propulsive industries as dominant economic actors. End-market industries in given clusters transmit growth pulses through the cluster through demand for intermediate and capital goods. In addition, because they are composed of internationally competitive, best-practice firms, they may play an important role as diffusers of process and product innovations. To the degree, for example, that large original equipment manufacturers (OEMs) can use their market power to dictate (or perhaps strongly encourage, even assist with) technology upgrades and improved manufacturing strategies to their suppliers, such end-market industries might be said to drive, at least in part, overall cluster competitiveness. On the other hand, one can also conceive of market power among some cluster members as exerting a detrimental influence on the overall cluster. For example, short-term, least-cost-focused contracting practices of OEMs with their suppliers may actually discourage strategic thinking and investment.

Perroux's macrolevel analysis is not unlike Porter's in another respect: the nature of the dynamics between economic actors is described only in general terms. Microlevel or mesolevel theories and studies of changing contracting practices, best-practice technology diffusion, and networking among related firms and industries are particularly important for the better specification of how clusters develop and grow (Helper, 1991; Imrie and Morris, 1992; Klier, 1994). Research on these topics has deepened in recent years and holds considerable promise for providing a clear and tangible picture of how enterprise transactions influence strategic choices, innovation, and competitiveness.

Perroux was particularly concerned with demonstrating that economies are characterised more by imbalance than by equilibrium. Technological change is a central feature of his framework. The introduction of a significant innovation can lead to the concentration of market power and influence in the hands of the innovating sector, which is then able to establish a lead or dominant position vis-à-vis other industries and firms, at least through the exhaustion of the economic life of the innovation. Thus, the introduction of new innovations establishes new or even competing poles, which may alter the structural relations among enterprises in economic space. The development trajectory of the economy is one of movement from imbalance to imbalance.

Schumpeter's (1934) theory of innovation and development is related to growth-pole theory in some of these respects, though not in terms of any focus on economic space. As pointed out by DeBresson (1996; DeBresson and Hu, 1997), Schumpeter suggested how innovations cluster *in time*: in particular, as a result of "reductions in uncertainty, entrepreneurial profits for rapid imitators, more ingenuity in times of recession, some periods of the business cycle being more conducive to entrepreneurial activity than others, and so on" (DeBresson, 1996, page 149). Schumpeter described the development process in terms of a series of waves. The entrepreneur who initially exploits an opportunity associated with a radical innovation initiates a flow of productive factors from consumption to investment activities (first wave) and is soon followed by imitating entrepreneurs. A general increase in prices, debt-financed increases in output, and a shift back toward consumption activities fuel a general expansion phase (second wave). As the innovation nears the end of its economic life, the economy, in effect, overshoots itself and enters a recessionary period. In this economy of fits and starts, we observe key and related innovations clustered in time.

But we may also think of innovations as clustered in economic space. Technological advances tend to establish paths of further innovation through the adoption of similar learning processes, as a result of the systemic nature of particular technologies, and through cumulative learning processes (Debresson, 1996). Clustering of innovations can increasingly be studied through recent national surveys of innovative activity, which trace flows of innovations among sectors in a traditional input–output framework. This is the focus of much of the research on clusters and national systems of innovation (Lundvall, 1996; Nelson, 1988; 1993).

Indeed, the study of industrial clusters in the context of national systems of innovation is an interesting example of how Porter's general framework has been received, embellished, and implemented differently in various international contexts. In the United States, industrial clusters are often simply a more sophisticated means of targeting traditional economic development programmes than the narrower sectoral schemes pursued in the 1970s and 1980s (see Sternberg, 1991), whereas in Europe they are increasingly viewed as an integral element in broader industrial innovation processes or systems. As a result, in the United States, industrial clusters are perhaps most frequently used for marketing purposes, whereas in Europe, they help identify and characterise the conduits through which learning, technology, and innovations diffuse (Lagendijk and Charles, 1997; Roelandt et al, 1997). These continental variations are a function of differences in the related literatures, theories, and concepts that inform regional economic policymaking in the two places. The systems-oriented view of economic processes is much more common in Europe than in the United States, which is still dominated by what might be described as an atomistic perspective of interdependence among economic actors.

3.2 Interdependence in geographic space
Of course, regional scientists and geographers are keenly interested in how and why clustering occurs in geographic space, and particularly in how such clustering influences regional development paths. It is important to note that Porter's framework implies that geographic clustering affects national as well as regional fortunes, that is, questions of location cannot simply be ignored or assumed away in any understanding of national competitiveness. But I defer my discussion of the linkage between geographic clustering and economic growth and development to the following section. Here, I focus on theories of why proximity matters to firms, that is, the determinants of spatial clustering. Relevant concepts include Marshallian spatial externalities, agglomeration economies, and new industrial districts (a resurgence of interest in Marshall's original analysis).

Why should firms seek to colocate in space? If we abstract from transportation or natural resource considerations [following Weber (1929) who defined these—along with agglomeration economies—as the three basic industrial location factors], firms presumably derive inherent benefits from locating near other enterprises. Two basic conceptual approaches to such benefits dominate the literature: (1) the industrial location theory perspective that builds on Weber and on Hoover (1937), where the benefits are called agglomeration economies, and (2) the Marshallian perspective that takes as its point of departure Marshall's (1890) original analysis of external scale economies and their typical presence in what he termed 'industrial districts'. In both cases, various specific types of externalities [or, more appropriately, *sources* of externalities (see Feser, 1998)] are cited as the reason why firms colocate. The literatures differ somewhat in their relative emphasis on static versus dynamic externalities. Neither perspective is particularly concerned with distinguishing between pecuniary and technological externalities, an underappreciated ambiguity that has direct implications for the need for, and utility of, some types of cluster strategies. Note that here the focus is mainly on externalities related to proximity among business enterprises (localisation effects), rather than on externalities associated with general urban advantages (urbanisation effects).

In agglomeration theory, the reasons why firms colocate are usually assumed or implied, rather than carefully specified. In his theory of industrial location, Weber (1929, page 126) assumes a known measure of the cost savings associated with spatial juxtaposition between producers (termed a 'function of economy of agglomeration'). He distinguishes between types of juxtaposition in terms of the (spatial) concentration of production within a single plant, the concentration of production across several plants in the same industry, and the concentration of production across multiple plants in multiple industries. [These are further refined by Hoover (1937) into the well-known distinction between localisation and urbanisation economies.] Weber is not particularly concerned with why such agglomeration economies arise, preferring to suggest that they are simply external varieties of the well-accepted notion of internal scale economies (see Weber, 1929, page 127). His primary aim is to demonstrate how the economies might lead to agglomeration. Weber therefore offers very little guidance as to what agglomeration economies really are.

To a limited degree, subsequent theorists in the industrial location tradition have tried to redress this particular problem. Hoover (1937, page 98) is also confident that the benefits associated with concentrated geographic production are self-evident enough to warrant little discussion. Interestingly, he does note that they likely "depend on the conditions of the particular industry, and also upon institutions (e.g., the way in which wages are determined, the presence or absence or labour unions, the industrial-promotion and tax policies of the communities involved, etc)". But this is the closest he comes to going beyond general references to industry and urban size as primary determinants of such economies. Other researchers cite particular advantages of proximity between firms, including increased market power through brokered buying and selling, the better availability and use of specialised repair facilities, shared infrastructure, reduced risk and uncertainty for aspiring entrepreneurs, and better information (Carlino, 1978; Isard, 1956; Lichtenberg, 1960; Vernon, 1960).

For the most part, one has to look outside the traditional agglomeration economies literature for any sophisticated discussion of firm interdependence and geographic clustering. One important exception is Chinitz's (1961) paper on market structure as a key determinant of agglomeration economies. In a brief but rich discussion that essentially anticipates the present-day focus on how firm and industry organisation influences regional development paths, Chinitz essentially draws a direct link between what Porter calls 'firm structure and rivalry' and regional economic fortunes. Critiquing the focus in the agglomeration-economies literature on urban and industry size, Chinitz argues that industrial structure particularly influences learning, innovation, and entrepreneurship, giving diverse and small-firm-rich places such as New York a leg up over large-firm single-industry towns such as Pittsburgh. This has become an important theme in the Marshallian new industrial district theory.

But before we consider the Marshallian perspective, it is worth noting that in his explication of localisation and urbanisation economies, Hoover relies partly on Robinson (1931), who makes an important distinction between mobile and immobile external economies. Immobile external economies are localised, that is, dependent on the growth of an industry in a given place. Mobile external economies are, in principle, global in scope. Firms may benefit from the worldwide development of the industry (usually through diffused technological advances). Interestingly, just as the global nature of some types of firm interdependence contributes to the ambiguity in the cluster concept today, early agglomeration theorists also faced the problem of disentangling local and nonlocal external economies. It is worth noting that Robinson's distinction is rarely, if ever, referenced in the agglomeration-economies literature after the 1960s [for early discussions of mobile and immobile economies in the regional economics literature see Guthrie (1955) and Nourse (1968)].

Marshall (1890) defines external scale economies as cost savings accruing to the firm because of size or growth of output in industry generally. Such economies contrast directly with internal scale economies, which are the source of increasing returns from growth in the size of plant. Such external economies are essentially spatial externalities, which may be defined generally as economic side-effects of proximity between economic actors. They can be either negative or positive, static or dynamic, pecuniary or technological. The static variety are reversible, whereas dynamic externalities are those associated with the technological advances, increased specialisation, and division of labour that accompanies and/or drives growth and development (Young, 1928). For the most part, regional scientists are interested in dynamic external economies, though this is not always explicitly stated. A static external economy enjoyed by a firm in a given industrial district might be the lower costs it enjoys for intermediate inputs because of proximity to its suppliers (for example, as a result of reduced shipping costs). This economy is also pecuniary and imposes no market failure because it is fully reflected in the price mechanism. There is certainly no role for government to encourage geographic clustering in this context. To the degree that such benefits outweigh any costs associated with agglomeration (congestion), enterprises will be inclined to cluster.

Of most relevance for understanding industry clusters are dynamic external economies associated with learning, innovation, and increased specialisation. Marshall illustrates the workings of (largely dynamic) external economies with reference to concentrated industrial districts, places where firms enjoy the benefits of large skilled pools of labour, greater opportunities for intensive specialisation (a finer social division of labour), and heightened diffusion of industry-specific knowledge and information (knowledge spillovers). Behind these dynamics is not just the size of the district alone, but social, cultural, and political factors, including trust, business customs, social ties, and other institutional considerations (Bellandi, 1989). Much of Marshall's analysis is relevant to Porter's (1990) discussion of firm structure, strategy, and rivalry as one of the four determinants of competitiveness (Peneder, 1995). In effect, Marshall provides some of the first hints as to how microlevel business relationships might influence regional growth and development. But Marshall also emphasises how important industrial districts are for small firms, which, through a social division of labour, may enjoy the same types of benefits large firms earn through internal scale.

Marshall's basic ideas are subjected to scrutiny and elaboration in the recent explosion of literature on new industrial districts (particularly the 'Third Italy') and small and medium-sized enterprises (see Asheim, 1996; Park, 1997; Park and Markusen, 1995). This literature, in turn, draws on theories of flexible specialisation (Asheim and Isaksen, 1997; Heidenreich, 1996; Isaksen, 1997), though the latter's focus on substantiating a basic sea change in the organisation of production is less important for understanding the specific microlevel relationships that link firm interdependence to geographic clustering. In other words, to understand why firms might cluster geographically, it is not necessary to demonstrate a general shift from the dominance of mass or 'Fordist' production methods to more flexible production regimes characterised by networks of smaller firms, a deeper social division of labour, and more cooperative business relations.

Of more importance are recent efforts to clarify the general relationships between scale and scope economies (for example, Bellandi, 1996), as well as the many case studies of particular industrial districts that identify not only basic economic trends in agglomerations of smaller firms, but also social and cultural behavioural codes that govern relationships between firms in these dynamic regions (see Humphrey, 1995 and related articles). The study of the 'social embeddedness' of economic transactions constitutes a principle contribution of the new industrial district literature (Harrison, 1992), and holds promise for making clearer the broad institutional factors Porter cites in his work.

Unfortunately, from an empirical perspective, drawing conclusions based on highly stylised case studies is difficult (Bellini, 1996). More attention to the task of assembling the many disparate studies and comparing findings is sorely needed. Such an effort would greatly facilitate more systematic theory-building.

3.3 Interdependence, agglomeration, and regional growth
What does economic and geographical clustering mean for explaining observed differential regional rates of growth and development? Porter suggests that a nation's competitiveness depends on its economic clusters, which themselves are likely to be spatially concentrated. But his is not a study of regional growth and change. Although many regional policymakers clearly perceive a linkage between the promotion of clusters and positive regional outcomes, the recent cluster literature offers few explicit clues as to what this linkage looks like or how it works. Instead, we must turn to a host of mainstream theories.

The discussion of economic interdependence began with a review of Perroux's (1950) conception of economic space and poles of growth. Here it is appropriate to start with the concept of growth centres, the regional extension of Perroux's nonspatial ideas. Though writing with only limited reference to Perroux's work, Hirschman (1958) offers one of the first justifications for a growth-centre strategy for underdeveloped areas. Indeed, Hirschman effectively provides a comprehensive study of economic interdependence (backward and forward linkages), geographic interdependence ('growth points'), and the implications for regional growth disparities ('trickling down of progress' and 'polarisation' effects). Granted, Hirschman (along with subsequent growth-centre advocates) is concerned with jump-starting stagnant underdeveloped areas with significant public directed capital investments in a few key sectors. Most of the current cluster debate is taking place in industrialised countries with already diverse economies and relatively strong effective demand (domestic and/or international). Also, most policy applications focus on the establishment of proper framework conditions for clusters to succeed, rather than strategic capital investments in favoured industries.

But at the same time, other observers have noted how commonplace it has become for local and regional agencies within industrialised countries to designate clusters for policy attention that are actually very poorly developed or that constitute the only viable industry in the given region [a designation of clusters motivated more by a limited choice set or by politics than by any economic justification (Held, 1996)]. In this context, given limited resources, the decision to concentrate policy attention on key industries—even if it means groups of related sectors—instead of focusing on basic infrastructural needs or other strategies that would serve best a broad array of industries cannot be taken lightly. And from a national policy perspective, a cluster-promotion strategy will benefit some regions over others. Those hurt by cluster policies might be peripheral areas subject to backwash effects from strong growth in neighbouring regions. A possible counterargument is that no one is advocating a significant diversion of development resources to a cluster-focused strategy. But if this is the case, all of this focus on clusters is unnecessary anyway.

The attempt to develop propulsive sectors with strong backward linkages that would start a process of cumulative regional advance was a hallmark of the growth-centre strategies of the 1960s and 1970s (Higgins, 1983). The history of failure in growth-centre policy is legendary. One of the most important reasons why such strategies often misfired is that too little attention was paid to the economic and social prerequisites that are necessary—at least as hypothesised in the vast theoretical literature—for growth centres to work (Malizia and Feser, 1998). Just as political and equity considerations often dictated, through a criterion of need rather than potential, the designation of very small and peripheral towns as growth centres, so it is the case that clusters identified in practice often bear little resemblance to Porter's (1990) ideal type.

Thus an important issue for industrial cluster applications is how cluster strategies can be developed while also addressing a traditional goal of regional policy: to reduce geographic disparities in income and employment. By some accounts, this is a new era of the city-state, marked by the ascendance of metropolitan regions as the relevant geographic unit for organising social, economic, and political life (Ohmae, 1995). Industry clusters are likely to be concentrated and fare best in prosperous and powerful regions. As industrial and locational advantage begets advantage through increasing returns, economic activity is further concentrated in select places. And, of course, the object of cluster policy is to leverage these advantages to promote positive synergies and returns. But what about peripheral and rural areas? Does the cumulative advance of some regions become the cumulative decline of others? In his classic criticism of neoclassical growth and trade theory's faith in the equilibrating force of labour and capital mobility, Myrdal (1957) argues that the backwash effects suffered by peripheral regions as a result of proximity to growth centres likely outweigh any countervailing spread effects. Although the spread–backwash debate has subsided to a degree, recent developments in growth and trade theory show some important consistencies with the early cumulative causation theories.

By replacing the standard neoclassical growth framework and its focus on factor accumulation and exogenous technological change with models that combine perfect competition at the level of the firm with industry-wide externalities, new growth theory opened the door for geography in mainstream economic models. The externalities highlight the role of human capital in generating long-run growth. Operationalised initially by Romer (1986), endogenous growth theory represents a technical refinement of Young's (1928) important thesis. Increasing returns yield growth models that predict perpetual growth. Building on Arrow's (1962) work, Romer identified knowledge accumulation (learning) as a form of investment that generates social externalities. According to Griliches (1992, page S29), the new approach emphasises two points: "(i) technical change is the result of conscious economic investments and explicit decisions by many different economic units and (ii) unless there are significant externalities, spillovers, or other sources of social increasing returns, it is unlikely that economic growth can proceed at a constant, undiminished rate into the future". One of the most important analysts of knowledge spillovers was Marshall (1890) in his study of industrial districts (Krugman, 1991).

As a practical matter, the focus in new growth theory on technology spillovers, information, and knowledge about process and product technologies as technological externalities, presented a solution to the problem of the incompatibility of increasing returns with perfect competition. The intuition behind the spillovers is that they permit the necessary productivity improvements that ensure continued investment. But importantly for regional analysis, the idea that financially uncompensated information exchanged between firms may be an important source of growth leads logically to the proposition that physical proximity is an important determinant of such externalities. This linkage between geographic proximity, externalities, and increasing returns brings mainstream economic growth theory closely into line with many regional models of growth and change, particularly those of the industrial-districts and agglomeration-economies literatures, and is also consistent with the industry-cluster perspective.

For example, in drawing a direct parallel between the new advances in growth theory and the role of cities in national and global economies, Lucas (1988) emphasised the linkage between proximity, externalities, and growth. In his survey of 'new regional economics' Glaeser (1994, page 13) writes: "Lucas followed [Jane] Jacobs and argued that when we are thinking about human capital, knowledge and growth, thinking about cities is almost inescapable. Ideas move quickly in cities.... Lucas brought into growth economics the idea that cities may be playing a major role in facilitating the accumulation of knowledge spillovers in the growth process." The geographically concentrated nature of knowledge spillovers has subsequently been the subject of empirical work based on patent data (Jaffe, 1989; Jaffe et al, 1993).

Lucas also noted the importance of grounding new growth theory by identifying spillover mechanisms outside the bounds of the generalised theoretical model: "The engine of growth in the [new growth] models... is *human capital*. Within the context of [these models], human capital is simply an unobservable magnitude or force, with certain assumed properties, that I have postulated in order to account for some observed features of aggregative behavior" (1988, page 35). He argued that, unless human capital is better defined, it "would make little difference if we simply re-named this force, say, the Protestant ethic or the Spirit of History or just 'factor X'" (1988, page 35). He charged growth theorists with coming up with a clear and testable explanation for such externalities. In this respect, the 'engine of growth' in new endogenous growth theory as 'factor X' is not unlike Weber's (1929) 'function of economy of agglomeration'; the latter is also something of a black box, an assumed behavioural phenomenon in a model serving other purposes.

Lucas's external effects of human capital have to do with influences economic actors have on the productivity of each other. The scope of such effects depends on the "ways various groups of people interact" (1988, page 37). In particular, these effects could be regarded either as global in nature or as purely localised at the level of family or firm. If this were the case, Lucas writes, "a model that incorporated internal human capital effects only plus other effects treated as exogenous technical change would be adequate". However, there is more likely some middle (geographic) ground, because individuals and firms both typically interact at a larger social scale, that is, the community or neighbourhood, city, and industrial district or complex. If knowledge spillovers are prevalent in industry clusters, then the linkage between clusters and long-run economic growth becomes clearer. The proof is in the identification of technological externalities in industry clusters.

Like the new growth theory, 'new international economics' also holds important implications for regional analysis. It is not that trade theory now admits a geographic dimension; trade theory has always been spatial theory (Krugman, 1991; Ohlin, 1933). Rather, the incorporation of increasing returns in models of trade implies the prospect of a highly concentrated geographic pattern of development (Krugman, 1990), including sustained disparities in regional income and employment. Again, the focus is on knowledge-related externalities as sources of increasing returns, particularly in advanced technology industries (Krugman, 1996). The process of cumulative advance in regions whose industries have established a competitive lead in given markets has been described as a 'lock-in effect' (Arthur, 1989; 1990a; 1990b). In principle, the initial lead may be as much a result of luck or historical accident as business acumen. But either way, particular 'locational clusters' may be able to establish a type of monopoly advantage over industries in other places. How likely or sustained such a process would be is an empirical matter. Nevertheless, because industry clusters, by definition, are not ubiquitous, industry-cluster policy would seem to imply at least acceptance of a potential worsening of regional economic disparities. Whether this is desirable is a matter of debate, and brings to the fore a host of other difficult questions, including the social costs of concentrated versus dispersed development and the inherent importance of place (the shoring up of perpetually lagging regions). Answers to these questions turn as much on issues of distribution as of efficiency.

4 Summary

Although the industry-cluster concept does not constitute a new self-contained model of regional development, it does represent a comprehensive description of how economic and geographic interdependences are integral to regional growth and development processes. But, although clusters provide a provocative and holistic perspective on interdependence, the research and policy activity on clusters to date has probably raised more questions than it has answered. Some of the most important of these are (1) the sources of technological externalities that drive increasing returns in industry clusters;

(2) the role of social and cultural versus economic factors in determining such externalities; (3) the role of proximity as an influence on externalities; (4) the prospects for leveraging technological externalities through policy interventions; and (5) the implications of spatially targeted development policy for the growth prospects of lagging regions. Undoubtedly there are others, but these are central to the cluster debate, particularly the question of policy efficacy. Most of them are also long-standing issues in regional analysis.

Answers to many of the difficult questions related to industry clusters will be possible only through continued careful and theoretically grounded empirical work. This invariably implies some (at least temporary) narrowing of terms and definitions, as well as the better integration of cluster ideas with traditional but still relevant regional theories and models. Failure to base industry-cluster initiatives on theory and empirical evidence increases the risk of ineffectual policy, wasted resources, and unintended consequences. But the empirical work must also strive for greater generalisability than has heretofore characterised the literature, or the industry-cluster concept faces the same problem identified by Bellini (1996, page 3) for the notion of industrial districts: without generalisation, "historic phenomena turn into concepts and one ends up talking about ideal-types, with a more and more vague connection with reality".

References
Arrow K J, 1962, "The economic implications of learning by doing" *Review of Economic Studies* **29** 155 – 173
Arthur B, 1989, "Competing technologies, increasing returns, and lock-in by historical events" *Economic Journal* **99** 116 – 131
Arthur B, 1990a, "Positive feedbacks in the economy" *Scientific American* February, 92 – 99
Arthur B, 1990b, "'Silicon Valley' locational clusters: when do increasing returns imply monopoly?" *Mathematical Social Sciences* **19** 235 – 251
Asheim B T, 1996, "Industrial districts as 'learning regions': a condition for prosperity" *European Planning Studies* **4** 379 – 400
Asheim B T, Isaksen A, 1997, "Location, agglomeration, and innovation: towards regional innovation systems in Norway?" *European Planning Studies* **5** 299 – 330
Bellandi M, 1989, "The industrial district in Marshall", in *Small Firms and Industrial Districts in Italy* Eds E Goodman, J Bamford (Routledge, London)
Bellandi M, 1996, "On entrepreneurship, region, and the constitution of scale and scope economies" *European Planning Studies* **4** 421 – 438
Bellini N, 1996, "Italian industrial districts: evolution and change" *European Planning Studies* **4** 3 – 4
Bergman E M, Feser E J, Sweeney S H, 1996 *Targeting North Carolina Manufacturing; Understanding the State's Economy Through Industrial Cluster Analysis* (UNC Institute for Economic Development, Chapel Hill, NC)
Best M, 1990 *The New Competition: Institutions of Industrial Restructuring* (Polity Press, Cambridge)
Boekholt P, 1997, "The public sector at arms length or in charge? Towards a typology of cluster policies", paper presented at OECD Workshop on Cluster Analysis and Cluster Policies, Amsterdam, 9 – 10 October; direct inquiries to Dr. T J A Roelandt, Ministry of Economic Affairs, Research Unit, Economic Policy Directorate, PO Box 20101, 2500 EC, The Hague

Carlino G A, 1978 *Economies of Scale in Manufacturing Location* (Martinus Nijhoff, Boston, MA)
Chinitz B, 1961, "Contrasts in agglomeration: New York and Pittsburgh" *American Economic Review* **51** 279 – 289
Dahmén E, 1984, "Schumpeterian dynamics" *Journal of Economic Behavior and Organizations* **5** 25 – 34
Dahmén E, 1988, "'Development blocks' in industrial economics" *Scandinavian Economic History Review* **36** 3 – 14
Darwent D, 1969, "Growth poles and growth centres in regional planning: a review" *Environment and Planning* **1** 5 – 31
DeBresson C (Ed.), 1996 *Economic Interdependence and Innovative Activity* (Edward Elgar, Cheltenham, Glos)
DeBresson C, Hu X, 1997, "Techniques to identify innovative clusters: a method and 8 instruments", paper presented at OECD Workshop on Cluster Analysis and Cluster Policies, Amsterdam, 9 – 10 October; see Boekholt, 1997
Feser E J, 1998, "Enterprises, external economies, and economic development" *Journal of Planning Literature* **12** 283 – 302
Gilmour J M, 1974, "External economies of scale, inter-industrial linkages and decision making in manufacturing", in *Spatial Perspectives on Industrial Organization and Decision-Making* Ed. F E I Hamilton (John Wiley, Chichester, Sussex) pp 335 – 362
Glaeser E L, 1994, "Cities, information, and economic growth" *Cityscape* **1** 9 – 47
Griliches Z, 1992, "The search for R&D spillovers" *Scandinavian Journal of Economics* **94** (supplement) S29 – S47
Guthrie J A, 1955, "Economies of scale and regional development" *Papers and Proceedings of the Regional Science Association* **1** J1 – J10
Harrison B, 1992, "Industrial districts: old wine in new bottles?" *Regional Studies* **26** 469 – 483
Hassink R, 1997, "Localized industrial learning and innovation policies" *European Planning Studies* **5** 279 – 282
Heidenreich M, 1996, "Beyond flexible specialization: the rearrangement of regional production orders in Emilia-Romagna and Baden-Württemberg" *European Planning Studies* **4** 401 – 419
Held J R, 1996, "Clusters as an economic development tool: beyond the pitfalls" *Economic Development Quarterly* **10** 249 – 261
Helper S R, 1991, "Strategy and irreversibility in supplier relations: the case of the U.S. automobile industry" *Business History Review* **65** 781 – 824
Higgins B, 1983, "From growth poles to systems of interactions in space" *Growth and Change* **14** 3 – 13
Hirschman A O, 1958 *The Strategy of Economic Development* (Yale University Press, New Haven, CT)
Hoover E M, 1937 *Location Theory and the Shoe and Leather Industries* (Harvard University Press, Cambridge, MA)
Humphrey J, 1995, "Introduction" *World Development* **23** 1 – 7
Imrie R, Morris J, 1992, "A review of recent changes in buyer-supplier relations" *Omega* **20** 641 – 652
Isaksen A, 1997, "Regional clusters and competitiveness: the Norwegian case" *European Planning Studies* **5** 65 – 76
Isard W, 1956 *Location and Space Economy* (John Wiley, New York)
Jacobs D, de Jong M W, 1992, "Industrial clusters and the competitiveness of the Netherlands" *De Economist* **140** 233 – 252
Jacobs D, de Man A-P, 1996, "Clusters, industrial policy and firm strategy: a menu approach" *Technology Analysis and Strategic Management* **8** 425 – 437

Jaffe A B, 1989, "Real effects of academic research" *American Economic Review* **79** 957 – 970
Jaffe A B, Trajtenberg M, Henderson R, 1993, "Geographic localization of knowledge spillovers as evidenced by patent citations" *Quarterly Journal of Economics* **108** 577 – 598
Kaufman A, Gittell R, Merenda M, Naumes W, Wood C, 1994, "Porter's model for geographic competitive advantage: the case of New Hampshire" *Economic Development Quarterly* **8** 43 – 66
Kirkpatrick D A, Gavaghan K, 1996 *North Carolina Environmental Business Study* prepared by Kirkworks for the North Carolina Environmental Technologies Consortium; http://www.ncacts.state.nc.us/NCAC/Ts
Klier T H, 1994, "The impact of lean manufacturing on sourcing relationships" *Economic Perspectives* **18** (4) 8 – 18
Krugman P, 1990 *Rethinking International Trade* (MIT Press, Cambridge, MA)
Krugman P, 1991 *Geography and Trade* (MIT Press, Cambridge, MA)
Krugman P, 1996 *Pop Internationalism* (MIT Press, Cambridge, MA)
Lagendijk A, Charles D, 1997, "Clustering as new growth strategy for regional economies? A discussion of new forms of regional industrial policy in the UK", paper presented at OECD Workshop on Cluster Analysis and Cluster Policies, Amsterdam, 9 – 10 October; see Boekholt, 1997
Lichtenberg R M, 1960 *One-tenth of a Nation* (Harvard University Press, Cambridge, MA)
Lucas R E, Jr, 1988, "On the mechanics of economic development" *Journal of Monetary Economics* **22** 3 – 42
Lundvall B, 1996, "National systems of innovation and input – output analysis", in *Economic Interdependence and Innovative Activity* Ed. C DeBresson (Edward Elgar, Cheltenham, Glos) pp 356 – 363
Malizia E E, Feser E J, 1998 *Understanding Local Economic Development* (Rutgers University Press, Brunswick, NJ) forthcoming
Malmberg A, Maskell P, 1997, "Towards an explanation of regional specialization and industry agglomeration" *European Planning Studies* **5** 25 – 41
Marceau J, 1997, "The disappearing trick: clusters in the Australian economy", paper presented at OECD Workshop on Cluster Analysis and Cluster Policies, Amsterdam, 9 – 10 October; see Boekholt, 1997
Marshall A, 1890 *Principles of Economics: An Introductory Volume* 9th edition published in 1961 (Macmillan, London)
Martin R, Sunley P, 1996, "Paul Krugman's geographical economics and its implications for regional development theory: a critical assessment" *Economic Geography* **72** 259 – 292
Myrdal G, 1957 *Economic Theory and Underdeveloped Regions* (Harper and Row, New York)
Nelson R R, 1988, "Institutions supporting technical change in the United States", in *Technical Change and Economic Theory* Eds G Dosi, C Freeman, R Nelson, G Silverberg, L Sote (Frances Pinter, London) pp 312 – 329
Nelson R R, 1993 *National Innovation Systems: A Comparative Study* (Oxford University Press, New York)
Nourse H O, 1968 *Regional Economics: A Study of the Economic Structure, Stability, and Growth of Regions* (McGraw Hill, New York)
Ohlin B, 1933 *Interregional and International Trade* (Harvard University Press, Cambridge, MA)
Ohmae K, 1995 *The End of the Nation State: The Rise of Regional Economies* (Free Press, New York)

Park S O, 1997, "Dynamics of new industrial districts and regional economic development", paper presented at the International Symposium on Industrial Park Development and Management, Taipei; copy obtainable from the Graduate Institute of Building and Planning, National Taiwan University, Taipei

Park S O, Markusen A, 1995, "Generalizing new industrial districts: a theoretical agenda and an application from a non-Western economy" *Environment and Planning A* **27** 81–104

Peneder M, 1995, "Cluster techniques as a method to analyze industrial competitiveness" *International Advances in Economic Research* **1** 295–303

Peneder M, Warta K, 1997, "Cluster analysis and cluster oriented policies in Austria", paper presented at OECD Workshop on Cluster Analysis and Cluster Policies, Amsterdam, 9–10 October; see Boekholt, 1997

Perroux F, 1950, "Economic space: theory and applications" *Quarterly Journal of Economics* **64** 89–104

Perroux F, 1988, "The pole of development's new place in a general theory of economic activity", in *Regional Economic Development: Essays in Honour Francois Perroux* Eds B Higgins, D J Savoie (Unwin Hyman, Boston, MA) pp 48–76

Porter M E, 1990 *The Competitive Advantage of Nations* (Free Press, New York)

Robinson E A G, 1931 *The Structure of Competitive Industry* (James Nesbit, Digswell Place, Welwyn, Herts)

Roelandt T, den Hertog P, van Sinderen J, Vollaard B, 1997, "Cluster analysis and cluster policy in the Netherlands", paper presented at OECD Workshop on Cluster Analysis and Cluster Policies, Amsterdam, 9–10 October; see Boekholt, 1997

Romer P M, 1986, "Increasing returns and long-run growth" *Journal of Political Economy* **94** 1002–1037

Rosenfeld S A, 1995 *Industrial Strength Strategies: Business Clusters and Public Policy* (Aspen Institute, Washington, DC)

Rosenfeld S A, 1996 *Overachievers: Business Clusters that Work* Regional Technology Strategies, Inc., Chapel Hill, NC

Rosenfeld S A, 1997, "Bringing business clusters into the mainstream of economic development" *European Planning Studies* **5** 3–23

Rouvinen P, Ylä-Anttila P, 1997, "A few notes on Finnish cluster studies", paper presented at the OECD Workshop on Cluster Analysis and Cluster Policies, Amsterdam, 9–10 October; see Boekholt, 1997

Schumpeter J, 1934 *The Theory of Economic Development* (Harvard University Press, Cambridge, MA)

Sternberg E, 1991, "The sectoral cluster in economic development policy: lessons from Rochester and Buffalo, New York" *Economic Development Quarterly* **5** 342–356

van der Laan H B M, 1997, "Everything you always wanted to know about clusters, but were afraid to ask", opening address, OECD Workshop on Cluster Analysis and Cluster Policies, Amsterdam, 9–10 October; see Boekholt, 1997

Vernon R, 1960 *Metropolis* (Harvard University Press, Cambridge, MA)

Weber A, 1929 *Theory of the Location of Industries* translated by C J Friedrich (University of Chicago Press, Chicago, IL)

Young A, 1928, "Increasing returns and economic progress" *Economic Journal* **38** 527–542

On Geography and Technology: Proximity Relations in Localised Innovation Networks

A Rallet
University of Bourgogne and IRIS, University of Paris-Dauphine
A Torre
Institut National de la Recherche Agronomique, Paris

1 Introduction
The concern for the relation between geography and technology is nowadays one of the major focal points in spatial economics and has given rise to a number of valuable areas of research. During the 1980s and 1990s, many studies have been devoted to such topics as innovative milieux (Ratti et al, 1997), technological districts, technopoles and science parks and, more generally, to localised systems of production and innovation. More recently, certain work in the field of economic geography stressed the spatial consequences of technological spillovers and proposed new measures concerning the relation between academic research and location of R&D expenditure at the level of the firm (Audretsch and Feldman, 1996).

Work performed in the so-called domain of the geography of innovation stressed the role played by geographical proximity [1] in the process of technological transfer, be it between private firms or public bodies. This work is based on the idea that geography provides organisation for the diverse types of knowledge needed for production and commercialisation, but also on the assumption that "knowledge transverses corridors and streets more easily than continents and oceans" (Feldman, 1994, page 4). Whereas information may be transmitted across distances, the transfer of knowledge needs communication and repeated interaction. It is assumed that this process of trial and feedback is facilitated by face-to-face interaction, which permits reciprocal exchanges, negotiations, and deep communication during the complex process of innovation. In most of these papers innovation is regarded as a cognitive process, which implies, in particular during its early stages, a great uncertainty requiring the building of common codes. The spatial dimension occurs here, for it is assumed that this process is enhanced by face-to-face interaction and thus by geographical proximity.

This assumption is supported by theoretical arguments borrowed from "the economics of information and knowledge" (Foray and Lundvall, 1996, page 11). As a result, proximity and location matter because of the specific nature of exchange between agents involved in R&D or innovative activities.

---

[1] The term 'geographical proximity' can cause confusion as distance is a relative notion and can be measured in different ways. In this chapter, a conventional definition of 'geographical proximity' is adopted: economic agents or individuals are considered geographically close when they can have daily face-to-face relations.

Indeed, if information and knowledge are mainly what is exchanged between these agents they are not public goods freely diffused in the economy as suggested by Arrow (1962). More precisely a crucial distinction has to be made between two kinds of knowledge, namely tacit and codified knowledge (Polanyi, 1966). Usually, "tacit knowledge refers to knowledge which cannot be easily transferred because it has not been stated in an explicit form" (Foray and Lundvall, 1996, page 11) whereas codified knowledge—or 'information'—is reduced to messages which can be easily transferred between economic agents through nonhuman supports. It is assumed, then, that codified knowledge can be exchanged regardless of distance by using technologies of communication, be they old (postal mail) or new (electronic mail, computer conferencing). At the opposite end, the transfer of tacit knowledge requires the sharing of a common work experience through face-to-face relations. As a consequence, geographical proximity appears to be a necessary condition for the efficient sharing of knowledge, especially in the case of tacit-knowledge-intensive activities such as innovation creation and diffusion.

In this chapter we intend to begin to discuss this argument, which can be regarded as the conventional explanation of the high degree of concentration of innovative activities. This major theoretical assumption—because of the need to transfer tacit knowledge—has to be put under closer examination. We follow two main directions.

In the next section, we ask if advances in information and communication technologies (ICTs) change the need for geographical proximity between knowledge users. It is claimed that recent advances in ICT alleviate this constraint to some extent even if it remains rather strong. More important is the argument that there are supporters of tacit knowledge exchange other than permanent location at the same place. Another kind of proximity—which we call organisational proximity—combined with the mobility of human resource is an alternative basis for knowledge exchange that no longer requires permanent colocation.

In the third section we enhance this thesis by examining the role of geographical proximity in networks of innovation. Localised networks of innovation are usually regarded as an efficient framework of transfer and innovation diffusion. This is why local development policies often focus on the networking of local producers and users of technology (firms, universities, public research laboratories, etc). In this section, lessons are drawn from case studies of localised networks of innovation supported by public institutions in three French regions (Corsica, Aquitaine, Rhône–Alpes). The conclusion is the same as that in the second section: organisational proximity appears to be a stronger supporter of technology transfer and innovation diffusion than does geographical proximity. As a conclusion, new directions for local development policies are suggested.

2 Coordination mechanisms in innovative and research activities, information and communication technologies, and the geographical proximity constraint

2.1 The traditional thesis: ICT does not call into question the need to be closely located
Let us remind ourselves of the arguments which justify the role played by geographical proximity in the joint development of innovative and research activities:
(a) these activities are characterised by the important weight of tacit knowledge,
(b) the more tacit knowledge is, the more face-to-face relations are necessary,
(c) the higher the frequency of face-to-face relations, the greater the need for permanent physical proximity.

Two of these criteria (the weight of tacit knowledge and the frequency of face-to-face relations) vary according to the nature of the activities and the stage of their development. Most studies show that the need for close location is stronger in the early stages of the development process of research activities because of the importance of tacit knowledge during these stages.

The question is whether ICT changes this situation. It is usually suggested that ICT does not basically modify the need for geographical proximity. According to this point of view, ICT is only increasing long-distance exchanges of codified knowledge. From a remote location, it is easy now to be connected to databases, to read technical instructions or working papers, or to send texts, data, or pictures. Consequently, it is expected that ICT will increase the scope of remote coordination insofar as it is based mainly on exchange of codified knowledge. But as research activities also involve intensive exchanges of tacit knowledge, the geographical proximity constraint remains very strong.

However, it could be argued that ICT increases the possibilities of remote coordination insofar as it is a powerful means to turn tacit knowledge into codified knowledge (for instance, conversion of tacit knowledge into expert systems and know-how databases, storage of organisational knowledge on CD-ROM, and automation of routines by means of workflow software). If such were the case, the geographical proximity constraint would become less and less strong.

It should therefore be supposed that ICT can gradually reduce tacit knowledge considered as a given stock. However, that is impossible for four reasons:
(1) The process of coding knowledge involves a cost which is an increasing function of the tacit degree of the knowledge. It is often more efficient and less expensive to rely on tacit knowledge exchanges than to codify knowledge in order to transfer it easily [compare the limits of expert systems (Hatchuel and Weil, 1995)].
(2) Advances in science and technology constantly rebuild new tacit knowledge. The development of science and technology takes the form of emergent knowledge which cannot be immediately codified. The domination of tacit knowledge in the early stages of development explains why knowledge cannot be easily transferred from one individual to another or from one team to another.

That is why invention and innovation are so concentrated in some places.
(3) Tacit and codified knowledge are complementary. As Nonaka (1994) underlines, the transmission of codified knowledge supposes the use and sharing of common tacit knowledge. Conversely, the transfer of tacit knowledge is based on the use of codified knowledge.
(4) The use of ICT tools requires the sharing of common codes and practices of communication which are tacit. This is why the tools of remote communication are especially used by individuals who meet frequently.
For all these reasons, tacit knowledge will always be used in research and innovative activities. Consequently, face-to-face relations prove to be necessary to these kinds of activity. The geographical proximity constraint thus remains strong.

Conclusion: One has to expect an extension of the geographical scale of the coordination process in research and innovative activities thanks to the possibility of remotely sharing codified knowledge. But the development of ICT does not basically modify the need for face-to-face relations because of the important weight of tacit knowledge in these activities. Geographical proximity remains a necessary and important tool of coordination.

2.2 Geographical proximity as a relative and less and less strong constraint

Two arguments moderate the above thesis. The first underlines the possibility of satisfying the need for physical proximity by the temporary mobility of individuals (that is, travel) and not by permanent colocation. The second stresses another kind of proximity (organisational proximity) which allows the sharing of tacit knowledge between remote locations.

The need for a face-to-face relation to exchange tacit knowledge does not imply that individuals are closely located. It implies only that individuals meet often. In certain circumstances, the problem can be solved by the mobility of individuals. This is the case when the frequency of tacit knowledge exchange is not very high. Then individuals can move when this exchange must be carried out. Their locations continue to be determined by other factors: proximity of production centres, marketplaces, specific scientific or technological resources, or historical and institutional factors. This case is frequent: to design and to develop a product, firms constitute project-oriented task forces based on teams gathered temporarily together and belonging to different plants. Individuals of the task force meet at the beginning of the process and then only at defined moments (to make a synthesis, to pass on to a new stage, to redefine the project).

Alternation between moments of proximity coordination and moments of remote coordination is supported by the decrease in transport costs and the development of high-speed means of transport. Temporary mobility thus appears as an effective way to coordinate individuals who have to share tacit knowledge.

Geographical proximity is not the only kind of proximity which makes it possible to share and exchange tacit knowledge. There is also a kind of proximity created by membership of the same organisation or professional community, which we will call *organisational proximity*. Organisations are characterised by collective value systems and representations of the world ('corporate culture') which tend to homogenise individual behaviours in given situations. They develop in the same way as a homogeneous technical culture, that is, common ways to think and solve production problems. This collective and technical culture guarantees that employees will spontaneously give the same interpretation to exchanged data or text, even if they are located in different places.

One can even suggest that organisational proximity is a much more effective supporter of tacit knowledge exchange than is geographical proximity. Indeed, it is well known that individuals can be closely located and nevertheless behave like foreigners. Geographical proximity is effective only if it coincides with the existence of organisational relationships. Whereas at the opposite extreme, one can imagine individuals sharing common tacit knowledge without being physically close.

Geographical proximity is not the only supporter of coordination, especially for research and innovative activities. This argument is strengthened by the use of ICT to coordinate individuals and teams. As we saw, the traditional thesis claims that ICT support codified knowledge exchange. For this reason, ICT considerably widened the potential scope for remote cooperative work or activities (search for new partners, greater access to knowledge databases, teleconferencing, codification of cooperative work procedures, etc). But, on the other hand, ICT is supposed to have a weak impact upon the exchange of tacit knowledge (unless it can be codified by ICT). The need for tacit knowledge exchange continues to lock the door against extensive remote cooperation.

But this argument does not take into account one of the most important changes brought about by ICT during these last few years, that is, its ability to support exchanges of tacit knowledge. However, we have to be careful with this assertion. We must keep in mind that ICT is not a simple substitute for in-person contact. ICT and especially computer-mediated communication are characterised by limited social presence. Indeed, social presence cannot be easily recreated by ICT. Many studies have shown this by comparing face-to-face relationships with mediated communications in laboratory simulations (the psychobehaviourist approach) or by analysing these two situations within the framework of an ethnological approach (on this subject, see the surveys by Cardon and Licoppe, 1997; Garton and Wellman, 1996; Wellman et al, 1996). Face-to-face meetings and computer-mediated communication are never equivalent. As a result, the need for geographical proximity cannot be totally eliminated by the use of ICT.

Nevertheless, ICT can be used to support the exchange of tacit knowledge and informal relationships. For instance, the practice of computer conferencing or e-mail does not replace face-to-face meetings but creates new kinds of social contact and even interpersonal relationships between persons who are physically distant. ICT increases access to new people, provides individuals with new opportunities of contact, and facilitates social networking by weakening social, spatial, and temporal barriers (see the use of e-mail, newsgroups, forums, discussion lists, etc, on the Internet). It is no longer possible to differentiate between ICT and formalised communication as was usual before the development of Internet. Other examples of ICT supporting the exchange of tacit knowledge can be quoted. For instance, ICT generates redundant information which is generally presented as one of the main advantages of geographical proximity because it provides the capacity to build up social ties as bridges between informal sources of information. Some ICT tools such as hypertext are based on cognitive processes similar to those which characterise tacit knowledge, for instance, the use of metaphors (see Nonaka, 1994) or the analogical way of reasoning.

Consequently ICT is used not only to support strong ties by codified relationships but also to support weak ties by informal interactions. We know the important role of weak ties in setting up, regulating, and widening social networks and professional communities in the field of research and innovation. So ICT raises the capacity to develop new ways of tacit knowledge exchange between physically distant individuals or teams. The possibilities of coordination through space are improved thanks to this capacity. This could be one of the major impacts of ICT on location patterns.

However, it must be repeated that ICT does not eliminate the need for face-to-face meetings. It generates a dynamic complementarity between face-to-face meetings and distant coordination. It is well known that the development of distant coordination by means of ICT increases incentives for people to travel in order to have face-to-face meetings. This rule is particularly true in the field of research and innovation: in many cases of telecooperation, airplane tickets are the main item of the team's budget. ICT thus reinforces the probability for coordination to be supported in an alternative way by mobility and distant coordination. In this perspective, the crucial location factor for individuals or firms engaged in cooperation is not to be physically close to partners but to be located close to high-speed transport infrastructure which allows them to meet when ever needed. This is what our case studies show us.

2.3 A few lessons drawn from case studies

Two case studies are related to development and research projects. The first is a graphic data-processing company (Silicon Graphics) whose R&D centres are distributed on five world sites. The purpose of their cooperation is to conceive and develop graphic animation software. The second case study is the design and development of a videoconferencing system by CNET (National Centre of Studies for Telecommunications, France). This project needs the cooperation of four research centres located at different sites. A third case study is on the development by a Corsican Studies Centre (CIRVAL) of an expert database on specific agricultural products. The database is fed and consulted by research and study centres located around the Mediterranean Basin. Fourth, a questionnaire was addressed to academic researchers and teachers at two French universities, the University of Bordeaux I (physics and chemistry) and the University of Bordeaux II (biology and medicine). The questionnaire focused on the communication practices of academic people in the framework of their research projects.

Some conclusions can be drawn from these four case studies.

1. *The geographical constraint of proximity is especially strong for research projects carried out within the university community.*
The need for frequent interaction is important throughout the whole process, not only for specific stages such as literature searches, the definition of a common framework, or the conclusion of the process, but also for the implementation stage for which the solution of a short, medium, or long-term stay at the same workplace is often used. This is because of the importance of tacit knowledge used through the frequent mutual adjustments between researchers at all the stages of research projects. The high weight of tacit knowledge can of course be explained by the importance of basic research in these projects but also by the organisational characteristics of academic communities. The weak division of labour which characterises them leads to many overlapping tasks and as a result to the need for partners to carry out frequent mutual adjustment over all the project. This need is reinforced by the absence of a strong authority able to solve the problems of coordination. Whether they are important or not, these problems must be regulated by a direct and consensual dialogue between researchers.

ICT does not basically change this situation. The need for frequent mutual adjustment explains why communication is mainly supported by the use of 'rich media' such as face-to-face meetings, telephone, fax, electronic mail, and electronic forums. 'Poor media' (that is, those which imply formalised and codified relationships) are hardly used. Such is the case of groupware tools whose diffusion is strongly limited by the need to formalise the organisational framework of cooperation.

2. *The more informal the organisation of the project, the more difficult the remote coordination.*
The CIRVAL example shows the difficulty in cooperating remotely through the sharing of knowledge databases within rather informal communities. The reciprocal and decentralised basis of the Internet—I put on the network information in exchange for other information I can find there (network externalities)—is adapted to information and knowledge which is already compiled and available. Difficulties appear when the network is used as a decentralised means of knowledge production. Such an attempt immediately highlights organisational problems. Who are the actors who will produce knowledge for the network? What are the incentives to do it? Is there a sufficiently strong common interest to prevent free-rider behaviour?

These problems are not technical but organisational. They are hard to solve when the community concerned in the network is not well organised. This is the situation in the CIRVAL project.

3. *In the case of R&D projects within firms, the need for geographical proximity is relative and can mainly be satisfied by periodic meetings.*

When research projects are developed within organisations characterised by well-defined objectives and strong central authority, geographical proximity is necessary only for specific and limited stages (the case of both Silicon Graphic and CNET). It is especially needed for the launch of projects. In the upstream stage of a project, teams engage in brainstorming. During this stage, they are occupied in confronting arguments, convincing others, and finally converging towards the same position. At this stage, face-to-face meetings are required because consensus is obtained much more quickly than through remote coordination even supported by ICT. Remote coordination by telephone can be used during this stage but in particular to discuss points of view related to a precise point within the framework of a bilateral relationship and not for long and multilateral discussions. Videoconferencing is more appropriate for technical meetings but cannot replace face-to-face meetings for complex discussions.

The face-to-face constraint is much less strong during the technical development stages. During these periods, tasks are defined and distributed prior to their being carried out, so that remote coordination becomes easier to manage. Adjustments between individuals or teams can be performed through remote coordination by using in a complementary way the whole range of ICT tools, from telephone to specific cooperative software (groupware) and by travel when an important difficulty must be solved.

In conclusion, the case studies show that:

(a) The need for being closely located remains strong for specific stages of innovation and research activities. This is the result of some characteristics of these activities, namely the importance of tacit knowledge whose exchange implies frequent face-to-face contacts between partners but the weight of this constraint also depends on organisational characteristics. When work is divided into precise tasks, when the coordination of these tasks is under the control of a central authority, and when partners share the same cognitive maps, the possibility of remote coordination increases.

(b) The need for face-to-face contacts is not permanent and can be satisfied by periodic travel and short stays combined with the use of ICT to transmit codified knowledge and to develop new kinds of social networking.

Geographical proximity always plays a role, but some of the needs for physical proximity can be satisfied more and more by the mobility of people and the use of ICT. One result is that functional needs for coordination are no longer a sufficient explanation of the high degree of geographical concentration of innovation and research activities.

3 The place of local networks in the process of technological development

We have seen that localised actors do not always need to be closely located to take part in a process of innovation and that organisational proximity is as important as geographical proximity. This result was obtained by starting from the analysis of the role of information and communication technologies. One obtains an identical conclusion by analysing another geographical dimension of the innovation process, namely the role of local networks in the diffusion of knowledge and technologies. In this section, we will try to assess local technological policies which aim to support the diffusion of knowledge and technology by easing contacts between geographically close actors. The weight of geographical proximity will be relativised there too.

3.1 From the importance of local networks of innovation to institutional support for these networks

It is nowadays widely assumed that local networks play a major role in regional economic and technological development. This idea gives birth in the economic literature to an increasing use of concepts such as local networks, localised systems of production, local systems of innovation, etc. All these concepts rest on the importance of geographical proximity relations in the network setting of the actors in the innovation.

Because they are convinced of the importance of these local networks, public actors develop regional technological policies directed towards the support or installation of collective processes of research and innovation. These policies, which have as their objective the sharing of knowledge or competencies within a local framework, from now on supplement traditional policies dealing with material infrastructure. The development of collective networks of actors is thus supported by local or regional institutions. One must, however, wonder about the relevance and the limits of these local technological policies. Their implementation reveals several obstacles that we will examine. The major one relates to the connection between spontaneously created local networks and local networks developed by institutions.

Spontaneous local networks are a grouping of local actors around one or several joint economic projects, according to a nonmarket form of organisation. The links are generally not based on contracts or completely explicit agreements, but rather on processes of cooperation or collective learning. Their main purpose is a common interest in the production of a good, in sharing a technique, or in the search for information needed by all the members. The exchanges mainly relate to the transfer or sharing of knowledge and are made through trust relations.

Institutional local networks correspond to structures settled by public bodies in order to support the firms. They concern flexible organisation forms, founded on a common acceptance of the rules which engage the participants, among which one can make a distinction between the producers and the users of information and technological knowledge. The link between the participants in the network is materialised by an adherence to as well as a utilisation of the services offered by an organising cell which also plays a part in animating the whole network. One can find general or specialist networks.

There is a difference between technological policies according to whether spontaneous local networks already exist. When they do not exist or are poorly developed, the policy aims to promote them, even to create them, by means of incentives or voluntarist policies. When they already exist, the objective is to support their development by, in particular, supporting transverse cooperation between partners belonging to different worlds (industry, research, universities, technical centres, etc). In both cases, the objective is to connect a spontaneous network of economic actors and an institutional network promoted by the local authorities. The relations between these two categories of networks are illustrated in the following examples, based on the French experience.

3.2 Two examples of regional technological policies based on institutional networks
3.2.1 *Corsica*

The first example is that of a region, Corsica, where spontaneous local networks are poorly structured.

In Corsica, the objective of the public authorities was to set up an institutional network to support the diffusion of knowledge and technology. In French regions there have been for twenty years so-called networks of technological diffusion (NTDs) whose mission is to diffuse innovation and to support technology transfer. Their main objective is to help small and medium-sized enterprises (SMEs) solve their innovation problems, whether related to the internal organisation of the firms or to their relations with external partners (laboratories, universities, other firms, or public bodies).

The network of innovation of the Corsican region is most of all characterised by cooperation between local and 'continental' firms. The connections between local firms lack coherence because of the very narrow local market and the weak development of science–industry relations. This weakness of local interactions shows that the search for competencies is the most important factor of interfirm cooperation. In this case such a search is done outside the region when competencies do not exist or are very weak at the local level. Geographical proximity is not a sufficient condition for the existence of a system of innovating enterprises as local firms are obliged to seek competencies outside the region. In this context, the support brought by the public institutions for the local firms is a major one and can take the form of government aid for the development of innovative firms. But institutional support can further try to support the formation of a local network of innovators. This is the objective of the NTD.

Created in September 1995, the Corsican NTD was founded on the following idea: not enough local firms have access to available technological competencies and use the device of innovation support. Its creation follows pressure from by the Local Authority of Corsica (Collectivité Territoriale de Corse) with the purpose of pushing NTD as one of the major pieces of strategy of technological development. The Corsican NTD was created to make it possible for SMEs to reach external innovation and technology competencies, to mobilise and regroup local actors, but also to match their actions and to promote consultation. It aims to improve the efficiency of the regional devices of development aid and support, by easing collaboration and exchange between the various operators in order to obtain synergetic effects. To fulfill its mission, it is thus based on 'the network effect' as well as on the formation of human resource and on the grant from the technological network service, intended to encourage the firms to launch technological innovation.

3.2.2 Aquitaine and Rhône–Alpes

The case of the Aquitaine and Rhône–Alpes regions is different because strongly structured spontaneous networks already exist. Consequently, the role of the institutions is less generic and answers the need for supporting specific projects or poles, in this case biological and medical poles (BMP).

The Aquitaine region is characterised by an old system of high and average technology industries such as pharmaceuticals and medical equipment goods (surgery and medical imagery). Concurrently with this industrial pole, there are solid scientific and technological competencies in the fields of health and life sciences. But these competencies are insufficiently developed and not articulated enough with the industry. In addition to the development of the infrastructure and human resources of scientific research, the industrial and academic actors thus sought to cooperate to facilitate technology transfer. However, the setting of these networks appeared insufficient and was often poorly connected with local industrial competencies.

The innovation network of the Rhône–Alpes region occupies a foreground position in activities related to sectors of health. At the end of the 1970s, the region was characterised by a strong presence of large pharmaceutical companies but also of firms specialising in medical goods or medical engineering activities. As regards hospitals and R&D, the region had an international reputation but cooperation between the local actors was considered insufficient. Relations between firms, hospitals and researchers were limited to traditional fields such as the drug industry. Construction of inter-relations appeared essential in emergent sectors such as activities related to biomedical technologies. But the formation of these relations was complex because of the heterogeneity of the sector; thus the networks could not be led by industrial actors alone.

BMP poles of the two regions (created in 1979 in Rhône–Alpes and in 1987 in Aquitaine) are specialised networks, organised around techniques and specific products. Constituted, directed, and coordinated by a local institution, they are intended to support the creation of a local technological milieu. Their actions consist in encouraging relations between research, industry, and public authorities in order to support innovation and encouraging the creation of internationally dynamic market-oriented firms.

These poles were in the beginning intended to intervene upstream of the chain of innovation, that is, to support relations between public scientific laboratories and firms. But they quickly changed and instead became providers of services in response to the needs expressed by local industries. Their intervention is now downstream. In 1993 the Rhône–Alpes pole was not considered to be efficient enough: the projects were associated mainly with public laboratories and not industries. The region then created an Agency for Biomedical Technologies (ARTEB) which gives the priority to industrial firms. The BMP Aquitaine followed the same evolutionary course, as revealed by the formation of a Health Strategic Organisation whose objective is to shorten the delays between the R&D process and industrial application. This reorientation testifies to the desire to listen carefully to the needs of the industrial leaders.

3.3 From the initial objectives of the policies to the actual local networks
It is worth emphasising the fact that, in the preceding experiments, the policies do not always achieve the stated goals. In particular, one can wonder whether institutional networks, which promote geographical proximity while considering that it supports the process of technological development, are able to encourage synergetic effects at the local level. From this point of view, the gap between the initial objectives and the actual networks shows the limits of the power of geographical proximity.

The investigations carried out in the three regions indeed show that there are few links between spontaneous and institutional networks and that, very often, the objectives originally defined by the public authorities have not been reached or have changed on the way.

The Corsican NTD primarily gathers together the regional public actors who intervene in the field of technology. After two years of existence, it succeeded particularly in coordinating the practices and research operations better between these partners. Their image, their competencies, and their fields of intervention also became more readable for local economic actors. However, several firms remain outside this institutional network even if this is not caused by a lack of information. The institutional network is poorly articulated with the private actor networks because they are mainly nonlocal.

This evidence reveals that, in the absence of strongly organised spontaneous local networks, the intervention of the public authorities to support local cooperation involves two main steps. The first is to set up institutional networks to cope with the absence of a dynamic economic environment and to organise assistance for local firms. These networks then tend to privilege an institutional function, that is, to develop coordination between the public organisations specialising in support for innovation without really articulating themselves with the local actors. As shown by Corsica, the logic of partnership then remains largely centrifugal.

The cases of the Aquitaine and Rhône–Alpes regions exhibit different relations between spontaneous and institutional networks. The problem is to produce synergy between various actors of the processes of innovation and production of knowledge.

In Aquitaine the institutional networks suffer from an asymmetry between actual academic competencies and insufficient industrial activities. The BMP appears at the same time to have shifted compared with the spontaneous networks where they existed and to be in a permanent search for a more solid anchoring. That condemns it to widen the field of its missions to all medical activities and to play on the connection with foreground institutional actors. In the Rhône–Alpes, the existence of a strongly developed industrial system confers on the firms a prior role. The institutional networks then have two problems: to define their place and to define their role. On the one hand, firms' relations largely extend beyond the regional area. On the other hand, the industrial environment is heterogeneous with regard to the types of activities and sizes of firms. Questioned on their relationships to the proximity networks set up to facilitate technology transfers, the firms, especially the smallest ones, consider them interesting but deplore a disconnection between collective interventions which are too general and their very specialised needs in terms of activities and market knowledge.

We moreover stressed that attempts to establish relations between various actors situated upstream of the chain of innovation failed and that the institutional networks attempted to offer services to downstream firms (case of the BMP).

All this shows that it is difficult to connect, in a voluntarist way, local actors belonging to different worlds. For example, the fact that doctors and entrepreneurs are located close together is not sufficient to make them work together and constitute themselves into a network, in spite of the efforts of the institutional actors. It is even observed that these efforts often lead to the opposite result. They end up supporting homogeneous but distinct networks (doctors, industrialists, chemists) and thus give up on their initial objective in spite of partial successes (a certain number of contracts concluded). The weakness of the relations between local actors is then likely to be reinforced and lead to the maintenance of a centrifugal logic of cooperation. Once more, geographical proximity fails as a form of proximity organised for activities of innovation.

3.4 Cognitive logic and importance of organisational proximity

The case studies reveal the uneasy installation of local networks of innovation supported by public policies and show at which point it is difficult to promote in a voluntarist way localised synergies in terms of innovation and technology.

The fact that these policies had to deviate from their initial objectives and to return to less ambitious goals shows that the analogy often made between geographical proximity and easy diffusion of techniques or knowledge must be seriously questioned, in particular if one connects it to the tacit or codified character of knowledge. It is clear that one of the aims of regional policies is to promote technological development on the basis of voluntarist networks of local actors. It is not so simple, however, to break with organisational or cognitive logics which have functioned for several years.

Our results reveal two main obstacles for the installation of local networks of innovation supported by public policies (these obstacles make clear why these policies had to deviate from their initial objectives).

3.4.1 *Cognitive logic differences, or the importance of organisational proximity*
One of the obstacles faced by local technological policies is to set up transverse cooperation between local actors of various kinds (entrepreneurs, researchers, trainers), as revealed by the experience of the BMP poles. Work practices and cognitive logic are very different from one world to another. Moreover, tacit knowledge is more easily transmissible within a professional world (even at distance) than between different worlds (even in proximity). Although it is supported by the voluntarist development of institutional networks, geographical proximity alone is not sufficient to break down these barriers. Consequently, the diffusion of knowledge and technology assumes that an organisational proximity exists between the actors, that is, that these are previous relations founded on professional links, whether or not these links have a local content.

Disjunction between researchers and firms whose cognitive logics are strongly differentiated is striking in the Aquitaine and the Rhône–Alpes regions. It blocks the transmission of knowledge. Visions and expectations remain unmatched; knowledge and fields of application are heterogeneous. As a consequence, each group of actors trusts only partners with whom it is accustomed to work, even if they are located outside the region and even if there are sometimes more skilled actors within the region. The actors privilege organisational proximity which is based on a long common experience of interaction and reciprocal learning and which enables them to overcome differences in cognitive logic. Moreover, frequently the required partner exists only in another region, or even in another country, which limits interest in the search for local relations, with or without the BMP pole. Between the 'territory' of spontaneous networks (which goes from the region to contacts on a worldwide scale) and the regional level of the institutional network, the interrelationships slowly build themselves.

3.4.2 *The weight of the past*
Thus, organisational proximity does not necessarily have local foundations. For reasons linked with the way in which the local systems were constituted, actors often engage in cooperation with partners external to the region. They are accustomed to cooperate, a practice which results in mutual knowledge of the people and organisations, and they have common work procedures which have proved reliable. Putting actors in contact who are geographically close to each other is not sufficient if they did not have organisational relations before.

The history of local relations counts ('proximity matters') but also the history of the nonlocal relations ('distance matters'). By forgetting that, voluntarist technological policies often end up reproducing the situations which they previously proposed to solve. The example of the Corsican NTD reveals that it is unrealistic to seek to impose fast technological development on an interventionist basis, and even more to support in a voluntarist way local interactions to the detriment of external contacts. This is why the current stage is that of an appropriation of new knowledge by the members of the institutional network. This stage is essential to build shared skills between them and because the development of projects and their realisation start a second phase, of recombining of the previous relations on the basis of now well-defined coordination. Once again, the permanence of the organisational and professional trajectories is very impressive. Previous relations appear strongest and it is only if they imply at the same time a geographical and an organisational proximity that it is possible to promote or support them within the framework of a regional technological policy.

4 Conclusion

The aim of this chapter was to engage in a discussion about the role played by geographical proximity in the process of technological transfer, a conventional argument often regarded as the explanation of the high degree of geographical concentration of innovative activities. Our starting point was to examine closely the theoretical assumption that geographical proximity is a necessary condition for an efficient share of knowledge, especially in the case of tacit-knowledge-intensive activities such as innovation. We explored this discussion in two main directions.

First, we asked if advances in information and communication technologies change the need for geographical proximity between knowledge users. We showed that another kind of proximity—organisational proximity—combined with the mobility of human resource and the use of ICT is an alternative basis for knowledge exchange that no longer requires permanent colocation.

Second, we developed this thesis by examining the role of geographical proximity in networks of innovation. Lessons drawn from case studies on localised networks of innovation supported by public institutions in three French regions reveal that organisational proximity appears to be a stronger supporter of technology transfer and innovation diffusion than does geographical proximity.

It appears that the role played by geographical proximity to set up and develop networks of innovation has been overestimated in the economic literature. Consequences should be drawn for local development policies. It would be appropriate for them to diversify their orientation and to move gradually from the exclusive search for local synergies to more open strategies of development.

Acknowledgements. This research was partially supported by the Commissariat Général du Plan (grant number 17-95) as part of the programme "l'entreprise et l'économie de l'immatériel". This help and support is gratefully acknowledged.

References
Arrow K, 1962, "Economic welfare and the allocation of resources to invention", in *The Rate and Direction of Inventive Activity* Ed. R Nelson (Princeton University Press, Princeton, NJ) pp 56–72
Audretsch D B, Feldman M P, 1996, "R&D spillovers and the geography of innovation and production" *American Economic Review* **86** 630–640
Cardon D, Licoppe C, 1997, "Approches des usages en computer supported cooperative work (CSCW)", Centre National d'Études des Télécommunications, 38–40 rue du Général Leclerc, F 92131 Issy-les-Moulineaux
Feldman M P, 1994 *The Geography of Innovation* (Kluwer Academic, Boston, MA)
Foray D, Lundvall B A, 1996 *The Knowledge-based Economy: From the Economics of Knowledge to the Learning Economy* (OECD, Paris)
Garton L, Wellman B, 1996, "Social impacts of electronic mail in organizations: a review of literature" *Communication Yearbook* **18** 434–453
Hatchuel A, Weil B, 1995 *Experts in Organizations* (De Gruyter, New York)
Nonaka I, 1994, "A dynamic theory of organizational knowledge creation" *Organization Science* **5**(1) 14–37
Polanyi M, 1966 *The Tacit Dimension* (Routledge and Kegan Paul, London)
Ratti R, Bramanti A, Gordon R, 1997 *The Dynamics of Innovative Regions* (Ashgate, Aldershot, Hants)
Wellman B, Salaff J, Dimitrova D, Garton L, Gulia M, Haythornthwaite C, 1996, "Computer networks as social networks: collaborative work, telework and virtual community" *Annual Review of Sociology* **22** 213–238

The Transformation of Local Production Systems: International Networking and Territorial Competitiveness

R Cappellin
University of Rome "Tor Vergata"

1 Introduction

The European economy is subject to a widespread process of restructuring, in which production is reorganised and relocalised, as a consequence of the phenomenon of the shortening of time and space. In the last few decades, in most European countries (such as Italy, Germany, France, and Spain) the pattern of industrial development occurring in many once typical rural regions is rather similar, a pattern characterised by the existence of local production systems in which small and medium-sized firms (SMEs) represent a high share of total employment, the unemployment rate is lower than the national average, the share of industrial employment is particularly high, and sometimes even increasing, and the firms export a high share of their turnover. Typical examples of this model of local development are various regions in Italy, such as Lombardy, Veneto, Emilia – Romagna, Tuscany, and Marche, or Baden-Württemberg and Bayern in Germany, or Catalonia and Valencia regions in Spain, and Rhône-Alpes and Provence-Côte d'Azur in France. Similar patterns of regional development have also recently been identified in Nordic countries, such as Denmark, Sweden, and Finland.

In Europe SMEs (firms with fewer than 250 employees) represented 66% of employment in 1995. Their share was 56% ten years ago (*Il Sole 24 Ore* 13 May 1995). SMEs created 75% of the new jobs in Europe during the period 1993 – 95 (*Il Sole 24 Ore* 18 May 1996). In the USA the 4000 firms listed in the NASDAQ have increased their employment by 500 000 jobs during the period 1990 – 94, while the Fortune 500 firms have decreased their employment level by 850 000 jobs (*Il Sole 24 Ore* 13 May 1995). In Italy firms with fewer than 100 employees represent 59% of total employment in manufacturing industry, and their share is 39% in France and 23% in the USA (*Il Sole 24 Ore* 26 February 1996).

A local production and technological system may be described as a local labour market or as an area comprising both an urban centre and the surrounding rural areas. Local production systems can also be a locally concentrated network made up of industrial firms, often but not always specialising in a particular sector, such as in the well-known 'industrial districts'. Intense networking also exists between the various firms and their suppliers and clients, research institutions, universities, schools of higher education, modern consulting services, sectoral associations of producers, and public institutions active in economic policy, such as chambers of commerce and local public administrations (Becattini, 1979; 1990; Fuà, 1983; Garofoli, 1989; Putnam, 1993).

Local production systems are characterised by an endogenous development model (Cappellin, 1983a; 1983b; 1992), in which external investments play a minor or ancillary role with respect to the birth and growth of firms owned by local entrepreneurs. The basic characteristics of this model are, first, a decentralised decisionmaking mechanism based on the cooperation of the various local firms, each of which performs an individual phase of a complex production process or specialises in different products which are strictly complementary. Second, a local production system is characterised by the existence of specific production know-how which has been accumulated over a long time, cannot be easily transferred to other locations, and which is capable of promoting new forms of production, because of the existence of dynamic economies of scope within the existing production systems.

Hence a local production and technological system is not just a territorial concentration of specific firms working in the same sector or in closely related sectors, but is also a specific form of organisation of the close relationships among all the above indicated local actors or a specific 'governance structure' which minimises the transaction costs in these relationships (Bianchi, 1993; Cappellin, 1988; Scott and Storper, 1990; Williamson, 1981). Local production systems behave like a 'learning organisation' and their internal structure and external relationships have constantly evolved during the last few decades because of the increasing importance of technology and the internationalisation of national and local economies. This implies that the actual structure of many local production systems is quite different from that which was illustrated in earlier studies of these areas (Alessandrini and Canullo, 1997; Becattini and Rullani, 1993; Falzoni et al, 1992).

In many local production systems, notwithstanding a very dynamic growth in industrial production and exports, the growth of industrial employment has either stopped or has entered a slow decline, because of the effect of the crisis in specific sectors (such as the metallurgical industry), the transfer to other countries of the first phases of production, and especially because of reorganisation both in large and in small firms, which results in a reduction in the number of employees in the individual firms, the elimination of superfluous plants, and a search for higher productivity (Cappellin, 1995a).

The recent evolution indicates a shift to a more complex form of organisation and the birth of many new firms in nonindustrial sectors (Cappellin, 1989). Thus, although industry remains the single factor capable of mobilising the entire local economy, new development has concentrated in the field of production services, such as wholesale trade, logistic activities, banking and insurance, and marketing organisations. In various cases these new activities have been capable of ensuring not only the employment of the workers laid off from the industrial firms but also the fast reuse of the industrial estates vacated by the original industrial firms, thus avoiding further urbanisation of the rural areas. The evolution from the production of individual goods or machinery to the satisfaction of the overall needs of the consumer has led to a tighter integration of service activities with manufacturing activities, such as in the technical design of integrated production systems, in the research of new markets, and in the supply of after-sales services to customers. Thus both the technology and the service content of the products are increasing. These changes with respect to the informal methods of cooperation of the past do not imply a revival of the old hierarchical model of the large firms. On the contrary, they may be interpreted as the evolution from a specific organisational model—the 'industrial district'—to a new organisational model—the 'network' at the interregional and international level.

In this chapter I aim to illustrate the recent evolution of the endogenous model of development of many local production systems in the North–Central regions of Italy and in various intermediate and nonmetropolitan regions of Europe. In particular, I will illustrate some major factors affecting the competitive scenario of the international economy, which may represent an opportunity for further development of SMEs. Then I will show that local production systems are no longer specialised in traditional production mainly oriented to the regional and national market and that both innovation and internationalisation processes have assumed a great importance and explain the international competitiveness and economic success of these areas. Last I will focus on the new factors of attraction of external firms and of the economic performance of local firms and on the need for a tighter integration of the policies of territorial organisation with the traditional instruments of local industrial policies, such as financial incentives and provision of specialised producer services.

2 The changing pattern of industrial organisation in Europe

Industrial firms, both large and small, have to adapt their strategies in the face of the evolution of the macroeconomic and technological scenario. Recent changes, which seem to be of crucial importance, include the following. The constant increase in wages in the intermediate and economically lagging regions of Europe, brought about by increasing labour mobility and the working of 'demonstration effects' on household consumer behaviour, are constantly decreasing wage differentials with the most developed European regions. Thus, the competitiveness of intermediate and economically lagging regions in Europe with respect to non-European countries (where the wages are ten or more times lower) should not be based on costs but on other factors, which should be actively promoted by regional industrial policy. The economic development of an area does not depend on advantages in terms of cost of production factors, but on the dynamic of productivity and the speed of the process of innovation adoption.

The adoption of rigorous monetary policies and the creation of the European Monetary Union is leading to lower inflation rates, fixed exchange rates, and lower growth in aggregate demand. This will produce a radical change in pricing policies of industrial firms, because prices can no longer be dictated by a fixed markup on costs. Prices are to be considered as fixed and sometimes they may decrease not only in particular sectors (as they did during the last few years in metal, chemical, electronics, and agroindustrial productions) but even at the aggregate level. Profit margins, wages, and especially productivity levels should be considered the strategic instruments firms use in order to guarantee their competitive position in their respective markets. Thus innovation and conversion to new production become the key goals of a regional industrial policy.

The process of outsourcing is constantly increasing both in large and in medium-sized firms because the gradual removal of nontariff barriers and the decrease in transport and communication costs lead to a greater integration of markets and to a greater division of labour between the various firms (Stigler, 1951). Economies of scale and of specialisation may be exploited at each phase of the production process and vertical integration is replaced by the exploitation of complementarity and increasing cooperation between firms, both within each individual regional production system and also in an international framework. That leads to the adoption of an intraregional and interregional networking strategy in regional industrial policy.

Increasing competition, standards of living and of culture, and the increasing diversification of society in various groups with different consumer behaviour imply the end of the 'mass production' model and of the myth of a large 'middle class' with homogeneous preferences. Quality control, product diversification, and product innovation become the crucial instruments in a market-oriented or consumer-oriented regional industrial policy.

The technological effort of firms is less oriented to a decrease in variable costs in manufacturing production (that is, wages, intermediate inputs, and raw materials) and increasingly oriented to a decrease in fixed costs (capacity utilisation, overhead costs, stock levels, logistics costs) and to a decrease in production and distribution time (just-in-time methods, lead time of production processes, delivery time, and time to market in product development). Thus cost competition should be increasingly integrated into time competition and there is often a trade-off between lower costs and shorter times. Clearly the flexibility and smooth integration of the regional production system require a greater effort in transport infrastructure and logistic services.

The strategic power of large manufacturing firms is challenged by the appearance of large distribution firms, which represent an unavoidable interface between the producer and the consumer, can impose lower prices on manufacturers, and may compete directly with the sale of 'own-label' products. Moreover, competitiveness is to a large extent determined by the ability of various manufacturing firms to guarantee adequate services to clients. Thus the ancillary role of services with respect to manufacturing seems to have been reversed and productivity increases in manufacturing phases may be less strategic than productivity increases in the service and distribution phases. Moreover, excellence in services allows some distribution firms to decide freely how to reallocate production tasks to manufacturing firms and regions where the costs are lower. Thus regional industrial policy should extend its field of intervention to include new service activities and to promote a closer integration of manufacturing and distribution.

Finally, the complexity of a modern industrial economy requires a new role for government. A 'prescriptive' approach, based on dirigism or top-down planning is increasingly being replaced by a 'transactional' approach, where the government, at the national or regional level, defines general norms (that is, 'property rights' or 'rules of the game') and aims to remove the obstacles to a greater and more flexible integration among the various economic actors through the provision of 'public goods', such as information, infrastructure, services, and strategic initiatives based on public–private cooperation.

3 Innovation, conversion, and the process of job creation

Faced with these changes in the international economy, the response of SMEs differs from that of large firms, which is based on reengineering and the adoption of the well-known modern paradigms of managerial economics. Empirical analysis indicates that a close inverse correlation exists between the share of SMEs and the unemployment rate at the regional level. Various new industrialised regions, where industrial employment was particularly high at the beginning of the 1980s, have even indicated an increase in industrial employment: a trend which contrasts sharply with the decline in industrial employment in most of the old industrialised regions of Europe. Moreover, new industrialised regions indicate a high share of exports in total turnover, both in large and in small firms, and this share has continuously increased during the 1990s.

It can be demonstrated that these phenomena are closely related and they can be interpreted according to a model of industrial development which is sharply different in the case of regions characterised by SMEs from the development model of areas where the large firms prevail. First, the process of creation and adoption of technological innovation in SMEs follows patterns and trajectories of evolution which are different from those of large firms. Technological and organisational know-how develops through a heuristic-type process of 'learning by doing' or through a close integration between research, manufacturing, and marketing activities. In particular, the competitiveness of SMEs is often related to their ability, mainly in the production of specialised machinery or components, to combine in original ways various techniques which belong to different technological paradigms. That requires a highly skilled labour force, which has been trained either in other firms of the same specialised production sector or in other complementary sectors.

This model of the innovation process clearly differs from the rationalistic model which prevails in the case of large firms or research institutions, where the adoption of innovation follows the explicit planning of research investment. SMEs are often not able to follow the complex and formal procedures which are required in order to apply for research grants offered by national institutions, as these procedures have been designed according to the characteristics of the innovation process in large firms and do not take into account the less formalised or heuristic process of innovation occurring in small firms.

Technological advance is based on the interaction between firms rather than on investment in in-house autonomous R&D. Innovations are adopted after observation of other firms, rather than through formal learning and searching, as emulation of the other producers is enhanced by the competition mechanism. Thus technological advance is based on the interaction between the various firms in the same sector and, in particular, it is the result of close client–producer relationships, rather than of large in-house autonomous research investments.

Second, SMEs seem to adopt a defensive rather than a proactive approach in the adoption of process innovation, which leads to an increase in labour productivity. Entrepreneurs seem to identify the lack of demand for their products as the main constraint to their success, rather than the limits of their technological capabilities. Only the pressure of decreasing product prices or of increasing labour costs and the risk of crisis lead the often reluctant entrepreneurs to adopt new production technologies. This approach contrasts sharply with that of large firms, where technological change and 'restructuring' or 'reengineering' is a strategy which is proactively pursued. On the contrary, product innovation or product customisation are actively pursued by SMEs, as they represent the conditions which ensure the ability to respond to the needs of new customers and ensure the growth of the firm into new markets (Stigler, 1951).

All firms, both large and small, are highly specialised and believe that increasing competitiveness requires a greater focus on those productions in which the individual firm has a competitive advantage and the parallel outsourcing of any other function which could be better performed by specialised suppliers. The firms while aiming to decentralise nonstrategic phases, which could be performed with lower costs and higher quality by specialised producers, still perceive the need through agreements with other firms to control all the 'filière' or the chain of value-added creation. The relationships between medium-sized firms and small firms are much less conflictual in regions, such as the Three-Veneto region in Italy, where medium-sized firms are important, than in regions where large firms dominate, such as regions which specialise in automobile production. Relationships with suppliers are based on the model of 'co-makership', according to which each individual supplier has increasing autonomy and is required to ensure an original contribution to the quality of the final product or service.

Although firms aim to decentralise nonstrategic phases, they are aware that their competitiveness relies greatly on their ability to control all the phases of the production and distribution cycle through various formal and informal agreements with other local firms. This leads to a new process of indirect vertical integration through the creation of 'constellations' of subcontractors and specialised suppliers around a 'leader' firm, as this allows the leader firm to maintain a tight coordination of the various stages in the chain of value-added creation. Moreover, the subcontracting relationships induce SMEs to develop new relationships with more than one main contractor and the stimulus of the specific needs of the new clients induce the firms to convert to new productions, which are tightly linked with the know-how accumulated during various decades of experience in the original field of production. Thus subcontracting relationships play an important role in promoting product innovation within the firms and among the firms in the areas where SMEs prevail.

The process of the diversification of the local economy toward new production is stimulated by the fact that new products and new production techniques mainly emerge first as the result of a learning process and in response to the internal needs of existing firms, which need new specialised intermediate products or services, and as the result of the development toward new production areas of the actual technological capabilities existing in the local firms. Later, the growth of the new productions may induce the spin-off of new firms, often owned by the same entrepreneurial family, and these new firms can develop in an autonomous way with respect to the original firm. The organisational strategy of SMEs in local production systems is based on the concept of decentralisation. Entrepreneurs prefer to create a new firm, rather than enlarge the original firm. This approach emphasises the responsibility of individual family members who are assigned as managers to the new firms and allows the division of the investment risk of the overall entrepreneurial family. Thus the dynamics of employment is basically connected with the change in the number of firms rather than with the increase in the size of individual firms. The average employment size of firms has actually been decreasing during the last few decades.

The different capabilities of SMEs with respect to process and product innovation indicate that public policy should aim to stimulate the demand for technology by SMEs rather than to intervene in order to expand the supply of technological services, both public and private, which is often adequate in many intermediate regions. In order to accelerate the adoption of product innovation, for which there is a high demand by SMEs, it seems necessary to stimulate the research activity both internal and external to the SMEs through the development of cooperative research projects. For this purpose it may be useful to create adequate interface organisations between SMEs and existing local research institutions.

A further difference with respect to large firms is the fact that large firms explicitly follow a strategy of decreasing production costs and choose the regions or countries where these costs are lowest. This may help them to resist wage increases in the areas where they are located and thus keep the wage level down in these areas. For SMEs the cost of labour seems not to be a constraint, but rather a stimulus leading to an increase in the overall productivity of the firm. SMEs cannot choose locations very distant from the origin of their entrepreneur and cannot react to wage increases by relocating in different regions or countries.

When productivity growth decreases the labour requirement of SMEs for given production levels, SMEs cannot fire their workers, as they belong to the same entrepreneurial family or are endowed with a specific and not easily replaceable skill. In fact, firms fear that these workers may move to competing firms in the same local production system, transferring the specialised know-how acquired from years of work in the firm, and this induces the firms to pay higher (that is, 'efficiency') wages than those required by the official sectoral contracts. Moreover, the limited absolute number of employees, often fewer than 10, makes unfeasible a decrease of even a few employees without disrupting the overall organisation of an SME.

Thus, the response of SMEs is a continuous effort to expand their production levels by identifying new market outlets for their traditional products or alternative uses for their existing know-how through the development of new products and the diversification of the sectors of activity, in order to ensure full utilisation of their labour resources. These external and internal constraints oblige SMEs both to increase productivity and continuously adopt product innovation, rather than aiming to minimise wage costs through explicit strategies of international relocation, as happens in large firms.

In particular, this process of conversion can be interpreted as an endogenous growth model, where the material and immaterial resources, such as the labour capabilities of the entrepreneurial family, the local highly trained labour force, local organisational capabilities, local technical know-how, production capital, etc, which are made idle by technological process innovations, are smoothly reinvested into new products and new firms (Cappellin, 1983b). It is clear that this process is closely related to the prevalence of small firms coupled with low barriers to entry, the family character of entrepreneurship, and the high social cohesion in the relationships between management and labour. These factors ensure that firms are strongly embedded in the regional community and promote the responsibility of firms in ensuring the full employment of regional resources.

4 Labour-market flexibility and increasing wage levels

The model of the local production systems may also indicate important prospects for labour-market policies. In fact, although flexibility of local labour markets has recently been pinpointed as the most important policy priority for labour markets in European countries, this flexibility has long been a structural characteristic of the areas where SMEs prevail. The economic success of the regions characterised by SMEs seems to indicate that the decrease in wages is not a factor of development, but rather is a result of the failure of the development strategy in those regions, which have not been able to stimulate the growth of SMEs. The most dynamic areas are those where the wage growth rates are higher, whereas areas in deep crisis see inevitable decreases in incomes and standard of living.

The key factors in the growth and creation of new workplaces are the increase in competitiveness of firms, the ability to increase productivity, and innovation and diversification into new products and new markets. Clearly, a faster productivity dynamic allows an increase in wage levels and it is the only instrument which can allow a firm to stay competitive notwithstanding the inevitable increase in wage levels. The decrease in wage costs, through various fiscal incentives, has a transitory effect, which will disappear in a couple of years, because wage increases are inevitably homogenous within a country and the temporary advantage of some regions will be soon reabsorbed. Moreover, firms' productivity levels tend to adjust to the level of wages and a wage differential in favour of some regions will induce the firms in these regions to be less efficient than those of other regions.

Local production systems are usually characterised by cooperative behaviour in labour relationships. In particular, the success of all firms specialising in a particular field of production depends greatly on the special skills of the local labour force and this leads firms to limit labour turnover by paying higher wages to the most qualified workers. Indeed, they often help these workers to create a new firm, which becomes a subcontractor or a specialised supplier of the original firm.

Within local production systems, the high turnover of firms, because of the high birth and death rates of SMEs, automatically determines a high mobility of workers, who are often obliged to find a new occupation. Thus the high turnover of firms compensates for the apparent rigidity determined by labour-market regulations. In particular, the flexibility of the labour market in areas where SMEs prevail is underlined by the fact that the frequent crisis of many firms, coupled with the existence of an entrepreneurship capability in the local labour force to exploit new business opportunities, means that many workers change from the status of self-employed to that of employees and vice versa at various times in their lives.

Thus, in these areas labour-market flexibility seems to be the implicit result of the behaviour of the labour supply, rather than of new policies of the firms, which are increasingly advocating the removal of rigid regulations in the hiring and firing of workers. SMEs are often deeply concerned with the unstoppable phenomenon of the frequent 'zigzag' movements from one firm to another by the most experienced workers, who are enticed by other firms through higher wages, after they have been trained for many years in the original firm. In this perspective female workers are often preferred by firms, as they ensure a greater stability and that has certainly contributed to the high female activity rate in these areas.

Too high a mobility of workers represents a critical factor not only for the smallest and least organised firms, but also for those firms which intend to increase their investment in labour training and would like to consolidate their internal organisation in order to comply with the new requirements of higher quality and shorter production and distribution times. On the other hand, the mobility of workers and the turnover of firms are factors that stimulate the diffusion of technological and product innovation among local firms and may have a positive effect on the overall competitiveness of local production systems.

A further important change in the labour markets of the local production systems is represented by the rapid ageing of the population, which is important in Europe and in particular in those areas which industrialised rapidly during the 1960s and 1970s and had attracted important immigration flows by a young work force. This phenomenon is related to a gradual weakening of the effort traditionally devoted to work, to the resistance by younger and more educated workers to accept the traditional long work shifts or even an occupation in traditional industrial activities. This has led firms to attract increasing flows of immigrants from non-European countries. On the other hand, these changes contribute to stimulating the firms to increase their technological level and productivity.

5 The internationalisation process of SMEs

SMEs do not operate only in a national framework and they contribute substantially to national exports, as is shown by various recent studies on the territorial disaggregation of national exports (Balloni et al, 1998; Cavalieri, 1995; D'Antonio and Scarlato, 1997; Viesti, 1998). In 1985 their share of the Italian national total was 53% and it has been increasing since then. Clearly not all SMEs export, but in 1996 68% of firms with fewer than 500 employees did. When exporting SMEs only are considered, the average share of exports on turnover is 21%. This share increases in some sectors and it was 45% in the mechanical sector in 1994. In particular, for a high percentage of the exporting SMEs (21.5%) the share of exports is larger than 50% (*Il Sole 24 Ore* 12 February 1995). Also the exposure of SMEs to international competition is high, as 30% of SMEs consider that their major competitors are located abroad (*Il Sole 24 Ore* 23 April 1996).

In particular, handicraft firms (that is, firms with fewer than 5 employees), which represent 12% of GDP in Italy (9.6% in Germany), have 19% of national exports (2% in Germany) (*Il Sole 24 Ore* 18 May 1996). Moreover, 30% of handicraft firms can export between 11% and 30% of turnover (*Il Sole 24 Ore* 14 February 1996). On the other hand, export capability is lower among subcontractors and only 23% of the subcontractor firms are able to export. In particular, subcontractor firms export on average only 8.5% of turnover, but they certainly contribute indirectly to exports through their sales to exporting SMEs and large firms (*Il Sole 24 Ore* 21 May 1996). The geographic orientation of SME exports indicates the predominance of the nearest countries, such as those in the EU. However, SMEs are increasingly capable of exporting to distant non-European markets, such as the Far East.

SMEs have created production and commercial units abroad or have developed strategic alliances with foreign firms in the production, technological, and commercial fields, thus strengthening their traditional presence on the international markets through exports. Recent investigations, which have considered only production plants and not commercial offices, have indicated that employment in foreign-controlled firms by Italian firms which are not among the largest twenty Italian multinationals had increased by 3.5 times in the period 1986–92. But employment in foreign firms controlled by the top twenty Italian multinationals had increased only 2.1 times (Cominotti and Mariotti, 1994). The same survey indicated that 324 of the 450 Italian firms which invested abroad in 1994 had fewer than 500 employees and only 6 firms had more than 10 000 employees (*Il Sole 24 Ore* 4 June 1996). Various SMEs control subsidiaries abroad which have a larger number of employees than they employ themselves. In particular, 2% of SMEs (1800 firms) have foreign subsidiaries, 4% of SMEs (4000 firms) have technical-production agreements abroad and 14% of SMEs (9000 firms) have commercial agreements abroad (*Il Sole 24 Ore* 23 April 1996).

In recent years the increasing importance of technology and the process of internationalisation of national and local economies have been transforming relationships between firms, which have become more complex, risky, and need to be redesigned in the long term. This has compelled firms to devise new organisational forms and contractual arrangements to manage these new and more complex relationships.

In a network model of organisation, SMEs may also aim to perform a global role, by being closely integrated with other SMEs in foreign countries. In fact, internationalisation requires firms to work in a different environment and requires a greater decentralisation of functions and the creation of flexible alliances with foreign firms. A decentralised structure is more efficient than a concentrated one and an SME may be able to manage a specific field of business at the worldwide scale when it is integrated with other SMEs in an international network. Thus the increasing integration of the international and European industrial system and its structural transformation is creating new challenges but also new opportunities for SMEs (Cappellin, 1991; 1995b). The new forms of industrial organisation seem to indicate that large firms no longer have an automatic advantage over SMEs. Firms which are capable of surviving in an increasingly competitive international market are those which have been faster to innovate or to adapt to the specific tastes of local consumers: this seems a specific characteristic of SMEs.

An interesting phenomenon is the increased activity in the same foreign countries of many small entrepreneurs originating from the same region, such as the textile entrepreneurs of the Veneto region in Romania. This phenomenon is the result of a bandwagon effect, which looks rather similar to the concentration of emigration flows which occurred during the 1950s from the same regions to specific countries and regions. Moreover, the firms of various local production systems in Northern Italy have recently undertaken joint investments aimed at artificially creating kinds of 'industrial districts' in various less developed regions in Southern Italy and also abroad.

A major characteristic of the internationalisation process of SMEs is the fact that it is similar to a gradual 'learning process' in which the forms adopted by the individual firms vary continuously, trying to adapt pragmatically to the different environments of the various countries on the basis of experience. Thus the internationalisation process of SMEs can be interpreted as an extension in an international framework of the same model of specialisation and cooperation with other firms, which has long existed within a regional framework. In particular, SMEs adopt a very cautious approach in their internationalisation strategies and they seem to be aware of new market opportunities abroad, although they do not exploit them if this would require too large an investment in human resources and too high a risk. The internationalisation of SMEs through commercial agents or their own foreign commercial offices or subsidiaries is more important than that through foreign investment in production structures, although the development of international subcontracting agreements and the sale of licenses often represent a valid alternative to these productive investments.

As operative functions abroad and alliances with foreign firms need to be monitored through a direct presence in that foreign country, the major obstacle to the internationalisation of SMEs is internal to the firms themselves and is represented by the lack of qualified human resources or by the frequent existence of a too parochial culture, which is too different from that in the foreign country in question. Moreover, the lack within the SMEs of many qualified technical collaborators with the entrepreneur hinders the systematic effort which would be required to implement a medium-term strategy. In particular, the production capabilities of SMEs are largely incorporated in the qualifications of the local labour force, in the local network of suppliers, and in other factors characterising the local environment, which may hardly be reproduced abroad. Therefore SMEs often prefer to keep their production concentrated and are reluctant to decentralise it to less developed European regions and countries. This is especially true in the case of products, the demand for which is unstable and for which time and quality are crucial factors of competition, as is typical for SMEs operating in very specific segments of the international market.

SMEs usually prefer to invest in marketing structures especially in the most developed countries, aiming to integrate the value added created in the phases downstream of the manufacturing phase. The problem of controlling the distribution chain for their own products has become a factor of similar importance to technological innovation and it often requires investments which are larger than those needed to increase production capacity abroad, as indicated by the high amount paid in order to control foreign firms with a well-known trademark or a diffuse network of retailers.

Various dynamic SMEs seem to be inspired by the example of firms which have recently become multinational or very large industrial groups (such as Benetton, Luxottica, Riva, and Radici) capable of competing with the traditional Italian large industrial groups (such as Fiat, Pirelli, Montedison), but which during the 1970s were themselves SMEs. Since the 1990s, these SMEs have considered investing in the most developed countries, such as the USA and Germany, not only in order to consolidate their market presence, but also to acquire access to specialised know-how and complementary technological competencies.

A further indicator of the internationalisation of SMEs is the fact that, when they recognise the need to raise capital through quotation on the stock exchange in order to sustain their growth process, they aim to be quoted on a foreign stock market, such as that of New York, rather than on the domestic stock market of Milan, thus exploiting the reputation which they enjoy in those countries where they have been exporters for many years.

These trends in the internationalisation process of SMEs certainly do not deny the increasing importance of the process of outsourcing of some parts of production through subcontracting agreements with firms in less developed countries, in the Far East, in Eastern Europe, or in North Africa. However, the firms which are interested in creating foreign production plants in less developed countries are mainly those which have maintained their specialisation in traditional production, where competitiveness is mainly determined by cost. This may also lead to the closure of domestic plants and to the entrepreneur emigrating.

5.1 The role of foreign multinationals in local production systems

An opposite phenomenon is represented by the increasing investment of foreign multinational firms in the local production systems of Northern Italy. Highly specialised industrial districts, which have become world leaders in specific segments of production (for example, in white goods or sports shoes) are endowed with important external economies, in terms of specific know-how, a well-trained labour force, and specialised suppliers. This has led multinational groups to locate production plants in these areas or to acquire local SMEs and to concentrate within these firms productions which were previously scattered around the world. Various SMEs have become the target of acquisition by foreign firms, especially when these firms are technologically well endowed and may make a jump in quality through the transfer of external technology and especially through access to a modern and worldwide marketing network. These SMEs have benefited from a direct access to the large distribution organisation of the multinational groups, to higher financial resources, and to technological transfers, and this has often meant an increase in employment in these firms or they have avoided closure as the result of an internal crisis.

5.2 New challenges for the internationalisation of SMEs

There is wide agreement that there is a need to consolidate the traditional export activities, which have characterised many SMEs in Italy for 20–30 years, through a direct presence in various foreign countries. Thus export revenues have recently been invested in the creation of stable commercial and/or production activities in those same foreign markets, rather than in the expansion of production plants at home. However, many exporting SMEs are unable to use abroad commercial distribution channels which differ from those used in the domestic market, where traditional small-scale retail distribution still prevails, and they are unable to adapt to the existence in these markets of a very limited number of very large operators, such as buying consortia and large retail distribution companies. Cooperation in exporting activities among SMEs operating in the same sector has been widely attempted with only limited success, although an organised export activity would require much larger investments than could be afforded by individual SMEs.

Thus the development abroad of efficient logistic structures, of transport, warehousing, sorting, adapting products with special components, repair, and service to clients, is becoming a crucial condition for the continuation of the high flows of exports by SMEs. This requires considerable investments abroad, which are different from production plants but may be a preliminary condition for their creation. A valid alternative to the direct creation of autonomous commercial structures in foreign countries is the establishment of stable contracts with large distribution companies. However, a crucial constraint is that small firms often lack that scale of production and reliability in terms of quality and time which would be needed in order to work with modern distribution companies. In general, the development of international activities of SMEs points up the need for a tighter integration between manufacturing activities and service-type activities and for higher investment in the improvement of the technological background and of the international culture of local human resources.

5.3 The internationalisation process of the local environment
In the past the concept of market globalisation meant increasing international flows of products and services and the existence of a price-determining mechanism operating at the international level and not at the national level, as in markets for foreign currencies, bonds, and stocks. Nowadays, the concept of internationalisation means that the behaviour of the individual economic actors is also changing and becoming transnational. This implies an overall international integration not only of product markets but also of the supply side of national economies, which is made up of the production organisation of the firms and of other economic actors which have usually been assumed to be protected from international competition, such as trade unions, public institutions, research organisations, etc. The internationalisation process has a clear social and institutional dimension and implies a change in the institutional and social mechanisms which link the various economic and social actors not only in an international framework but also within each national or regional economy.

The internationalisation of economies is a broader process than that of market globalisation and it does not imply increasing homogeneity within a 'global village'. Instead, it implies increasing interdependence and the need for a policy of differentiation or of exploitation of the differences characterising each actor and of the original role which the actor may perform. These changes may also be defined as the shift of the industrial system of the advanced economies from an 'international' structure to an 'interregional' structure, as the economies of the various countries become integrated in a way similar to those of the regional economies within the same country (Cappellin, 1993; 1995a).

6 Territorial quality as a factor of competitiveness

The development of the abilities to compete at the international scale is the crucial objective of a development strategy for regions which are tightly integrated in the European Union and in the international economy. The competitiveness of firms is increasingly affected not only by the adoption of technological innovation within the firm, but also by the competitiveness of the overall regional economy in which the firms are located and by its ability to adopt broad organisational and institutional changes which may provide a favourable local environment.

In particular, no European region can be really competitive on labour costs alone with the various newly industrialised countries of the world. The development strategy of the local production systems in intermediate European regions should shift from competitiveness based on costs and on low salaries or on various forms of public financial support, as it was in the earlier phases of development, toward a 'market-oriented' development strategy. Competitiveness should be based on the growth of productivity, on product quality, on fast adjustment to the continuously changing needs of consumers, and on the continuous adoption of product and process innovation. Clearly, the competitiveness of a region depends on the integration between firms and their specialised suppliers, modern logistic, distribution, communication, and transportation services, other business services, and research and education institutions (Cappellin and Nijkamp, 1990). The international competitiveness of an SME is based largely on the spatial concentration of many SMEs in the same region and on their relationships with their territory.

The factors contributing to the competitiveness of a modern industrial economy are not only 'hard' factors, such as: transport costs; availability of nonqualified labour; the cost of labour and the cost of raw materials; the cost of industrial sites; the size of the local market; exploitation of internal economies of scale; and availability of public financial incentives. New qualitative or 'soft' location factors (Cappellin, 1997c), which may create a favourable environment for the development of local firms, seem to be increasing in importance, such as: the dynamism of the local economy; the increase in labour productivity; the exploitation of the industrial traditions or the technological know-how of the local economy, and the local entrepreneurial culture and its capabilities; the availability of specialised suppliers; forms of bank-industry relations; educational opportunities for specialised workers; forms of labour-management relations; conflicts between local industry and local residents; forms of social relations for entrepreneurs; the quality of the local social environment; the capability of responding to the new residential preferences of the entrepreneurs and of the labour force; the cultural resources, the development of a regional identity, and the sharing of a common development strategy; the quality of the organisation of the territory and the effectiveness of urban policies; the development of the local transportation networks and the availability of logistic services; local administrative capabilities; the level of autonomy of local institutions; the international image of the area and the extent of international relations; and the existence of an explicit local 'foreign policy' of the regional and urban institutions.

A modern industrial organisation, which looks to the quality of the products and of the processes, should be based on a carefully designed but flexible organisation of the relationships between the individual firms. As competitiveness is increasingly determined by the pace of response to external shocks and by the acceleration of production and distributions times and less by the costs of production alone, this requires that the flows of products, factors, and information are as smooth as possible. There is a close relation between the objective of 'total quality', at which modern just-in-time techniques aim, and the objective of better environmental quality or better quality organisation of the territory, because the first requires a better organised environment external to the firms and a careful territorial planning of the various local production and residential settlements. Modern logistics indicates that a decrease in the transport time for goods and people requires wider spaces devoted to the management of these flows, such as the areas of modal interchanges, parking, and deposit. Similarly, a more efficient work organisation requires wider working spaces and a more efficient layout of plants.

In contrast, low-quality territorial organisation may represent an obstacle to economic development. The highly congested traditional road network, which can hardly be expanded because of the already high density of the settlement structure, is causing continuous queues of thousands of commercial and passenger vehicles both in the internal transport of many local production systems and in the linkages between the various urban centres of the same region. Quite often three hours are needed to travel less than 100 km on a highway between two cities. Circulation is often slower along secondary roads. This results in frustration among the local population, as planned improvements have not materialised after many years. On the other hand, there are increasing tensions between firms and local institutions determined by the external diseconomies generated by the firms and by the new needs for a better environment for the local population. The environmental decay caused by existing production activities may result in hostility to the location of new production activities in these areas. Moreover, the inefficiencies within the external system of public services for the population may result in a claim for higher wages and increased labour costs in the firms.

Territorial quality does not only consist in planning constraints aiming to preserve the natural environment but more generally includes an active effort aimed at improving the standard of living of the citizen as well through economic and social policies, such as lower unemployment or higher security against crime and corruption. In particular, the decrease in working time and the increase in leisure time result in an increased demand for public services for sport and recreation in the smaller centres. On the other hand, population ageing and the accelerating pace of industrial restructuring is leading local institutions to increase the financial resources devoted to social programmes for the many people who are marginalised by an increasingly competitive labour market.

Thus, although sometimes economic development may have a negative impact on territorial quality, it is also true that territorial quality may represent a strategic factor for the economic success of a local production system. The territory is not the simple object of changes in production levels and in the internal organisation of large firms and it performs a more active role in the case of the SMEs as it represents the set of those external resources which to a large extent determine the long-term development of the individual local firms. The concept of quality may play a strategic role in the framework both of economic policies and of physical planning. The traditional approach in national regional policies, based on financial transfers, and the traditional approach of regional institutions, based on macrosectoral economic planning, are incapable by their very nature of taking into account the close relations between the economic development of an area and the organisation of its territory.

There is a need for horizontal policies capable of affecting the local environment in which the firms operate and of ensuring greater efficiency in the local system of public services. Thus infrastructure networks and territorial and urban policies will represent new key elements of industrial policies in local production systems.

7 The role of urban centres in a network model

A regional economic development strategy based on the network model of organisations has clear implications for regional industrial and territorial policies, as this model implies: an increasing importance for physical infrastructure both traditional and very innovative; an increasing importance for service activities, which may be considered a kind of 'soft infrastructure'; an important role within a regional economy for cities and not only large metropolitan areas but also intermediate and even minor urban centres.

A working model of industrial organisation emphasises the interdependence between economics and territory and the role of urban centres (Cappellin, 1988). In particular, an urban centre may be considered as a pool of infrastructures or a centre of 'soft' location factors. Urban centres act as nodes in the transport and communication network or as gateways between regions and the outside world. Cities are the centre of economies of agglomeration and allow a decrease in transaction costs between the various firms and stimulate their spatial concentration. They represent centres of business and public services and have power over their territory, as they exercise a sort of 'leadership' and fulfil incubation functions in the creation of new firms particularly in new service-related activities, which allow a move toward new sectors of the regional economy. Cities are cultural centres, centres of higher education and of other quality services to the population which determine the attractiveness of an area to foreign investors. They give identity to the surrounding regions and play a leading role in defining the development strategies of a regional economy and in the creation of consensus among the various local actors.

These different dimensions of the nature of a city indicate that urban policies have a specific regional and national dimension and do not necessarily coincide with the traditional field of 'urbanism' or of physical planning. On the other hand, urban policies also have a distinct economic dimension, which differs from that of traditional regional economic planning. A modern regional and urban policy should aim both at economic development and at territorial quality and it should be capable of integrating the spatial, economic, technological, social, and environmental dimensions. These different dimensions of a city may be useful in interpreting the changing role of small and intermediate cities within nonmetropolitan regions during the last few decades (Cappellin, 1997a). Cities played only a passive role in the early phases of industrial development of nonmetropolitan regions, as they performed the role of simple supply centres of basic commercial and public services for the populations of their respective hinterlands, whereas the development of industrial activities was mainly scattered in rural areas or concentrated in rather small country villages. Thus, the export base of the intermediate regions has for a long time been in rural areas rather than in the cities. Moreover, the industrial development based on SMEs in these regions has had a clear antiurban bias, as a reaction to the model of economic and social organisation and standard of living characteristic of the large metropolitan areas, where the large firms were concentrated.

Intermediate regions which during the early phases of their industrial development in the 1960s and 1970s did not consider urban centres to be important, now acknowledge the crucial role that various intermediate cities, such as their regional capital cities, may have in the actual process of transition of their industrial structure. The traditional conflict between rural areas and cities is changing into a more synergetic relationship, where the rural areas maintain their industrial specialisation and are the origin of manufacturing exports, while intermediate cities play a crucial role in the provision of strategic producer services, of fairs, of research and higher education centres, and of important commercial transport and communication infrastructures. Moreover, the increasing level of per capita income and the decrease in work time allow the development of new and more complex needs, other than simple housing needs, by the local labour force and this stimulates demand for a wider range of services. Thus the intermediate and small cities of local production systems are being transformed from an undifferentiated concentration of housing for the regional labour force to new complexes of qualified private and public services. Intermediate cities of nonmetropolitan regions have had a rather subordinate role with respect to large metropolitan cities and they have all been rather similar or characterised by a rather undifferentiated economic structure in every region.

Transaction costs are lower when distance is lower and they may also explain why a diffuse urban system made up of various intermediate and small cities may be more efficient than a centralised model, such as that of a large metropolis (Cappellin, 1988). Intermediate cities now reveal an increasing importance for interregional and international relationships with respect to internal regional relationships and they are transforming themselves into the gateways of their regions in their relations with the outside world. They are becoming interdependent between themselves and increasingly specialise in particular activities, which give them a particular identity. Thus, they may now compete at least in these specific activities with large metropolitan areas. Intermediate urban centres have also developed increasingly extensive international relations and are challenging the role of the large metropolis in their role of gateways in international transport and communication flows.

Recent European experience seems to indicate a decrease in the power of the large metropolis and a 'renaissance' or a substantial increase in the international standing of smaller urban centres and especially of intermediate cities, which are now linked with each other in a framework of increasingly larger regional and interregional or transnational networks. The ability of a smaller city to compete at the national and international level depends, first of all, on a greater specialisation in specific activities and the ability to create a specific image which may make one city distinct from another. Second, the creation of alliances with other cities in the same region or urban network allows them to reach a critical mass necessary for supporting innovative activities and to have access to the resources and the specialised know-how which may be available in the other cities of the same network. This may allow a small regional centre to become a national or international actor.

The increasing complexity of technologies and processes of internationalisation has increased the scope of interregional and international relations, which consist not only in ever-increasing flows of exports, but also in the exchanges of commercial, financial, and technological information. That can also mean that infrastructure again becomes a strategic factor or even a bottleneck in economic development, as the need for new modern infrastructure is coupled with the increasing congestion of traditional infrastructure. A typical case is the increase in international connections which is occurring in 'third-level' regional airports, which are not only the destination of tourist flows but are facilitating the international contacts of many SMEs which are active in international markets. For the last forty years airports have determined the hierarchy of cities, by determining their accessibility from medium and long distances. Intermediate cities, which were linked only to the major airports of the large metropolitan areas through hub-and-spoke connections have in the last few years developed various transversal or nonradial international and interregional connections with other intermediate cities. This has allowed an enlargement of the action area of business and better access to distant markets.

The new role of the urban centres as hubs in the internationalisation process is enhanced by the improvement in international and interregional transport. The development of the Trans European Networks and especially the creation of the high-speed trains may represent threats to the intermediate regions because they facilitate communications particularly between the largest European metropolises. However, their effects seem to be compensated by the increasing importance of air transport and the expanding international connections of regional airports. Moreover, a further technological innovation is represented in freight transport by the revolution in the organisation of the logistic chain from raw material sourcing to final product distribution, by the development of containers on sea–land transport, and by the development of rail–road combined transport. Further important innovations may be represented by the cable connections to be developed in many intermediate cities and by the spread of cellular phones which allow an increase in 'teleworking' or by the use of underground infrastructures for road transport to preserve residential or natural areas in the densely populated rural areas around the intermediate urban centres. A positive effect of these trends is the rediscovery of the importance of rail both for passengers and for freight transport. Thus many regions acknowledge the need to renew regional rail networks, which were left to decay and sometimes even dismantled, especially when that network may be used in the transverse relations between the various smaller centres of the same region, rather than only in the radial connections with the major urban centres, as it was in the past. An improvement in the transport infrastructure internal to the various regions may allow the various intermediate centres of that region to overcome a purely competitive strategy and aim at a greater integration in order to compete jointly with the much larger metropolis.

All these new trends imply a change in the traditional approaches and methods of policymaking. The spatial perspective of traditional urban planning should be enlarged to the organisation of larger areas encompassing entire provinces or the basins defined by various contiguous urban centres. This leads to the development of modern approaches of intermunicipal planning and cooperation, in order to overcome the risk that each local public administration will intervene in its own interest. The need to solve the traffic problems created by just-in-time underline the need to combine the creation of new transport infrastructures with a new planning policy aimed at bringing better order in the location of the various activities in the regional territory. The methods of environmental impact assessment or of the cost–benefit analysis of transport infrastructures may be integrated with other more complex types of analysis in the elaboration of 'area plans', aimed at the management and enhancement of the economic, social, and territorial effects of transport infrastructure and at designing the complex of accompanying measures which may tackle new opportunities and threats. These plans may be elaborated through the cooperation of many local public and private actors and the major transport operator.

New policies and projects may favour the development of public–private partnerships, which may concentrate especially in the provision of new advanced services complementary to the new transport infrastructures, in the creation of major transport terminals both for passengers and for freight, such as regional airports, old railway stations, and new intermodal logistic nodes, and in the conversion of large redundant industrial and transport areas within urban centres and rural areas.

8 Conclusions

The wide literature on 'industrial districts' has had the great merit of having clarified the existence of an industrialisation pattern which is very different from that which has historically characterised the metropolitan regions. However, the results of these studies could be considered as just interesting isolated cases, with limited relevance in explaining the structure and performance of the overall European industrial system, unless an effort is made to extend this different industrialisation model to encompass the more general case of the many intermediate regions characterised by the existence of various local production systems based on SMEs, but which have an economic structure which is more diversified than in the traditional industrial districts.

Since the 1970s, many intermediate regions, where SMEs prevail within the industrial sector, have shown a virtuous process of development, characterised by low unemployment rates, a large increase in productivity and in per capita income, high export propensity, and a strong and sometimes even increasing industrial specialisation. The importance of this new pattern of industrial development is indicated by the fact that the European territory is characterised not only by a few large metropolitan areas, but also by a large number of intermediate and small urban centres, which are far more important in terms of surface area and also in terms of population and employment than the metropolitan areas. Thus, to ignore the existence of the specific pattern of industrial development in local production systems is to ignore the reality of a very large part of the European economy. Moreover, the experience of these intermediate regions may be very useful for development policies in the peripheral regions of Europe and may provide greater opportunity of success than a too ambitious attempt to create millions of jobs through the attraction of many plants of multinational firms or in the creation from scratch of many industrial districts in these regions.

However, local production systems are very different from those of twenty years ago. In particular, the model of the monosectoral industrial districts has evolved, at least in the North Italian region during the 1990s, toward a model of multisectoral and outward-oriented local production systems characterised by a network of horizontal and vertical relationships between the local SMEs. In this chapter I have tried to analyse three related aspects of the recent transformation of the local production systems.

First, the process of diversification and reconversion through the adoption of innovation and the exploitation of production know-how into new productions is a peculiar characteristic of local production systems. This process is tightly linked to the existence of networks of firms, to outsourcing to subcontractors, to the spin-off of new firms, and to the high mobility of qualified workers between firms. It is also the factor explaining the full use of local human resources and low unemployment rates in these areas. In the future a critical factor of the flexibility of local production systems seems to be a more explicit effort in human capital investment and in a tighter linkage between education and R&D policies.

Second, the internationalisation process is not only the result of the transformation in the organisation of large firms in the large metropolitan areas, but also of an active role of SMEs within the local production systems of the intermediate regions. Although the globalisation process is often indicated as an external constraint and a risk for the survival of the local production systems, the increasing internationalisation of these systems may be described as the gradual extension at the international level of the close relationships of co-makership which have traditionally existed at the local level and have then expanded at the interregional level between the various firms of these local production systems. In this respect, investments in the development of modern logistic and distribution services are a critical factor for a wider internationalisation of SMEs.

Third, the organisation and quality of the territory seem to be key factors in international competitiveness, implying a new relationships between urban centres and their rural hinterlands. Moreover, the increasing internationalisation of SMEs is accompanied by the increasing role of cities and regions in international relations. Thus, an efficient territorial policy seems to require a new role for rail versus road transport and municipal and regional government should develop a local 'foreign policy' aiming to promote various forms of cooperation between urban centres or regions in the framework of interregional and international networks.

The economic success of local production systems was largely determined by the almost spontaneous availability of a favourable environment. Explicit industrial public policies were largely absent, and an important role was played by an often informal but intense and effective networking, which has allowed implicit forms of coordination of the various local actors, both public and private, in the framework of various institutions such as chambers of commerce, trade fairs and other local consortia, industry associations, trade unions, local banks, and major political parties. Local and regional governments seem to have played a largely secondary role in this phase of development.

However, a major challenge to the continuation of the development process in local production systems is represented by the need for major decisions, in order to undertake large projects on the infrastructure network and on other modern collective services, which require new forms of explicit or formal coordination and a greater role for local institutions, such as the municipalities and the regional government. The claim for greater political autonomy and for federalist reform (Cappellin, 1997b; 1997c), which has characterised the political debate in the regions of North Italy during the 1990s, seems to be based on the wide consensus that economic development will increasingly depend on the ability to solve problems external to the individual firms and which could be tackled only through public policy measures taken by local and regional institutions.

References
Alessandrini P, Canullo G, 1997, "I distretti industriali marchigiani: evoluzione e prospettive" *Economia Marche* number 1, 3 – 27
Balloni V, Conti G, Cucculelli M, Menghinello S, 1998, "Modelli di impresa e di industria nei contesti di competizione globale" *L'Industria* **19** 315 – 319
Becattini G, 1979, "Dal settore industriale al distretto industriale" *Economia e Politica Industriale* number 1, 7 – 21
Becattini G, 1990, "The Marshallian industrial district as a socio-economic notion", in *Industrial Districts and Inter-firm Co-operation in Italy* Eds F Pyke, G Becattini, W Sengenberger (International Institute for Labour Studies, Geneva) pp 37 – 51
Becattini G, Rullani E, 1993, "Sistema locale e mercato globale" *Economia e Politica Industriale* number 80, 25 – 48
Bianchi P, 1993, "Industrial districts and industrial policy: the new European perspective" *Journal of Industry Studies* **1** 16 – 29
Cappellin R, 1983a, "Osservazioni sulla distribuzione inter ed intrareginale delle attività produttive", in *Industrializzazione senza Fratture* Eds G Fuà, C Zacchia (Il Mulino, Bologna) pp 241 – 271
Cappellin R, 1983b, "Productivity growth and technological change in a regional perspective" *Giornale degli Economisti e Annali di Economia* number 7 – 8, 459 – 482
Cappellin R, 1988, "Transaction costs and urban agglomeration" *Revue d'Économie Régionale et Urbaine* number 2, 261 – 278
Cappellin R, 1989, "The diffusion of producer services in the urban system" *Revue d'Économie Régionale et Urbaine* number 4, 641 – 661
Cappellin R, 1991, "The European Internal Market and the internationalisation of small and medium size enterprises", in *Regional Science: Retrospect and Prospect* Eds D E Boyce, P Nijkamp, D Shefer (Springer, Berlin) pp 317 – 338
Cappellin R, 1992, "Theories of local endogenous development and international co-operation", in *Development Issues and Strategies in the New Europe* Ed. M Tykkylainen (Avebury, Aldershot, Hants) pp 1 – 19
Cappellin R, 1993, "Interregional cooperation in Europe: an introduction", in *European Research in Regional Science 3. Regional Networks, Border Regions and European Integration* Eds R Cappellin, P Batey (Pion, London) pp 1 – 20
Cappellin R, 1995a, "Regional development, federalism and interregional cooperation", in *Competitive European Peripheries* Eds H Eskelinen, F Snickars (Springer, Berlin) pp 41 – 57
Cappellin R, 1995b, "Subsidiarity as a new economic strategy", in *Regionale Innovation: Durch Technologiepolitik zu neuen Strukturen* Ed. M Steiner (Leykam, Graz) pp 27 – 51
Cappellin R, 1997a, "The economy of small and medium-size towns in non metropolitan regions", document prepared for OECD, Programme of Dialogue and Cooperation with China, SG/CHINA/RUR/UA(97)3, OECD, Paris
Cappellin R, 1997b, "Regional policy and federalism in the process of international integration", in *Regional Growth and Regional Policy Within the Framework of European Integration* Ed. K Peschel (Springer, Berlin) pp 111 – 141

Cappellin R, 1997c, "Federalism and the network paradigm: guidelines for a new approach in national regional policy", in *European Research in Regional Science 7. Regional Governance and Economic Development* Ed. M Danson (Pion, London) pp 47–67

Cappellin R, Nijkamp P (Eds), 1990 *The Spatial Context of Technological Development* (Avebury, Aldershot, Hants)

Cavalieri A (Ed.), 1995 *L'Internazionalizzazione del Processo Produttivo nei Sistemi Locali di Piccola Impresa in Toscana* (Franco Angeli, Milano)

Cominotti R, Mariotti S, 1994 *Italia Multinazionale 1994: Le Nuove Frontiere dell'Internazionalizzazione Produttiva* (Etas Libri, Milano)

D'Antonio M, Scarlato M, 1997, "Struttura economica e commercio estero: un'analisi per le province italiane" *Economia Marche* number 2, 209–270

Falzoni A, Onida F, Viesti G (Eds), 1992 *I Distretti Industriali: Crisi o Evoluzione?* (EGEA, Milano)

Fuà G L, 1983, "L'industrializzazione de Nord-Est e nel Centro", in *Industrializzazione senza Fratture* Eds G Fuà, C Zacchia (Il Mulino, Bologna) pp 7–46

Garofoli G, 1989, "Modelli locali di sviluppo: i sistemi di piccola impresa", in *Modelli Locali di Sviluppo* Ed. G Becattini (Il Mulino, Bologna) pp 75–90

Il Sole 24 Ore various issues, Quotidiano Economico Politico e Finanziario, Milano

Putnam R D, 1993 *Making Democracy Work: Civic Tradition in Modern Italy* (Princeton University Press, Princeton, NJ)

Scott A J, Storper M, 1990, "Regional development reconsidered", The Lewis Center for Regional Policy Studies, University of California, Los Angeles, CA

Stigler G, 1951, "The division of labour is limited by the extent of the market" *Journal of Political Economy* **29** 185–193

Viesti G, 1998, "Le strategie di internazionalizzazione delle piccole imprese italiane nel nuovo quadro internazionale" *L'Industria* **19** 320–335

Williamson O E, 1981, "The modern corporation: origin, evolution, attributes" *Journal of Economic Literature* **19** 1537–1568

Regional Cooperation and Innovative Industries: Game-theoretical Aspects and Policy Implications

S Mizrahi
Ben-Gurion University of the Negev

1 Introduction
In the last decade we have witnessed a twofold transformation in the economic structure both of developed and of underdeveloped countries. First, there has been a continuous trend of privatisation and decentralisation of the public sector. Second, economies needed to transform their industrial basis from traditional (Fordist) industries into innovative (post-Fordist) and high-tech industries. Although traditional industries have operated at the national level both in terms of infrastructure and in terms of financial support, post-Fordist industries are more independent and compact in their structure.

The combination of these two trends focuses attention on the regional factor (Hilpert, 1991a). In the process of decentralisation local regions get a certain level of autonomy while, in parallel, innovative and high-tech industries are usually developed at the regional level. Various aspects of these parallel processes are discussed in the edited volume *Regional Innovation and Decentralization* (Hilpert, 1991a).

Existing studies suggest that a public policy based on decentralisation of innovative and high-tech industries (for example, through the establishment of industrial or technological parks), can accelerate economic growth and social welfare both regionally and nationally (Berra and Gastaldo, 1991; Gibb, 1985; Goldstein and Luger, 1991; Hilpert and Ruffieux, 1991; Jowitt, 1991; Oakey, 1984). Such studies also investigate the economic and structural conditions for successful decentralisation of such industries, thus offering outlines for national policy towards regionalisation (Hilpert, 1991b; Moulaert and Swyngedouw, 1991).

However, there has not been a systematic analysis of the various conflicts which may arise through such decentralisation processes at national and regional levels. Such conflicts may slow down or completely stop regionalisation or alternatively make the process inefficient. More specifically, there are three sources of conflict and inefficiency in the process of regionalisation of innovative industries.

First, decentralisation and regionalisation processes may be objected to and slowed down by the bureaucrats in the central government who will lose power and authority. In order to overcome this obstacle, a balancing pressure from the local level is required. Yet, in order to exert such pressure there is a need to mobilise the local population to act for a collective goal. This may be a difficult task if we assume that people are directed by self-interest.

Second, private firms are also guided by self-interest and therefore may not see the advantages of contributing to the development and welfare of the region. In that case, the population of the region will not benefit from the regionalisation of industries and conflicts may soon arise.

Third, in order to motivate private firms to operate in peripheral areas and contribute to the prosperity of the region, politicians who make short-term calculations in addition to private firms may prefer direct and immediate material incentives rather than the creation of endogenous and long-term incentives. Such a policy may create inefficiency.

In this chapter I address these points by using the framework of public choice and game theory. Within this framework rational players attempt to maximise their self-interest under given structural conditions. More specifically, I model the various conflicts related to regional cooperation as a collective action problem and analyse the possible strategies to solve it.

Solutions for collective action problems are usually induced by external factors or specific players who alter the conditions of the situation. Such factors either externally change the players' utility function or internally influence the players' beliefs and incentives structure. Solutions of the first type require enforcement and therefore also call for regional political organisations which specify the hierarchy of power. On the other hand, solutions of the second type are based on a tradeoff between interests and resources, thus creating internal incentives for mutual cooperation. I discuss the different aspects of these solutions in the context of innovative and high-tech industry, arguing that solutions of the second type are more efficient than enforcement for achieving regional development and growth.

2 Modeling collective action problems and political interests in regional sciences

The dynamic of regional cooperation involves both political and economic aspects. Although economic models may show the advantages of cooperation, the fact that individual behaviour is motivated by self-interest creates a strong potential for conflict in any human interaction. Indeed, most of the time self-interest appears to be a stronger motivation than considerations of the collective good. In order to explain the ways in which conflicts evolve and are solved we apply game theory.

2.1 Game theory and the regional sciences

Game theory is the study of multiperson decision problems. A game is defined as any interaction between agents that is governed by a set of rules specifying the possible moves for each participant and a set of outcomes for each possible combination of moves. By applying this theory to the regional sciences the wide range of issues related to regional development and growth can be understood from a strategic point of view rather than one looking solely at economic indicators.

Game theory assumes that rational players who act under certain structural factors strive to maximise their self-interest whilst considering the possible actions of others. It follows that the explanation of their actions concentrates on revealing the factors that might have led them to their particular decisions (Dowding, 1994). In this respect, the 'structure of the situation', modeled by a game, represents actions of other individuals, the rules of the game, the information set, and external factors.

Within a game each player chooses a strategy, and the combination of all the chosen strategies leads to the equilibrium—that is, the outcome—of the game. We choose the equilibrium concept to be used in a particular game on the basis of our assumptions and knowledge about the situation. The most common solution concept in game theory is the Nash equilibrium (Nash, 1951; Osborne and Rubinstein, 1994). Put simply, a combination of strategies is in Nash equilibrium if and only if there is no player who has an incentive to deviate from his or her strategy when all the others do not deviate. Such an equilibrium is very stable because no player will individually change strategy. When the Nash equilibrium is not socially efficient there is a certain level of conflict between the players. A game-theoretical model therefore reveals possible ways to solve conflicts and increase cooperation.

2.2 The political economy of regional cooperation

The ways in which self-interest influences the potential and dynamic of regional cooperation may be gathered in two groups of models which constitute the core of public choice theory: (1) noncooperative games which model the problem of collective action over the supply of public goods; (2) models which explain the involvement of various players in public policy: politicians, bureaucrats, and interest groups. I begin the analysis with the first group of models.

Assuming that individuals follow their cost–benefit calculations in order to achieve their private goals, participation in collective efforts to provide public goods can be beneficial for them. Yet, when there is neither a binding agreement nor coordination mechanisms, individuals will prefer not to pay the cost of participation and yet want to benefit from the outcome (Olson, 1965, pages 13 – 14). Under these conditions a collective action problem is likely to emerge.

This rationale characterises, for example, environmental problems. Because an individual may benefit from improved environmental conditions without participating in the collective efforts which brought them about, a collective action problem may arise. This problem exists at any level of interaction between unitary players—whether they are people, groups, cities, regions, or states.

The particular type of collective action problem and the game that best describes it differ according to the particular empirical situation. Among many possible models, public choice literature concentrates on prisoner's dilemma, chicken games, and assurance games (Chong, 1991, page 6; Hardin, 1982;

Taylor, 1987). A prisoner's dilemma describes a situation with neither binding agreement nor communication. There are benefits from the supply of the public good (b) but there are also costs of cooperation (c). That decision situation is described by the payoff matrix in figure 1 where the entries are the payoffs of player A.

In this game a self-interested individual can benefit more from mutual cooperation than from mutual defection ($b > c$). Yet, under the given conditions player A tries to minimise the risk of losing everything and prefers to defect regardless of what the others do. Thus, if we apply the Nash solution concept all players in the game have a dominant strategy of defection. Collective action is unlikely to emerge and the desired public good will not be supplied. This results in a unique Nash equilibrium with suboptimal outcomes. Moreover, the potential for conflict is very deep even if there are mechanisms of cooperation. Therefore, any strategy of regional cooperation should not only offer ways to solve this collective action problem, but also suggest mechanisms to neutralise the potential for conflict, thus stabilising the solution.

	Group Cooperate	Defect
Player A Cooperate	$b + c$	$-c$
Player A Defect	b	0

Figure 1. A simple prisoner's dilemma.

As mentioned, this rationale characterises cooperation dynamics at all levels of human interaction. Yet, at the macrolevel, where groups or organisations act as unitary players, the fact that aggregation of the members' interests is required worsens the cooperation problem. The leaders of such a group, organisation, city, or state must have the support of their public for cooperative policies as well as that of other intragroup players. Otherwise, there will be great internal obstacles to any policy of regional cooperation.

Indeed, within the second group of models two subgroups explain the difficulties in reaching an agreement over a common goal. One has been developed on the lines of Arrow's impossibility theorem (Arrow, 1963). These models show that, when individual preferences vary, there is no social welfare function which will satisfy both efficiency and fairness (Arrow, 1963; McKelvey, 1976). Moreover, in any policy equilibrium there are dissatisfied players who will try to change the status quo in the long run (Riker, 1980). Hence, aggregating the interests of group members to constitute a clear common goal is a complex and slow process, if it is possible at all. This aggregation problem reduces the ability of local or national leaders to set decentralisation and regionalisation of innovative industries as their policy goal.

A second subgroup of models refers to the ways in which specific players—that is, interest groups and bureaucrats—influence public policy. Both are characterised by rent-seeking behaviour, attempting to maximise privileges or budget (Bendor, 1990; Mitchell and Munger, 1991; Niskanen, 1971). The interactions between these players, the public, and government create great difficulties in reaching an agreement over local or national interests. In particular, these players may consider initiatives of decentralisation and regionalisation as a threat, because their privileges may be abolished and competition will grow. For example, cooperation between municipalities to advance regional interests threatens bureaucrats in central government, because they may lose power and control. Similarly, regionalisation of innovative industries may increase competition, thus raising objections from interest groups which may lose their privileges and favourite conditions. On the basis of these groups of models, we may conclude that a successful strategy of regional cooperation should offer ways to bridge such conflicting interests, thus creating incentives for central players to participate in initiatives of regional cooperation.

Hence, a strategy of regional cooperation should include two stages: one leading players to internalise the advantages of regional cooperation and a second overcoming conflicts between specific organised interests. In the following section I discuss the theoretical aspects of such a strategy, concentrating on ways to transform the prisoner's dilemma.

3 Solutions to collective action problems and regional cooperation

Public choice literature suggests several rationales to explain cooperative solutions to collective action problems. Taylor suggests that there are two types of solutions: 'internal' solutions that do not involve changing the payoff structure of the prisoner's dilemma and 'external' solutions that work by changing individual preferences and expectations (Taylor, 1987, page 22; Sened, 1991).

An internal factor which may explain cooperation under the conditions of a prisoner's dilemma is the fact that social players are involved in an open-ended n-person continuous prisoner's dilemma supergame (Axelrod, 1984; Taylor, 1987). When players in such a game adopt a tit-for-tat strategy, in which each player begins with cooperation and continues to cooperate as long as the others do, mutual cooperation will evolve if the players do not discount future payoffs too much. In other words, when players do not know when the game will end, they calculate their moves to the long term and attach a certain value to future payoffs for cooperating in the present.

A similar uncertainty is created by incomplete information (Kreps et al, 1982). Cooperation may therefore also emerge when the game is finite but the players have incomplete information.

Nevertheless, Fundenberg and Maskin (1986) prove by a formal model of the folk theorem that under the above conditions of the prisoner's dilemma supergame there is an indefinite number of possible equilibria. Therefore, the tit-for-tat strategy or incomplete information explain how different outcomes are possible in addition to mutual defection; but in order to explain a particular outcome we need to specify more parameters.

Cooperation may be also explained by group size as an internal factor (Chong, 1991, page 41; Olson, 1965, pages 53 – 55; Ostrom, 1990). In a small group, where the contribution of each individual is more apparent than in large groups, factors such as social pressure play a major role. Furthermore, because large groups are composed of small subgroups, we can also use the parameter of social pressure to explain some forms of mass collective action (Dowding, 1994). Thus, another criterion that may explain cooperation in a large group is a high degree of internal interaction within it.

Although internal solutions are not supposed to lead by themselves to spontaneous collective action, external solutions explain particular outcomes. The set of external solutions assumes the existence of a third party—the government or social entrepreneurs—who are able to change either the payoffs associated with each strategy or the players' expectations (Taylor, 1987, pages 22 – 23). The payoff matrix is altered by negative and/or positive incentives. That is, penalties for those who do not cooperate and rewards for those who cooperate. If adequately applied, such a direct method of changing the payoff function transforms the prisoner's dilemma into a game with a dominant strategy of cooperation. Yet, because such solutions are based on external incentives, players hardly internalise the advantages of cooperation, thus destabilising the solution in the long term. Moreover, direct incentives strongly increase enforcement and transaction costs, and economic efficiency declines (North, 1990). It follows that a direct transformation of players' utility functions through positive or negative incentives is likely to produce results in the short term but is most likely to be unstable in the long term.

Another strategy to transform the prisoner's dilemma is to include structural changes which lead players to learn through experience the advantages of cooperation. Such changes are termed 'indirect external solutions'. Chong (1991, page 104) suggests that the first step towards the solution of a collective action problem is a transformation of the payoff matrix which represents the situation of the prisoner's dilemma into an assurance game. Political entrepreneurs provide 'selective social and expressive benefits' to achieve this.

"In order to coax an individual to cooperate under these more difficult conditions [of the prisoner's dilemma], it is necessary to provide him with an added inducement to participate over and above the benefits he will receive from the public good itself. Selective incentives are private benefits that can be enjoyed only by those who cooperate; therefore cooperators are rewarded with these additional benefits, whereas noncooperators are denied them" (Chong, 1991, page 31).

Such alternative incentives or by-products may be material (Olson, 1965, page 60), social, or psychological (Chong, 1991, page 32). The transformed game is presented by the payoff matrix in figure 2 where, for player A, there

	Group	
	Cooperate	Defect
Player A Cooperate	$b + b' - c$	$-c$
Player A Defect	b	0

Figure 2. A simple assurance game.

are costs of participation (c), benefits from the public good (b) and benefits from the selective incentives (b').

In this game an individual will participate if the rest of the group does and $b' > c$. If all the others do not participate, he or she will not. If we assume symmetric individuals, the players have no dominant strategies and, therefore, multiple equilibria are possible. Because formally no communication exists, they will all either cooperate or defect.

The transformation of the prisoner's dilemma into an assurance game creates mutual dependence between players and therefore strengthens group identity. The fact that players can have the positive incentive only if all other players participate motivates players to act as a coherent group. In that way the level of conflict is considerably reduced. Yet, in order to guarantee cooperation, a coordination mechanism is required. As will be explained later, this kind of transformation can be a very efficient strategy for regional cooperation.

Another indirect external solution may be transforming the prisoner's dilemma into an n-person chicken game by fixing a number of required participants (ω) for the supply of a certain public good. There are costs of participation (c) and benefits from the public good (b). In deciding whether to cooperate or defect an individual has to consider three possibilities—that is, the number (N) of participants excluding A is either less than $\omega - 1$, equal to $\omega - 1$, or greater than $\omega - 1$. The payoff matrix is presented in figure 3, where entries are the payoffs of player A.

In this game, there are two possible equilibria: (1) none will participate and the public good will not be supplied; (2) the number of participants will be exactly as required and the public good will be supplied through partial participation. The condition for a cooperative equilibrium is that players will

	$N < \omega - 1$	$N = \omega - 1$	$N > \omega - 1$
Player A Cooperate	$-c$	$b - c$	$b - c$
Player A Defect	0	0	b

Figure 3. An n-person chicken game or participation game.

believe that they are decisive. A decisive player faces a situation where he or she prefers to cooperate if the others defect and to defect if someone else carries out the task (Taylor, 1987, pages 45 – 48). To understand the structural conditions under which an individual is most likely to believe subjectively that he or she is decisive, Taylor proposes that $P > c/b$, where P is the objective probability of a player being decisive, c is the cost of participation, and b is the benefit from the public good. Because the value of P grows as the group gets smaller and the threshold, ω, gets higher, we may argue that the subjective probability that players assign to being decisive grows as the group gets smaller, the costs of participation are lower, and the benefits from the public good are significantly higher.

This rationale characterises 'red-line' policies for air pollution or water resources. They fix a threshold under which defection may cause great damage. Therefore, the benefits from the public good are very high. When individuals internalise the danger embodied in reaching the red line, and they actually face such a danger, they will probably participate in collective efforts to improve air or water quality. People will not participate, however, if air or water quality is above the threshold. As will be elaborated in the next section, this rationale can be used in developing a strategy of regional cooperation, thus reducing the potential of conflict between the various layers involved in decentralisation and regionalisation of industries.

4 Discussion: strategies of regional cooperation

The analytical framework developed so far points to two aspects of the conflicts embodied in dynamics of decentralisation and regionalisation of innovative industries. First, guided by self-interest regionalised industries may reach a noncooperative equilibrium between themselves as well as towards the region. When private firms have to cooperate between themselves in order to solve collective-good problems such as pollution or to contribute to the welfare of the region, the free-rider problem may arise. Second, other players such as interest groups and bureaucrats may prefer noncooperation rather than cooperating with decentralisation policies. They then intensify the conflict between them and the government, creating the conditions of the prisoner's dilemma.

Hence, the two types of problems can be basically described by the prisoner's dilemma. In analysing the solutions to this dilemma in the previous section I concentrated on the first type of problem, though the same principles of solution apply to the second type as well.

Realising those problems is the first stage in forming decentralisation and regionalisation policies. In order to solve them and neutralise the three sources of conflict which were mentioned in the introduction, one should adopt a long-term strategy with various components based on the rationales suggested here. At the core of such a strategy stands the transformation of the prisoner's dilemma into another game with a lesser degree of conflict. When the transformation is not entirely external but, rather, creates the conditions for players to learn for themselves the advantages of cooperation, long-term regional cooperation is more likely to evolve. Therefore, certain structural changes which will trigger learning processes among the players are crucial for sustainable regional cooperation both between firms and between firms and the population of the region. Positive or negative incentives alone will hardly achieve long-term cooperation in these dimensions.

Concentrating on indirect external solutions, the analysis suggests several guiding principles for a long-term strategy of regional cooperation. First, selective incentives can neutralise conflicts at any level. By eliminating the motivation to free ride and transforming the prisoner's dilemma into an assurance game the conflict of interest is eliminated as well. Moreover, because players need to mobilise others in order to gain the selective incentive, they are most likely to internalise the advantages of regional cooperation. This method can be used at all levels to neutralise the objections of interest groups and bureaucrats. Such players will object to decentralisation and regionalisation policies which may endanger their power and privileges. To avoid that possibility they have to be compensated through other channels. The rationale of selective incentives can then be used to form a cumulative compensation scheme based on tradeoffs. That is to say, when all privileges are abolished and compensation is given only if a certain level of cooperation is achieved, not only will those players have an interest in participating but they will also try to involve others as well.

Second, red-line policies will probably create a certain level of regional cooperation. Such policies can be combined with the rationale of selective incentives, meaning that incentives will be given only if a certain level of cooperation is achieved. Thus, when a certain level of cooperation within a region is presented as a condition for significant incentives, the benefits from the public good are extremely high and players are likely to cooperate.

Strict regulations which fix the required level of public goods supplied within a particular region differ among regions according to geographical location and power struggles. This may lead to asymmetric interests between firms with regard to certain public goods such as air or water quality. The fact that some players are more interested than others in the public good guarantees that a certain level of the public good will be supplied by those groups or regions which attach high value to its supply. On the other hand, those players who consider the specific public good less beneficial are most likely to free ride.

Hence, red-line policies will produce an asymmetry of interests between players, leading to partial supply of public goods and a potential for conflict between those who cooperate and those who do not. For example, because of asymmetric interests and different geographical conditions, regions operate under different regulations with regard to public goods such as preservation of the environment and contribution to the welfare of the region. This means that some regions contribute more than others to the collective good and therefore the asymmetry of interests may turn into a conflict about the distribution of natural resources. It follows that the success of any strategy of regional cooperation depends on organisational mechanisms which coordinate policies on natural resources.

Another component which was indicated in the analysis is group size. Different models reach a similar conclusion—that is, cooperation is more stable in small groups. Taking organisations, cities, or states as unitary players we may argue that regional cooperation at any level should not include too many players. Clearly, the exact number of players which will optimise the outcomes of regional cooperation depends on the parameters of the situation, but, in principle, when a group of firms, cities, or states which constitute a region grows above a certain critical mass, the motivation to free ride increases as well. In such a case, it may be more efficient to form several subregional blocs within one region—each with common characteristics—and to encourage competition between them. In that way we create controlled competition between relatively small groups of firms, cities, or states.

In this chapter I have analysed the various conflicts which may arise through decentralisation and regionalisation of innovative industries at the national and regional levels, the possible solutions to the problems, and the size limit for any regional cooperation. However, the general framework can be applied to many aspects and dimensions of the regional sciences.

Acknowledgements. This chapter was presented as a paper at the International Conference on: "Mediterranean Regions Economics and Sustainable Development", Aix-en-Provence, 19 – 20 June 1997.

References
Arrow K, 1963 *Social Choice and Individual Values* (John Wiley, New York)
Axelrod R, 1984 *The Evolution of Cooperation* (Basic Books, New York)
Bendor J, 1990, "Formal models of bureaucracy: a review", in *Public Administration: The State of the Discipline* Eds N B Lynn, A Wildavsky (Chatham House, Chatham, NJ) pp 373 – 403
Berra M, Gastaldo P, 1991, "Science parks and local innovation policies in Italy", in *Regional Innovation and Decentralization* Ed. U Hilpert (Routledge, London) pp 89 – 111
Chong D, 1991 *Collective Action and the Civil Rights Movement* (Chicago University Press, Chicago, IL)
Dowding K M, 1994, "The compatibility of behaviouralism, rational choice and 'new institutionalism'" *Journal of Theoretical Politics* **6** 105 – 117
Fundenberg D, Maskin E, 1986, "The folk theorem in repeated games with discounting or with incomplete information" *Econometrica* **54** 533 – 554
Gibb J M (Ed.), 1985 *Science Parks and Innovation Centers: Their Economic and Social Impact* (Elsevier, Amsterdam)
Goldstein H A, Luger M I, 1991, "Science/technology parks and regional development: prospects for the United States", in *Regional Innovation and Decentralization* Ed. U Hilpert (Routledge, London) pp 133 – 153
Hardin R, 1982 *Collective Action* (Johns Hopkins University Press, Baltimore, MD)
Hilpert U (Ed.), 1991a *Regional Innovation and Decentralization* (Routledge, London)

Hilpert U, 1991b, "Regional policy in the process of industrial modernization: the decentralization of innovation by regionalization of high tech?", in *Regional Innovation and Decentralization* Ed. U Hilpert (Routledge, London) pp 3–34

Hilpert U, Ruffieux B, 1991, "Innovation, politics and regional development: technology parks and regional participation in high tech in France and West Germany", in *Regional Innovation and Decentralization* Ed. U Hilpert (Routledge, London) pp 61–87

Jowitt A, 1991, "Science parks, academic research and economic regeneration: Bradford and Massachusetts in comparison", in *Regional Innovation and Decentralization* Ed. U Hilpert (Routledge, London) pp 113–131

Kreps D M, Milgrom P, Roberts J, Wilson R, 1982, "Rational cooperation in the finitely repeated prisoner's dilemma" *Journal of Economic Theory* **27** 245–252

McKelvey R D, 1976, "Intransitives in multidimensional voting models and some implications for agenda control" *Journal of Economic Theory* **12** 472–482

Mitchell W, Munger M, 1991, "Economic models of interest groups: an introductory survey" *American Journal of Political Science* **35** 512–546

Moulaert F, Swyngedouw E, 1991, "Regional development and the geography of the flexible production system: theoretical arguments and empirical evidence", in *Regional Innovation and Decentralization* Ed. U Hilpert (Routledge, London) pp 239–265

Nash J, 1951, "Non-cooperative games" *Annals of Mathematics* **54** 286–295

Niskanen W A, 1971 *Bureaucracy and Representative Government* (Aldine-Atherton, New York)

North D C, 1990 *Institutions, Institutional Change and Economic Performance* (Cambridge University Press, Cambridge)

Oakey R P, 1984, "Innovation and regional growth in small high technology firms: evidence from Britain and the USA" *Regional Studies* **18** 237–251

Olson M, 1965 *The Logic of Collective Action* (Harvard University Press, Cambridge, MA)

Osborne M J, Rubinstein A, 1994 *A Course in Game Theory* (MIT Press, Cambridge, MA)

Ostrom E, 1990 *Governing the Commons* (Cambridge University Press, Cambridge)

Riker W, 1980, "Implications from the disequilibrium of majority rule for the study of institutions" *American Political Science Review* **74** 432–447

Sened I, 1991, "Contemporary theory of institutions in perspective" *Journal of Theoretical Politics* **3** 372–402

Taylor M, 1987 *The Possibility of Cooperation* (Cambridge University Press, Cambridge)

Industrial Trade Clusters in Action: Seeing Regional Economies Whole

E M Bergman
University of Economics and Business Administration, Vienna

1 Introduction

Industrial clusters, which consist of firms linked actively together and in close spatial proximity, have become a very popular concept, one that is now embraced widely. It is difficult to identify another equally obscure concept that appeals to such a broad spectrum of academic disciplines, professions, and even lay people. By breadth, I refer first to the fitful and diverse theoretical traditions upon which the concept freely draws, from the seminal work of Alfred Marshall in the late 19th century, through mid-20th-century elaborations of agglomeration economies and externalities advanced by numerous economists, economic geographers, and regional scientists, and propelled now into the 21st century by a widening corps of enthusiasts drawn from nearly every social science, plus newly interested business strategists, international trade theorists, and economic development specialists (see Feser, 1998; this volume).

But its breadth, some might say its present ambiguity, results also from so-called 'Rashomon effects' (named after Kurosawa's film): different perspectives are brought to the cluster concept by the swelling ranks of its many diverse adherents. Certain adherents favor only one particular version of what constitutes the *essence* of a 'cluster,' often reserving—sometimes actively defending—exclusive use of the term for their version alone, no matter what the application or purpose. In this chapter I acknowledge such ongoing theoretical-cum-taxonomic disputes, mainly as the point of departure for first comparing approaches and then presenting one essential variant that is useful for framing analyses of regional economies.

2 Micro versus macro study approaches

Investigations of industrial clustering behaviour among firms proceed initially from either micro or macro perspectives. Both approaches account in different ways for many of the same strong, frequent, and often striking interdependencies firms enjoy with one another. Interdependent firms are typically colocated within fairly close proximity, the reasons for which are adduced as resulting from shared access to one or more uniquely supplied local resources. These include the availability of valuable infrastructure services that supply indirect (and *underpriced*) inputs, various economic or technological spillovers from other firms and regional institutions, or embedded social advantages that work almost invisibly ('in the air') to reduce transaction costs and improve the efficiency of trust-based market and governance systems. These may arise in very subtle ways, perhaps initially from historical accidents of 'path-dependent' advantage, where the sustained velocity of initial regional advantage continues to propel long-term success of firms in one or several linked industries. To these explanations of colocation must be added the more typical traded interdependencies, such as factor supplies of shared direct inputs [principally shared labour supplies, according to Dumais et al (1997)] and typical contractual relations between local suppliers, including service sector firms. All factors mentioned to this point are classic elements of agglomeration and localisation economies, many of which have been employed in earlier studies to document the presence and composition of industrial clusters and complexes.

Microstudies begin with some of the same broad insights of why firms successfully colocate and then redirect their principal focus to how groups of similar-sector firms cooperatively share production capacities, markets, labour, and technologies, reserving for such *Italianate* arrangements the term 'cluster'. The underlying cooperative behaviour is seen as a channel-following current, where:

> "The 'current' of a working production system [is] less easily detected, often embedded in professional, trade, and civic associations, and in informal socialization processes ... [such] that a cluster is a 'geographically bounded concentration of interdependent businesses with active channels for business transactions, dialogue, and communications, and that collectively shares common opportunities and threats'" (Rosenfeld, 1997, page 10).

Rosenfeld also describes such collectivities as "... *like and related businesses* [that] cluster geographically and become increasingly interdependent" (page 3).

Beyond the policy questions that arise concerning the cyclical stability of single industry massing (Fritz et al, this volume) is the equally relevant point that such clusters consist of very similar types of firms selling similar consumer or design-intensive products. The supplier chains often consist simply of commodity or raw material inputs that are transformed by cooperating producers employing similar production technologies and cooperative cultures; being relatively shallow, supply chains are of comparatively little importance to this definition of clusters.

The richly detailed accounts of these uniquely successful industrial groupings[1] are instantly familiar and compelling[2], particularly to politicians and policymakers desperately seeking immediate solutions to regional economic

---

[1] *Unsuccessful* groupings of similar industries, lacking inherent interest to study sponsors, remain relatively unresearched, therefore leading to selection bias in available scholarship. In the absence of studies that investigate *why* certain firm clusters are unsuccessful, we cannot be sure which factors are responsible for cluster success and which are simply found in clusters everywhere.

[2] "As I mentioned at the beginning of this lecture, in 1895 the teenaged Miss Evans made a bedspread as a gift. The recipients and their neighbors were delighted with the gift, and over the next few years Miss Evans made a number of tufted items, discovering in 1900 a trick of locking the tufts into the backing ... [2-paragraph expansion traces origins of carpet cluster] ... And so the little Georgia City [of Dalton] emerged as America's carpet capital" (Krugman, 1991, pages 60–61).

problems, while they are also of occasional use to theorists who wish to illustrate far more complex concepts. As in other ethnographic inquiries, these studies are so *uniquely* etched that enduring lessons and generalisations prove difficult to distil or to apply in other precincts.

Further, such studies by definition limit attention to evidence of 'currents' flowing among similar sector firms that are best detected up close and at fairly small geographic scales. This approach restricts its view to a *single visible collection* of similar-sector firms, thereby overlooking linkages that some of its members may have with regionally colocated firms from very different sectors, or robust clustering of other sectors. It is therefore not surprising that microstudies document only *one cluster per region*. The apparent indifference to the presence of additional clusters, particularly those based on alternative criteria or detectable only from a wider spatial view, is the result mainly of micro-oriented investigations of a priori cluster definition. An implication is that significant instances of regionwide industrial clustering go unrecognised by microstudies.

Macrobased definitions start instead by analysing transaction networks of universally documented flows that channel various interactions among all active industrial members of a trade-based economy, increasingly one of the multinational trading blocs (for example, NAFTA, EU). These networks are to be distinguished from the sole concern for internal layerings of a national economy that animated much early analysis of French *filière* (Jacobson and Andréosso-O'Callaghan, 1996, pages 118–120). Network flows consist of communication and transportation links, technology or patent diffusion and exchange, or interindustry trade, as primary examples. Various analytic techniques are applied to observed flow data as a means of detecting repeated transaction tendencies among subsets of all firms and industries, which then defines a flow-based 'cluster'. The most common flows are intersectoral factor purchases and product sales, as captured by industrial input–output tables. This approach to traded interdependencies represents what one OECD research team calls a 'value-chain' cluster (Roelandt et al, 1997, page 3):

"Clusters can be characterised as being economic networks of strongly interdependent firms, knowledge producing agents and (demanding) customers, linked to one another in a value-adding production chain ... (which) focuses on the linkages and interdependence between actors in a production chain when producing products and services and creating innovations."

Value-chain macrostudies capture highly probable interdependencies, mainly the interindustry trade of advanced industrial goods that Krugman (1991) and other trade theorists now stress, thus drawing attention to *industrial trade clusters*. Comparative production advantages brought about by reduced trade barriers, increasing returns, declining real costs of transportation, and the revitalised significance of agglomeration economies increase a region's (or nation's) focus on what can be produced competitively for export trade, and to purchase at lower cost from others that which cannot be made competitively.

Regions can become very significant nodes in expanding trading regimes, particularly as they specialise in production of a narrower, trade-linked range of products, and to produce them in ever larger volumes for sale in national-to-global markets. Increasing returns permit the growth of larger scale firms and industries in specialised regions to produce more efficiently and cheaply, and to ship production easily throughout open markets. Industries dependent upon strong supplier ties colocate in the same or nearby regions to take advantage of jointly valued production factors, creating afresh the familiar preconditions for trade-driven industrial clusters.

The 'currents of cooperation' observed in microclusters by Rosenfeld are instead present as 'collaborative arrangements' (Polenkse, 1998) among the majority of industrial trade cluster firms, because even fiercely competitive firms seldom prosper nor long survive in absolute isolation. Competitive firms rely utterly upon supplier agreements to make timely deliveries of high-quality intermediate inputs and services. They also learn about and continually incorporate key technological innovations and product improvements available through collaborative contacts with nearby suppliers of inputs, production machinery, and other forms of capital equipment. Finally, firms rely upon informal exchange of information, technology, and specialised knowledge that routinely flows between trading partners, a practice many acknowledge widely as an important source of unpriced externalities that improve individual and collective competitiveness, extending even to distant trade partners (Keller, 1997).

Macroclusters based on trade flows remain for some mainly abstractions with little recognisable content, even to business managers and procurement or sales officers whose daily commerce is presumably modeled. Rosenfeld cites Doeringer and Terkla (1995) approvingly along these lines:

"Although interindustry transactions incorporated within production channels can sometimes be detected in input–output tables, *neither the character of relationships among firms nor the benefits of clustering* can be discerned in this way" (1997, page 10, emphasis added).

Analysts comfortable with macro-evidence and approaches appreciate the comprehensive scope of clustering transactions studied, but also warn against direct application of insights to particular regions, because industrial trade clusters necessarily remain *probabilistic artifacts* that represent data extracted from a weighted sample of firms in widely varying regional economies.

3 Regional mesolevel clusters

In this chapter I present a regional mesolevel resolution of the two extremes. I do so partly in the interest of finding some common ground between extreme-scale approaches, but principally to advance our understanding of regions that host various industrial cluster constellations. Seeing a regional economy 'whole' requires that we do not accept macrolevel cluster definitions, that is, derived at national strategic or trading bloc levels, as sufficient evidence of their presence in regions. At the same time, 'wholeness' means that we include cluster links and additional whole clusters that remain unrevealed at microscales, which means we do not restrict our attention to the single most visible cluster of a region. Ideally, we want to know all the major clusters present in a region, their sectoral composition, their spatial concentration, and their essential dynamics.

One approach could involve the simple scaling down of strategic level analyses from national to city or regions, often using 'Porterian' or similar industry-policy screening principles[3]. Another would retain macrodefinitions of industrial clusters as working hypotheses that are then 'elaborated' through representative case-study inquiries. One such approach compiled several macrocluster definitions from a review of existing studies of Austrian industry clusters (Hutschenreiter, 1994; Peneder, 1994; Weiss, 1994), then mapped their relative presence in representative regions, and conducted several in-depth case studies designed to inform UNIDO's developing-nation clients of various cluster dynamics and governance (Bergman and Lehner, 1998). In this chapter, I report a more comprehensive variant of the national-to-regional approach; it is based on previous work in which *all* major industrial trade clusters in the United States were estimated and integrated them with highly detailed firm-level evidence in one US state.

Important distinctions should now be made between macro, micro, and regional levels of cluster analysis. First, industrial trade clusters are estimated, based on average US macro input–output purchasing patterns observed among groups of producers engaging in trade. Macrobased industrial clusters that result are then considered as hypothetical 'templates', which constitute the fullest possible pattern of interindustry trade that occurs among cluster members. Microdata for firms and industries of specific subnational areas (in this case, manufacturing firms for North Carolina and its many spatial subdivisions) are subsequently entered in each cluster's template of industries, and the resultant configuration constitutes a state or regional trade cluster. Rather than a true hypothesis, the template is used more as a 'counterfactual', in the sense that state or regional configurations are compared by using national cluster templates to detect analytically useful variations or key segments that permit strategic policy opportunities.[4]

---

[3] A series of such studies have been performed in many OECD countries, some of which have been recently documented (Drejer et al, 1997; Rouvinen and Ylä-Antilla, 1997; Strenberg and Strandell, 1997), and many US cities, regions and states have also adopted variants of this approach to identify potential industry clusters (http://www.hhh.umn.edu/Centers/SLP/edweb/example.htm).

[4] Microlevel firm-panel data for specific regions of North Carolina were available to apply national industry trade clusters including detailed information about every manufacturing firm (a unique identification number, total employment, total wages, branch versus single firm status, and 4-digit SIC). The panel data apply to postal codes (1100 total) for two observation years (1989 and 1994). Wage and SIC detail permit further derived inferences about cluster members, for example, estimates of value added, technology intensiveness, etc; the two-year panel observations permit an understanding of underlying processes of cluster formation and change for whole regions.

4 Manufacturing base economy: twenty-three industrial trade clusters

Analysis of the US manufacturing economy revealed twenty-three industrial *clusters* or extended input–output chains. These clusters represent aggregations of closely related individual *sectors* (industries). The 1987 US input–output tables were the basic source of information on industrial interdependencies. These tables (1) identify all significant sales of one industry's output to all other industries; (2) show how much of all other industries' outputs are purchased by each industry as production inputs; and (3) with some manipulation, reveal implied sales and purchases between sectors that occur through intermediate industries. Thus both direct and indirect relationships between buyers and suppliers are considered when grouping sectors. A cluster, for example, may include seemingly unrelated industries (for example, vacuums and vehicles) that purchase (sell) similar input (output) mixes, though they may not trade significantly with each other. A summary of the methodology is presented in the appendix.

An important distinction in this chapter is made between *industrial trade* and *regional trade* clusters. Industrial trade clusters are based on national input–output patterns observed among groups of producers that do, in fact,

Table 1. Twenty-three US benchmark manufacturing clusters (source: Bergman and Feser, 1997; Bergman et al, 1997).

Cluster	Primary only		All sectors	
	no. of sectors	2-digit SIC sectors	no. of sectors	2-digit SIC sectors
1 Metalworking	93	9	116	10
2 Vehicle manufacturing	35	13	58	16
3 Chemicals and rubber	20	6	48	14
4 Electronics and computers	25	6	38	8
5 Packaged foods	21	1	44	5
6 Printing and publishing	21	5	32	8
7 Wood products	16	2	23	6
8 Knitted goods	13	3	23	5
9 Fabricated textile products	12	7	22	9
10 Nonferrous metals	8	3	14	4
11 Canned and bottled goods	6	2	12	2
12 Leather goods	6	1	9	1
13 Aerospace	5	2	10	6
14 Feed products	5	1	10	2
15 Platemaking and typesetting	4	3	14	7
16 Aluminum	4	3	9	4
17 Brake products	4	3	9	4
18 Concrete, cement, and brick	3	1	8	2
19 Earthenware products	5	1	8	1
20 Tobacco products	4	1	4	1
21 Dairy products	3	1	6	1
22 Petroleum	3	2	5	2
23 Meat products	2	1	5	2

currently engage directly or indirectly in trade. These national clusters provide essential templates to benchmark the extent of clustering and likely trade occurring among regional manufacturers. The full list of national clusters and their sectoral complexity are depicted in table 1.

Regions to which such templates are applied will be shown to have key segments of industrial trade clusters in their economic base, but regions are quite unlikely to include the full set of industries that constitute any single national trade cluster. A substantial regional representation of firms in key trade-cluster segments implies that local producers are also engaged *presently* in trade; however, substantial representation alone is not conclusive evidence that regional trade necessarily occurs. Trade also increases the likelihood that firms in a given regional cluster interact through informal channels with other local firms in that cluster, more than with firms in wholly different regional clusters.

5 Microcluster singularity versus regional cluster wholeness

Microclusters include the highly localised examples of the 'Third Italy', where nearly all economic activity takes place in small, tightly bounded territories—usually within 'industrial districts'—devoted to the production of one or very few related cluster products. The truly homogenous nature of an 'Italianate' cluster means that the surrounding territory is 'wholly' devoted to it, and this implies that the regional economy is essentially captured in a comprehensive view of the industrial district cluster. Indeed, much of local governance, society, and culture effectively nurture its principal cluster.

Single microclusters consisting of one or few sectors have also been extended to more territorially diffused locations, such as sections of the US Appalachian Region, which extends from New York to Alabama (RTS, 1997), or of several large US states and cities. However, microclusters based on industrial district assumptions are far less likely to afford an adequate strategic view in the USA and other Western economies, simply because of the far more heterogenous mix of industries that constitute their regions. The handful of sectors included in a typical microcluster reduces drastically one's view of the full regional economy and how it functions.

These ideas are perhaps best illustrated with contrasting approaches applied to approximately the same region: thirteen counties in North Carolina and Virginia that represent part of the Appalachian Regional Commission (ARC) region [Regional Technology Strategies (RTS) study] versus Western Economic Development Partnership (WEDP) region [North Carolina Alliance for Competitive Studies (NCACTS) study], which consists of twenty-two NC counties. The apparel and textile industries are employing fewer workers over time, even though they still dominate the economies of their common counties (44% of NC counties from the RTS study are also in the NCACTS study). Each approach included the same general cluster: the knitting mills cluster in the ARC subregion and the knitted goods cluster in the WEDP region.

The RTS group (Liston, 1997) produced an in-depth case study that focused on four main industry sectors (firms): knit fabric (8), knit underwear and outerwear (30), hosiery (88), and miscellaneous knit (6). Suppliers of various key inputs and machinery are related sectors that also received attention, all of which together are said to employ about 20 000 workers (percentage of region is unstated), although the four main knitwear sectors are of principal interest. In addition, the case study documented the role of various service industries, distributors, industry membership organisations, and public services and infrastructure. The many and various connections, linkages, and cooperative agreements among all the contributing cluster elements were detected from face-to-face interviews and reviews of existing documents, as summarised in figure 1.

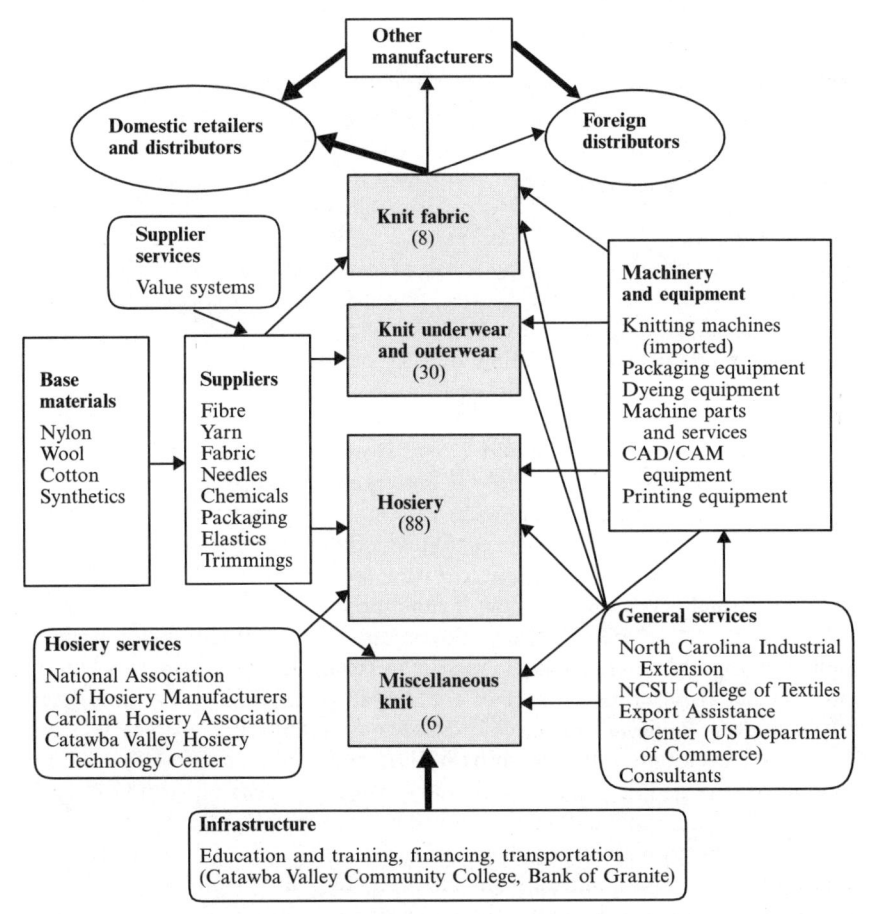

Figure 1. Knitting mills in North Carolina, Virginia, and Tennessee (figures in parentheses indicate number of establishments) (source: RTS, 1997).

Table 2. Basic data by detailed industry, by cluster, for the Western Economic Development Partnership region (source: Bergman et al, 1997).

SIC	Industry description[a]	Load[b]	Firms[c]			
			89	94	ind.[d]	small[e]
2253	Knit outerwear mills	0.93	5	2	50	50
2257–8	Knit fabric mills	0.93	11	12	75	50
2254	Knit underwear and nightwear mills	0.93	1	1	0	0
2284	Thread mills	0.92	5	4	25	0
2259	Knitting mills, nec	0.91	2	2	100	50
2395	Pleating and stitching	0.88	3	4	100	50
2397	Schiffli machine embroideries	0.85	1	1	100	0
2252	Hosiery, nec	0.85	16	21	90	62
224	Narrow fabric mills	0.78	5	5	40	20
3965	Fasteners, buttons, needles, and pins	0.77	1	1	0	0
2251	Women's hosiery, except socks	0.71	6	7	86	29
231–8, 3999	Apparel made from purchased materials	0.70	75	74	62	41
2269, 2281–2	Yarn mills and finishing of textiles, nec	0.65	16	19	53	32
315	Leather gloves and mittens	0.58	1	1	0	0
221–3, 2261–2	Broadwoven fabric mills and fabric finishing plants	0.57	30	28	68	50
227	Carpets and rugs	0.52	15	14	93	86
2298	Cordage and twine	0.52	1	1	0	0
2296	Tire cord and fabrics	0.51	1	1	0	0
2299	Textile goods, nec	0.49	4	4	75	75
2297	Nonwoven fabrics	0.47	2	4	25	25
2824	Manmade organic fibres, except cellulosic	0.44	2	2	50	50
2295	Coated fabrics, not rubberised	0.36	1	1	0	0
2392	Housefurnishing, nec	0.36	7	11	73	55

[a] nec not elsewhere classified.
[b] Load is the principal component correlation of any specific industry with all other industries that constitute the knit goods trade cluster.
[c] Average value for the third quarter of the year.

Table 2 (continued)

Employment[c]				Wages		
89	94	change 89–94	% change 89–94	total[f]	weekly[g]	% change 89–94
737	244	−493	−66.8	835	263	8.9
1.176	1.526	350	29.8	6.225	314	6.3
0	0	0	0.0	0	0	0.0
1.491	1.658	167	11.2	6.158	286	−2.8
408	568	160	39.3	2.820	282	17.0
190	199	8	4.4	695	269	16.0
327	271	−56	−17.0	907	257	10.9
1.223	1.254	31	2.5	4.172	256	1.0
602	405	−197	−32.8	1.294	246	−0.8
0	0	0	0.0	0	0	0.0
2.979	1.995	−984	−33.0	6.907	266	4.5
9.411	8.818	−593	−6.3	29.133	254	11.4
3.443	3.051	−392	−11.4	12.691	320	8.1
0	0	0	0.0	0	0	0.0
9.007	9.056	49	0.5	42.832	364	−1.6
985	1.018	33	3.4	4.332	327	−1.6
38	0	−38	−100.0	0	0	−100.0
0	0	0	0.0	0	0	0.0
161	94	−67	−41.7	356	292	−10.6
589	802	212	36.0	5.413	519	1.5
2.489	1.000	−1.488	−59.8	7.394	569	−9.6
0	0	0	0.0	0	0	0.0
1.333	1.342	9	0.7	4.753	273	16.8

[d] Ind. independent firms are production establishments that are wholly owned and not branches of a parent firm.
[e] Small firms employing fewer than 50 workers.
[f] Total wage bill in 1994 in thousands of 1989 dollars.
[g] Weekly wage per worker ($)

We can now compare the regional trade cluster approaches to see how they differ in results and interpretation. First, there are some obvious similarities simply because firms (number) in the same general cluster are studied: knit fabric (14), knit outerwear and underwear (3), and hosiery (28) firms are equivalent sectors. Second, the NCACTS approach (Bergman et al, 1997a; 1997b; Feser and Bergman, 1998) includes a far more detailed group of additional sectors that trade with the main knitted goods sectors and with each other: apparel made from purchased knit fabric (74), yarn mills (19), narrow fabric mills (5), pleating and stitching (4), thread mills (4), and a scattering of other sectors that routinely trade with the WEDP region cluster.[5]

We see in figure 2 which *specific segments* of the full knitted goods cluster are concentrated in this region, and which are far less fully represented, relative to the national trade cluster. As we know all the sectors that trade with each other within the cluster, regional development policy might be focused on the most beneficial cluster segments, particularly those that deserve special attention to retain production in the face of overall cluster declines. Recruitment policies can also be adjusted to focus on missing or underrepresented cluster segments that would value existing comparative advantages of the region, assuming such segments are thought beneficial as well. Regional development officials can augment their knowledge of regional comparative advantages by considering additional relevant facts, particularly the relative strength of trade connections among all the sectors of this regional trade cluster (as shown in table 2).[6]

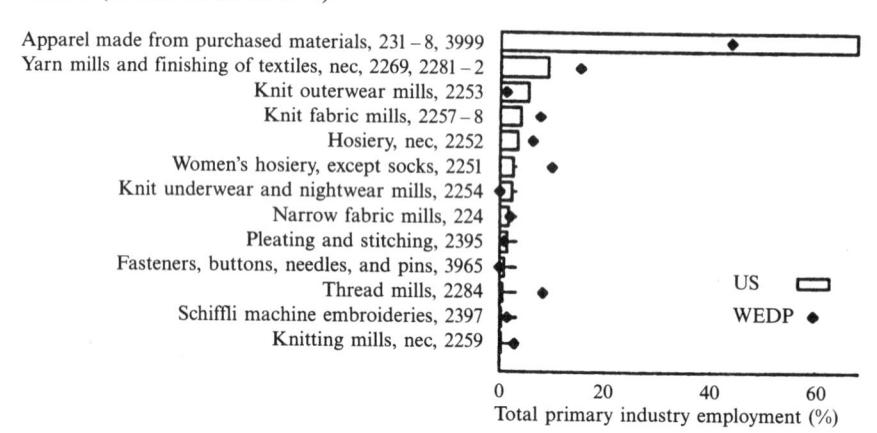

Figure 2. Estimated primary sector employment, 1993/94, in the knitted goods cluster of the Western Economic Development Partnership (WEDP) region compared with the United States as a whole (source: see Bergman et al, 1997).

[5] Differing data sources and regional definitions affect the number of firms included in constituent sectors of each region's cluster: the RTS group went outside ARC— and WEDP—region boundaries to include firms that were considered linked from neighbouring counties.

[6] All linkages between specific sectors are available in the full US input–output tables, including trade with service sector firms.

6 Strategic thinking: for whom?

Each of the two study approaches offer cluster details beyond the simple sectoral identification that characterises many studies, but they differ in type, quantity, and utility. There is no universally defined methodology for conducting microcluster studies, but there is a remarkable congruity in the types of results depicted in figure 1 and others like it produced for similar clients. Clients of these studies essentially determine which clusters are studied and how they are presented, and their interests are understandably similar. However, the very specificity of microstudies also limits their usefulness to apparently similar clusters located in different regions that may have quite distinct institutions and leadership dynamics.

Microstudies tend to revolve around the needs of the focal industries to survive or thrive in their settings, and most inquiries are therefore geared to learning what is needed to act decisively in their specific economic and regional environments. These studies are attempts to provide useful specificity, detail, and subtlety of how connections are made, networks are maintained, and interpersonal assets are translated into cluster advantages of utmost importance to the sponsoring clients. These interests may be at odds with host regions that wish to restructure their economies away from the most vulnerable to the most promising clusters.

Microstudies provide good coverage of the subtle and less-visible connections of cluster firms with their supporting environment, whereas regional trade clusters of the type described here offer deeper insights into the overall structure of *all* member industries and firms (for example, independent firm share, small firm share, average wages paid, average employment and estimated output by all categories, change over time in these categories and measures, precise spatial positioning of firms relative to each other and to key infrastructure). Because the information about regional trade clusters is usually relatively uniform in availability and quality, yet comparatively inexpensive to provide, its value may be greatest for use at the state or multistate policy levels where resources must be allocated both across *regions* and across *clusters*. These possibilities can be illustrated by comparing the presence of the knitted goods cluster in North Carolina's Economic Development Partnership regions, as shown in figure 3 (see over).

Figure 3(a) maps regions that are far more concentrated in some rather than other segments of this cluster. Because regional trade clusters also reveal these relative concentrations, as in figure 2, valuable information is available when specific policy mixes of targeted programmes for worker training, technology adoption, export assistance, infrastructure support, and so on are considered. Even stronger mixes of policies and resource allocations are entirely possible because *spillovers between clusters* can be optimised or otherwise taken into consideration. This is particularly important in designing policies that help diffuse the commonplace technological advances of some clusters to members of other regional clusters that would benefit; this possibility is one of the

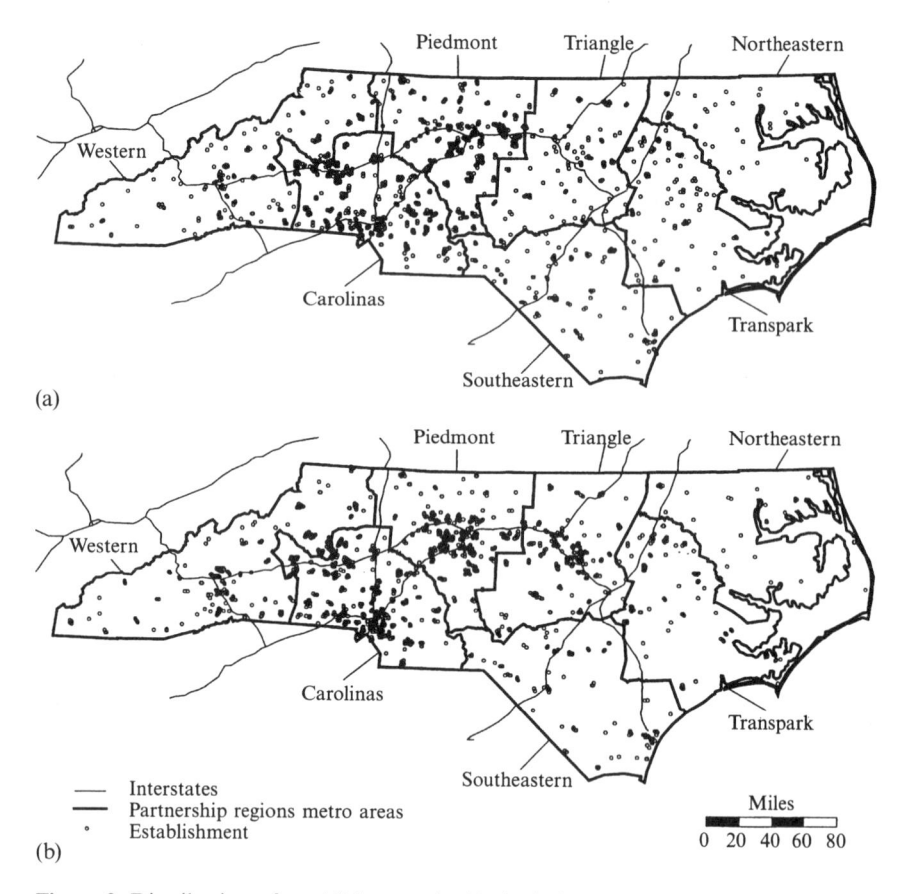

Figure 3. Distribution of establishments in (a) the knitted goods cluster (total = 1831) and (b) the vehicle manufacturing cluster (total = 2140), for the third quarter of 1994, primary and secondary industries (source: Bergman et al, 1997).

principal reasons why NCACTS supported developing the concept of regional trade clusters (Bergman et al, 1997).

Detailed regional trade cluster information remains of marginal use to microcluster members who seek to improve their relative position within their region. However, such details are of vital importance to regional officials who—as we shall see below—are responsible for balancing the competing claims of several cluster constituencies when designing and implementing regional development strategies. The option of designing and implementing overall policy for several industry clusters is itself novel: regional officials seldom have more than a lengthy list of individual sectors and their basic characteristics, neither of which permit a common policy approach. Regional trade clusters add strategic value to this basic information by depicting all parts of a regional economy in terms that permit customised policy approaches for *cluster groupings of specific sectors*.

7 Regional economic strategies

To appreciate the possibility of improved strategies, consider the following evidence for the WEDP region. First, the spatial concentration of the knitted goods cluster, as revealed by common GIS mapping procedures in figure 3, clearly distinguishes its spatial intensity, major highway systems (designated on maps as 'Interstates'), and proximity to nearby cluster concentrations of other regions. Knowing the spatial pattern of all firms in this and other significant clusters permits their alignment with existing or planned features of infrastructure in the region and neighbouring regions. Transportation improvements, essential public utilities, key public service areas, education and training facilities, and similar development policies may be fully reconsidered in light of the pattern of each cluster.

Second, although the knitted goods cluster is the largest in this region (19%), the wood product (18%), vehicle manufacturing (15%), and fabricated textile product (11%) clusters also account for more than 10% of total primary sector employment (figure 4). All are significant components of the regional economy whose combined needs require a portfolio of suitable policies. Larger regions or cities typically adopt policies suited to a continually *changing mix of industries* distributed across segments of several trade clusters, rather than to only one of twenty-three total possibilities. The four larger clusters mentioned above are joined at some threshold presence by electronics and computers (4%), printing and publishing (3%), chemicals and rubber (2%), plus slight traces of ten other clusters. Any of these may contain the seed of quite

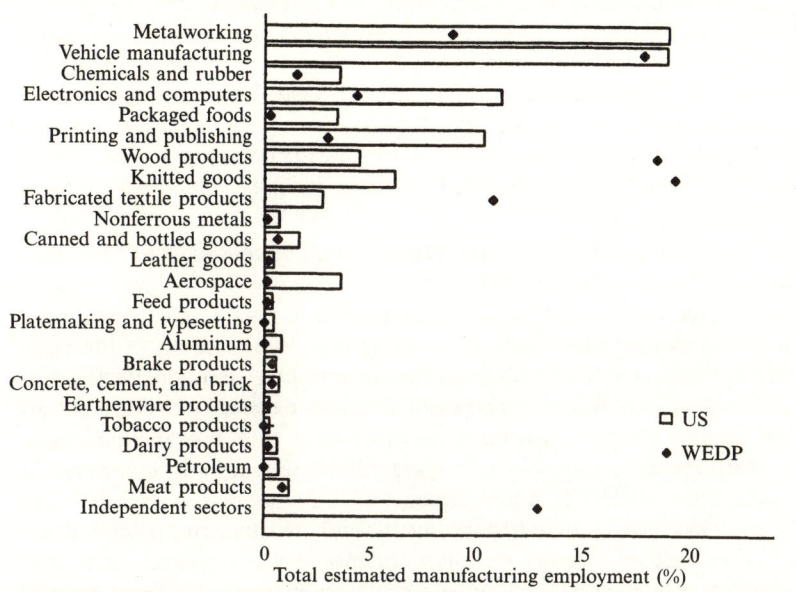

Figure 4. Estimated manufacturing employment by cluster: comparison of USA and Western Economic Development Partnership region, for primary industries only (source: Bergman et al, 1997).

dramatic economic transformations in future decades, or the linking agents that connect important common elements of larger clusters now active in a region. Accordingly, regional development officials and company managers who are committed to making interdependent investment decisions concerning the full regional economy may proceed with greater confidence, particularly when informed by readily envisioned clusterings and potential reconfigurations of a region's many firms and industries.

Joint spatial patterns of a region's principal clusters reveal possibilities for designing comprehensive policies, particularly because the mountainous landscape and widely dispersed towns and production centres of these regions require very careful infrastructure planning. The same logic also applies at higher levels of policy responsibility, for example, statewide policies, where spatially networked clusters of several regions are involved. For example, North Carolina's declared interest in promoting the vehicle manufacturing cluster *and* the present strength of this cluster in neighbouring regions [see figure 3(b)] readily suggest a strategic restructuring of the WEDP region away from its primary dependence on the knitted goods cluster.

A key question remains: does this analytic framework permit us to detect momentum that could be accelerated with further strategic initiatives? This is a relevant question, for if regional trade cluster dynamics can be detected in terms of direction and degree, then coordinated strategies to propel whole clusters in desired directions can be more easily designed and implemented. The portfolio quality of a cluster also implies that some segments of the cluster may respond better to strategic initiatives directed to the whole cluster than others or at different times, thereby revealing at an early stage whether such strategies are working.

The answer to this question depends upon evidence of cluster shifts in recent years. Detailed microdata for all cluster firms was analysed for 1989 and 1994 for all NC regions (see figure 5). The results for any particular region are read along its axis, where the shaded intersection measures the 1989 share of total NC cluster employment in the region, and the dark line intersection measures the regions share of total gains (vehicle manufacturing) or losses (knitted goods) in North Carolina.

As I am mainly illustrating how this applies to the WEDP region, note substantial shares of total knitted goods cluster losses there, at the same time it and the much larger Carolina Economic Development Partnership gained higher than initial 1989 shares of the motor vehicle cluster. Both are favoured regions for this growing cluster, because they both lie within the main north–south motor vehicle supply routes that now link the middle south with the great lakes automobile production regions (see Feser and Bergman, 1998). The detailed mapping of both clusters in figure 3 offers excellent evidence of where the restructuring will take place, as knitted goods continue to decline and vehicle production gains. The highest risk, least desirable sectors *within* this cluster can be further detected with the microdata presented in table 2. This example illustrates how similar evidence

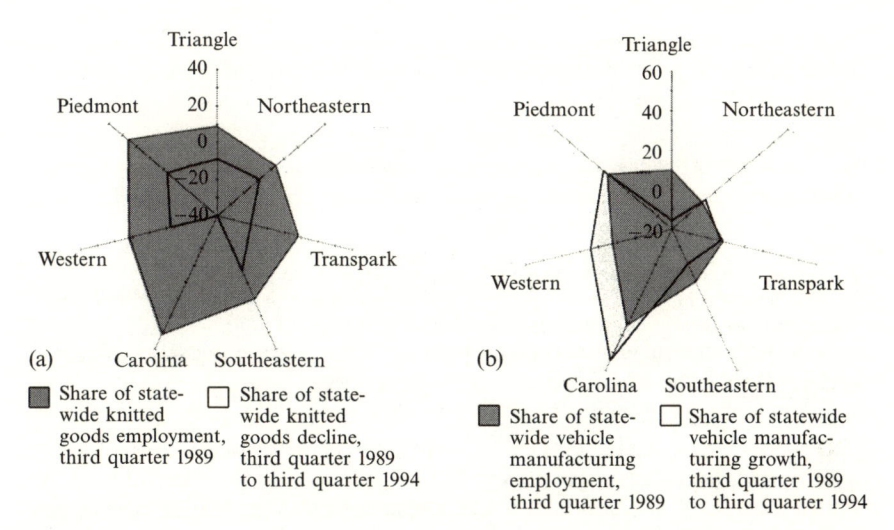

Figure 5. Trends in regional cluster distribution (includes primary and secondary industries): (a) knitted goods, (b) vehicle manufacturing (source: Bergman et al, 1997).

of restructuring for all clusters can be included in various infrastructure and other investment strategies established for the region as a whole.

8 Conclusions

There are very clear differences in the strengths and purposes of various cluster approaches. As I have demonstrated in this chapter, regional trade clusters have a potentially very significant role to play in how regional economies can be analysed as a framework for broad regional development strategies. This and similar cluster approaches may be expected to improve the ways that nations, regions, and localities integrate common concerns with the strength and competitiveness of all their economic assets and key industries.

References
Bergman E M, Feser E J, 1997, "Industrial, regional or spatial clustering?", paper presented at OECD Workshop on Cluster Analysis and Cluster Policies, Amsterdam, 9–10 October; direct inquiries to Dr T J A Roelandt, Ministry of Economic Affairs, Research Unit, Economic Policy Directorate, PO Box 20101, 2500 EC, The Hague
Bergman E M, Feser E J, Sweeney S, 1997, "Targeting North Carolina manufacturing: understanding a state economy through national industrial cluster analysis", RR 55, Institute for Urban and Regional Studies, University of Economics and Business, Vienna
Bergman E M, Lehner P, 1998, "Regional industrial clusters in Austria: mapping and documentation for UNIDO clients", working paper, Institute for Urban and Regional Studies, University of Economics and Business, Vienna
Doeringer P B, Terkla D G, 1995, "Business strategy and cross-industry clusters" *Economic Development Quarterly* **9**(6) 225–237

Drejer I, Kristensen F S, Laursen K, 1997, "Studies of clusters as a basis for industrial and technology policy in the Danish economy", position paper presented at OECD workshop on Cluster Analysis and Cluster Policies, Amsterdam, 9 – 10 October; see Bergman and Feser, 1997

Dumais D, Ellison G, Glaeser E L, 1997, "Geographic concentration as a dynamic process", WP 6270, National Bureau of Economic Research, Cambridge, MA

Feser E J, 1998, "Enterprises, external economies, and economic development" *Journal of Planning Literature* **12** 283 – 302

Feser E J, Bergman E M, 1998, "National industry clusters: frameworks for state and regional development strategy" *Journal of Regional Studies* forthcoming

Hutschenreiter G, 1994 *Cluster Innovativer Aktivitäten in der Österreichischen Industrie* Österreichisches Forschungszentrum, Siebersdorf

Jacobson D, Andréosso-O'Callaghan B, 1996 *Industrial Economics and Organization: A European Perspective* (McGraw-Hill, Reading, Berks)

Keller W, 1997, "Trade and the transmission of technology", WP 6113, National Bureau of Economic Research, Cambridge, MA

Krugman P, 1991 *Geography and Trade* (MIT Press, Cambridge, MA)

Liston C, 1997, "Knitting mills in North Carolina and Virginia", in *Exports, Competitiveness, and Synergy in Appalachian Industry Clusters* Regional Technology Strategies, Chapel Hill, NC, pp 150 – 174

Peneder M, 1994 *Clusteranalyse und Sektorale Wettbewerbsfähigkeit der Österreichischen Industrie* Österreichisches Forschungszentrum, Siebersdorf

Polenske K R, 1997, "Competition, collaboration, and cooperation: an uneasy triangle in networks of firms and regions", working paper, Multiregional Planning Research Project, Department of Urban Studies and Planning, Massachusetts Institute of Technology, Cambridge, MA

Roelandt T J A, den Hertog P, van Sinderen J, Vollaard B, 1997, "Cluster analysis and cluster policy in the Netherlands", position paper presented at OECD workshop on Cluster Analysis and Cluster Policies, Amsterdam, 9 – 10 October; see Bergman and Feser, 1997

Rosenfeld S A, 1997, "Bringing business clusters into the mainstream of economic development" *European Planning Studies* **5**(1) 3 – 23

Rouvinen P, Ylä-Antilla P, 1997, "A few notes on Finnish cluster studies", position paper presented at OECD workshop on Cluster Analysis and Cluster Policies, Amsterdam, 9 – 10 October; see Bergman and Feser, 1997

RTS, 1997 *Exports, Competitiveness, and Synergy in Appalachian Industry Clusters* Regional Technology Strategies, Chapel Hill, NC

Stenberg L, Strandell A-C, 1997, "An overview of cluster-related studies and policies in Sweden", position paper presented at OECD workshop on Cluster Analysis and Cluster Policies, Amsterdam, 9 – 10 October; see Bergman and Feser, 1997

Weiss A, 1994 *Österreich als Standort International Kompetitiver Cluster* Industriewissenschaftliches Institut, Vienna University of Economics and Business, Vienna

APPENDIX

Overview of technical methodology[7]

The basic methodology for clustering manufacturing industries consisted of factor analysis on a data matrix constructed from the 1987 US input–output coefficients. In factor analysis each given industry is treated as a variable, with a measure of the linkages between the industry and all other industries treated as observations. There is then an attempt to reduce the number of variables by exploiting the common variation among them, that is, industries are grouped together on the basis of similarities of interindustry trade, as revealed by their input–output coefficients. The result is a set of input–output-based industrial clusters. A detailed analysis of all the methods, as well as the criteria developed for identifying clusters from the statistical output, is available in other sources (Bergman et al, 1997; Feser and Bergman, 1998).

Input–output-based industrial clusters

Because interindustry trade linkages involve extremely complex networks, the task of aggregating industries into mutually exclusive clusters risks masking key input–output relationships. In reality, many industries trade with others in more than one cluster, and their trade linkages across clusters vary accordingly in degree or strength; that is, an industry may be tightly linked to one group of sectors and weakly or moderately linked to one or more additional groups. Such interrelationships can make interregional or intertemporal comparison of clusters difficult, because a significant amount of double counting may arise when calculating sectoral aggregations. Factor analysis provides a useful way around this problem by generating a set of 'loadings', which measure relative degrees or strength of linkage between a given industry and the cluster of which it is a part. Loadings closer to 1.0 indicate tighter linkages of the primary industries.

Secondary industries are defined as those sectors with cluster loadings of between 0.35 and 0.60. By focusing the overall analysis first on the primary industries, one obtains 23 mutually exclusive clusters that may be used for cross-comparison purposes. However, the full value of clusters requires the presence of both primary and secondary industries to provide the most complete picture of interindustry and interfirm trade. Linkages between clusters are best revealed through an examination of their secondary industries.

Primary versus secondary cluster industries

Loadings have been used to designate whether the sectors or a cluster are considered 'primary' or 'secondary'. In general, *primary* industries are those that are most tightly linked to a given cluster whereas *secondary* industries are those that are only moderately or weakly linked to the cluster. Specifically, primary industries for a given cluster are defined as those sectors (1) that achieved a loading of at least 0.60 on that cluster; *and* (2) that did not achieve a higher loading on any other cluster.

[7] Revised text from Bergman et al (1997).

Weakly clustered industries
Not all industries trade within sufficiently deep supply chains to exhibit distinct trade-clustering tendencies. Of 362 input–output sectors, 44 failed to achieve any loading of 0.60 or higher. These sectors are classified as purely secondary industries in their respective clusters. Although 22 sector loads did exceed 0.50 on at least one cluster, the 318 industries classified as primary industries achieved loadings of 0.80 or higher on one or more clusters. Three sectors (SICs 328—cut stone and stone products; 387—plumbing fixtures, fittings, and trim; and 3432—watches, clocks, watchcases, and parts) achieved maximum loadings below 0.35 and thus fall short of secondary industry thresholds, according to the criteria above. Nevertheless, to ensure that all manufacturing sectors are included in regional analyses, the weakest loadings are also included as secondary industries in the cluster where they attained their maximum loading.

In Search of Agglomeration Economies: The Adoption of Technological Innovations in the Automobile Industry

R A Dubin, S Helper
Case Western Reserve University

1 Introduction

Marshall (1898) argued that firms which are near other firms in similar industries benefit from three types of 'external economies'. First, an agglomerated industry benefits from a pool of specialised labor. Second, specialised services are more readily available. Third, information travels more rapidly. Marshall's work provides the theoretical foundation for the belief that geographical clustering of firms can be beneficial.

Marshall's three factors might also increase the likelihood of adoption of new innovations by spatially clustered firms (indeed, an advantage in adopting new technology would be an important contributor to the continued agglomeration of an industry). Firms in an agglomeration might find workers trained in the new technology more readily than firms located far from their rivals. Similarly, agglomerated firms would have faster access to specialised services in the new technology, such as marketing and maintenance. Most important would be the information advantages, because in a localised industry, "inventions and improvements in machinery... have their merits promptly discussed; if one man starts a new idea, it is taken up by others and combined with suggestions of their own; and thus it becomes the source of further new ideas" (Marshall, 1898, page 350).

An example of the importance of geographic proximity is given by the following anecdote. A systems analyst was having trouble with a Kodak machine designed to scan invoices. The Kodak salesperson explained that the performance of the machine would be improved if the company bought a newly designed (and expensive) attachment. The analyst was sceptical, but the salesperson said that a nearby firm had bought the attachment and, therefore, the analyst could see it in operation. After making the visit (which he did only because the distance was so small), the analyst became convinced that the product would solve his problem and adopted the new equipment (interview by Dubin, July 1995).

Evidence consistent with the hypothesis that information about new technology travels more rapidly in agglomerated areas is presented by Case (1992). She finds that Indonesian rice farmers are more likely to adopt new technology if their neighbours have characteristics that promote the adoption of the technology. Grilliches (1960) also found that adoption of hybrid seed corn in the USA was faster where a farmer's neighbours had previously adopted the technology. Proximity facilitates repeated dealings, which should reduce opportunism (Brusco, 1982).

However, in this age of telecommunications, geographic distance may not be the main barrier to information flow. Social distance may be a far more important barrier: people are more likely to communicate with those who have a closer tie to them than geographic proximity. For example, large firms have dense internal communications between their plants, even though they are spread throughout the country. Firms in the same industry also have specialised conferences, newsletters, trade shows, etc. Managers from firms of a similar size may also have more opportunities to meet. Large firms may jointly sponsor research consortia with firms outside their industry.[1] Small firms, on the other hand, may meet each other through events sponsored by regional chambers of commerce.

In this chapter, we present some preliminary findings about the nature of information flows. Our measure of information is based on observing the rate of adoption of relatively new technology. We hypothesise that the more information a firm has about the innovation, the more likely it is to adopt. Thus, by observing the pattern of new adoptions, we can draw conclusions about the flow of information. We further hypothesise that information flows between firms are positively related to their proximity. We use two measures of proximity. The first is based on the geographic closeness of the locations of the firms. The second is based on how similar the firms are to each other. Our results indicate that the similarity measure does a better job of explaining technology adoption than does the geographic measure.

2 Model

A firm's decision to adopt new technology is affected by firm-specific factors as well as its proximity to other adopters. Unfortunately, it is not obvious how to measure proximity. We take two approaches. In the first, we hypothesise that geographic space acts as a barrier to information exchange. Thus, firms which are physically closer together are more likely to communicate. This approach is consistent with the agglomeration economy literature.

In the second approach, we test the hypothesis that the more similar firms are, the more likely it is that they will communicate. We measure similarity in two dimensions. The first dimension is firm size. We posit that small firms are more likely to talk to other small firms, whereas large firms are more likely to talk to large firms. The second dimension has to do with the industry of the customer. Firms which are heavily involved in selling to a particular set of customers are more likely to share information. There may be newsletters, trade magazines, or professional meetings set up to aid this type of communication. Unfortunately, our data contain only one measure of customer industry, ORIGEQ. This variable is the percentage of the firm's sales that is original equipment for cars or light trucks. A high score on this variable means that the customer industry is primarily automobile. A low

---

[1] One example is CAMI, a consortium which sponsors research on computer-aided manufacturing, which is funded by large firms in the aerospace, automotive, and communications industries.

score means either that the firm primarily supplies to another industry or that the firm supplies to many industries.

Note that direct communication is not required. All that matters is that firms which are close in terms of the distance measure share the same set of information. This information may be provided by any number of mechanisms: salespeople, trade publications, or the owners having lunch at the same club. In the US automotive parts sector, from which the data for this paper are drawn, avenues for exchanging information are particularly rich. There are a large number of publications aimed at the industry, including *Metalworking News, Chilton's Automotive Industries,* and *Automotive News*. There are international technical associations, such as the Society of Automotive Engineers, which sponsor journals and major conferences. Engineers come from around the nation to hear technical papers presented, and to discuss their contents informally. Large firms participate in the Automotive Industry Action Group, which focuses on issues such as standardisation of bar codes and standards for computer-aided-design software. Smaller firms are more likely to belong to the Motor and Equipment Manufacturers Association, a more traditional trade association.

Because our data contain the longitude and latitude of the location of each firm, we are able to calculate the geographic distance separating each pair of firms.[2] Alternatively, we substitute firm employment and ORIGEQ into the Euclidean distance formula to obtain distance in 'similarity space'. Distance in similarity space is thus calculated as

$$[(\text{FEMP}_i - \text{FEMP}_j)^2 + (\text{ORIGEQ}_i - \text{ORIGEQ}_j)^2]^{1/2},$$

where FEMP_i is the employment of firm i and ORIGEQ_i is the percentage of firm i's sales that are original equipment.[3]

Regardless of whether we use geographic or similarity distance, the principle is the same: the smaller the distance measure, the more likely firms are to communicate. Thus, we model the influence that a prior adopter has on a potential adopter as a negative exponential function of the distance separating the two:

$$\rho_{ij} = b_1 \exp\left(\frac{-D_{ij}}{b_2}\right), \qquad (1)$$

where ρ_{ij} is the influence of prior adopter j on potential adopter i, b_1 and b_2 are parameters to be estimated, and D_{ij} is the distance (geographic or similarity) separating the two firms. The parameters b_1 and b_2 allow the

---

[2] Our geographic distance measure is distance 'as the crow flies'. That is, we use the distance along a straight line connecting the two firms, accounting for the curvature of the earth. The one exception is that, when firms are on opposite sides of the Great Lakes, the distance calculation breaks the trip into two straight lines by inserting a prespecified node, so that travel is primarily over land. Details of the geographic distance calculation can be found in appendix A.

[3] Because the scales of these variables are very different, we standardise them by subtracting the mean and dividing by the standard deviation.

researcher to estimate the influence of prior adopters. The parameter b_1 determines the effect that adjacent firms (small D_{ij}) have on each other, and b_2 determines the rate that influence attenuates with distance. If either b_1 or b_2 is insignificant, there is no interaction.

The model may now be formulated as follows. A firm's expected profit from adopting an innovation is a function of its own characteristics plus its distance from previous adopters.

$$Y_{it}^* = X_{it}\beta + \sum_{j=1}^{N} \rho_{ij} y_{j,t-1} + u_{it}, \qquad i \in N_t, \tag{2}$$

where Y_{it}^* is the unobserved expected profit from the innovation for firm i at time t. N is the total number of firms, and N_t is the set of potential adopters (that is, the set of firms that have *not* adopted prior to time t). X_{it} (a row vector) are the characteristics of firm i at time t. Term $y_{j,t-1}$ is one if firm j has adopted the technology at any time prior to t and zero otherwise. The error term is u_{it}, assumed here to be independently and identically distributed logit, conditional on X_{it} (a row vector) and the set of prior adopters. Last, β is a vector of coefficients (to be estimated) which represent the influence of the firm's own characteristics on the profitability of the innovation.

The dynamic nature of the model is clear from equation (2). As time progresses, more firms will adopt the technology. Thus, the probability that hold-out firms will adopt increases as nearby firms become adopters. After enough time has passed, virtually all firms will adopt the technology, giving rise to an S-shaped diffusion curve. However, the model is flexible, in that some firms' characteristics may so dispose them against the innovation that they never adopt.

The expected profit of the firm is not observed; instead we observe whether the firm adopts the innovation in time period t, as shown by term y_{it}. The firm will adopt the innovation if expected profits (Y^*) are positive. Thus,

$$y_{it} = \begin{cases} 1, & \text{if } Y_{it}^* > 0, \\ 0, & \text{otherwise}, \end{cases} \tag{3}$$

and

$$P(y_{it} = 1) = P\left[u_{it} > -\left(X_{it}\beta + \sum_{j=1}^{N} \rho_{ij} y_{j,t-1}\right)\right]$$

$$= P[u_{it} > -(X_{it}\beta + A_i)] = 1 - F[-(X_{it}\beta + A_i)]$$

$$= \frac{\exp(X_{it}\beta + A_i)}{1 + \exp(X_{it}\beta + A_i)}, \tag{4}$$

where F is the cumulative distribution function for u and $A_i = \sum_j \rho_{ij} y_{j,t-1}$. A_i represents the influence of prior adopters. The last equality in equation (4) is based on the assumption that the error terms come from a logistic distribution.

The log-likelihood function for this model is given by

$$\ln(L) = \sum_{i \in N_t} \left\{ y_{it} \ln\left[\frac{\exp(X_{it}\beta + A_i)}{1 + \exp(X_{it}\beta + A_i)}\right] + (1 - y_{it}) \ln\left[\frac{1}{1 + \exp(X_{it}\beta + A_i)}\right] \right\}$$

$$= \beta \sum_{i \in N_t} y_{it} X_{it} + \sum_{i \in N_t} y_{it} A_i - \sum_{i \in N_t} \ln[1 + \exp(X_{it}\beta + A_i)]. \tag{5}$$

Dubin (1995) suggests estimating the model by maximising equation (5) with respect to all parameters. The problem with this approach is that the absence of spatial effect may be indicated by a zero value for *either* b_1 or b_2. If the algorithm picks a small value for one of these parameters, changes in the value of the other will have only a small effect on the value of the likelihood function. Thus, when the spatial effects are small, the optimisation routine will converge slowly, if at all. However, both parameters are required for proper scaling of the spatial effects when they are present.

A better approach to estimation (and one that requires only standard software) is to choose values of b_2 which span its plausible range. The term, SUM,

$$\text{SUM} = \sum_{j=1}^{N} \exp\left(\frac{-D_{ij}}{b_2}\right) y_{j, t-1},$$

can be calculated for each chosen value of b_2. The model is then estimated as many times as there are values of b_2, with the appropriate SUM included as an explanatory variable. A standard logit routine can be used for this estimation. The maximum likelihood estimate of b_2 is the value that gives the highest value of the likelihood function. The maximum likelihood estimate of b_1 is the coefficient of SUM. The significance of the information term is determined with a likelihood ratio test, by using models with and without A_i (or, equivalently, SUM). A t-test cannot be used because the information term requires estimates of two coefficients, both of which are subject to sampling error.

Note that the summations in equation (5) are over the set N_t, which is the set of potential adopters. That is, the value of the likelihood function is determined by looking only at the firms which have not yet adopted. The influence of prior adopters enters through the term A_i.

3 Empirical results

The data used for the estimation are from a survey of automobile suppliers conducted by Helper. The survey was mailed to every automotive supplier and auto maker component division named in the *Elm Guide to Automotive Sourcing* (available from Elm Inc., East Lansing, MI). This guide lists the major first-tier suppliers (both domestic and foreign owned) to manufacturers of cars and light trucks in the United States and Canada. The target respondent was the divisional director of marketing at independent firms, and the divisional business manager or the director of strategic planning at automaker components divisions. These individuals were selected on the grounds that they would have the broadest knowledge about their firm's products, processes, and customer relationships. The response rate was 55%.

The model described above is dynamic: prior adopters influence potential adopters by increasing the information available to potential adopters. Therefore, to implement the model, the data must be divided into stages. The survey asks about current (1993) technology and technology four years ago (1989). In what follows, we take adoption prior to 1989 to be stage-one adoption. Firms which have not adopted prior to 1989 are potential stage-two adopters. The technology used in our analysis is robots.[4]

Before we discuss the maximum likelihood results, it is worthwhile examining the locations of the prior and potential adopters. These are presented in figure 1. Here, prior adopters are represented by white circles. Prior adopters are defined as firms which had any robots in period 1 (that is, prior to 1989). Thus, a firm with at least one robot prior to 1989 is classified as a prior adopter. If a firm does not have any robots by 1989, then it is a potential stage-two adopter. Triangles are used to identify these firms on the map. If the firm adopts the technology in the period 1989–93, it is a current adopter and is represented by a grey triangle. If the firm has no robots by the end of period two (1993), it is a nonadopter and is represented by a black triangle.

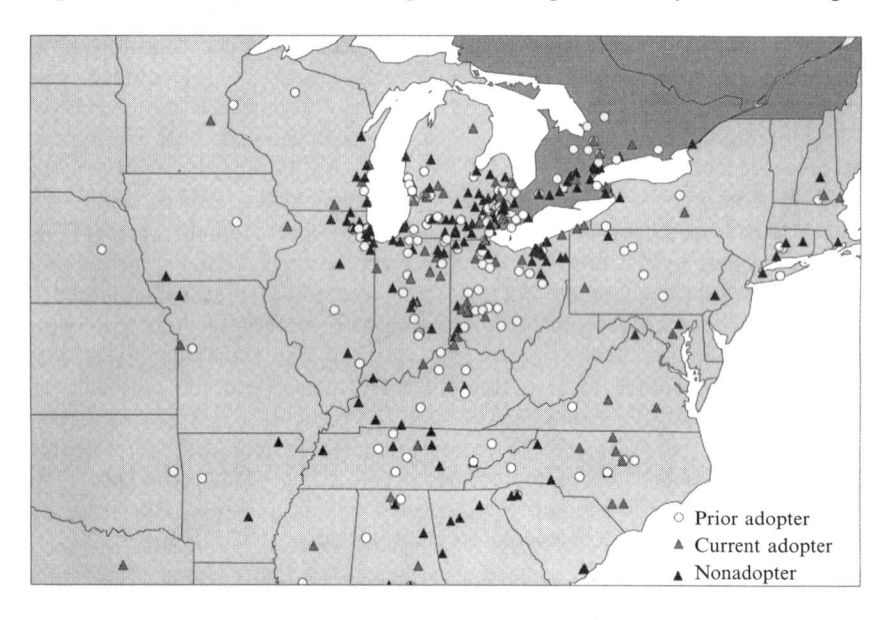

Figure 1. Adoption of robots in the study area.

[4] The survey also contains information about the adoption of programmable logic controller and computer numerically controlled machine tools. However, these technologies were highly diffused at the time of the survey, and therefore, their adoption patterns were not explained by any of the variables in the survey, including the information term.

If (the inverse of) geographic distance can be used to measure information flows, then current adopters should be clustered around prior adopters. Figure 1 shows that there are some areas which exhibit this pattern and others that do not. For example, North Carolina contains a cluster of prior and current adopters. Also, the Dayton area, in Southern Ohio, contains a large number of prior adopters and more current adopters than nonadopters. However, there are many areas that do not exhibit this pattern. The Chicago area has a large number of prior adopters and nonadopters, but no current adopters. The Cleveland area has prior adopters but many more nonadopters than current adopters. The Detroit area, which contains the largest concentration of firms in our sample, is very mixed: although there are many prior adopters, the current and nonadopters seem to be uniformly scattered throughout the area.

Figure 1 shows that it is unlikely that geographic space provides a barrier to information exchange. Although the statistical model is more sophisticated than the maps, in that it allows us to hold constant firm-specific variables affecting profitability, it seems likely that the spatial term will be insignificant.

The data contain a wide variety of measures concerning the characteristics of the supplier firms and their relations with their customers. Because the estimation is based on potential adopters, and because the number of potential adopters is relatively small, we attempted to keep the number of explanatory variables to a minimum.[5]

Based on the previous discussion, the independent variables should be the characteristics of the firm which affect the profitability of the investment. These include:

(a) Firm size. A proxy for a host of factors including risk of bankruptcy, and management depth. Our measure of firm size, FEMP, is firm employment in 1993.

(b) Market share. Presumably the more dominant the firm the better able it will be to appropriate the returns to its investment. The data contain a self-reported measure of the firm's market share, MKTSHR.

(c) Trust. Inability to trust the customer may result in highly variable future revenue. To measure the trust existing between customer and supplier we used two variables: FAIR and SWITCH. SWITCH is coded 1 if the firm believes its primary customer will switch to a competitor if that competitor is able to offer a product of equal quality at a lower price, and 0 otherwise. FAIR describes the firm's belief that the customer is fair in its dealings with the firm, with 5 being 'customer always treats us fairly' and 1 being 'customer never treats us fairly'.

(d) Location. We would like to test the hypothesis that being in an urban area makes adoption of robots more profitable. This might occur because urban areas contain workers who are skilled in the use of these technologies, the infrastructure required to service the machines, and firms outside of the auto industry to talk to (these are aspects of agglomeration economies).

[5] We further reduced our sample size by excluding firms with more than one plant, because between-plant communication is an option for these firms.

To determine whether the firm has an urban location we created a variable, URBAN, which is 1 if the firm is located in a metropolitan statistical area and 0 otherwise.

(e) Research and development. A firm which engages in a lot of R&D is likely to be well informed with respect to new technologies. PCTRD is the business unit's R&D, expressed as a percentage of sales.

(f) ORIGEQ. According to the previous discussion, ORIGEQ may affect information exchange. However, there is no reason to believe that it affects the profitability of the investment directly. Profitability is determined by the nature of the firm and its product, *not* by its customers. When a proper measure of information is included as an explanatory variable, ORIGEQ should be insignificant.

We also included two variables that measure the suitability of a firm's product line for use of flexible automation.

(g) Complexity of the product. COMPLEX is the response to the survey question asking the sales manager to estimate the technical complexity involved in manufacturing the product. The responses are coded 1 to 5, with 1 representing 'fairly simple' and 5 representing 'highly complex'.

(h) Changing product line. The survey asks, "What percentage of your business unit's sales come from products which it did not make four years ago?" A high value for this variable, NEWPROD, indicates that the firm's product line is changing rapidly.

Our variables thus include descriptors of the firm (size, market share, customer relations, R&D, and location) and descriptors of the product line (complexity of the product and changing product line). The product line variables are important because robots are well suited to producing complex products. They are also flexible so that they can save the firm money if its product line changes often. Thus we control for the characteristics of the product that make these technologies profitable, rather than attempting to control for the particular product that the firm manufactures.

Our hypothesis is that the distance from prior adopters influences the probability of adoption. However, in order for the prior adopter to have this influence, the adoption must be significant and successful. Thus, the prior adopter must have an installation that is capable of favourably influencing potential adopters and we therefore make a distinction between a prior adopter (a firm that has any of the equipment in stage one) and an influential firm (a firm whose installation is large enough to influence potential adopters). Unfortunately, we do not know how large an installation must be in order to have an influence. Therefore, we used all possibilities allowed by the survey instrument.

The survey asks how many robots were in use four years ago. The possible answers are (a) 0; (b) 1–2; (c) 3–5; (d) 6–10; and (e) more than 10. If the answer is (a) the firm is a potential adopter. The survey allows four possible definitions of influential adopter: (a) one or more robots (ROBOT1B); (b) more than two robots (ROBOT2B); (c) more than five robots (ROBOT3B); and (d) more than ten robots (ROBOT4b).

Our data set is relatively small, and the list of potential independent variables is long. To get around this problem, we estimated six different models.[6] Model 1 is our ideal model, that is, it is the one we believe to be correct. The other models are variations on model 1.

In model 1, FEMP, NEWPROD, and MKTSHR all appear as logs. FEMP ranges from 12 to 400 000. NEWPROD and MKTSHR are percentages. Because the maximum value of the remaining independent variables is no greater than 6, and because logit is somewhat sensitive to the scale of the variables, we used logs to scale these variables. The other independent variables in model 1 are COMPLEX, URBAN, and the trust variable, FAIR.

Model 2 is the same as model 1, except that the trust variable SWITCH is used in place of FAIR. Model 3 is model 1 with the addition of ORIGEQ. This variable is a percentage, and therefore is also scaled by taking its log. We believe that model 3 is misspecified in that the effect of information should be picked up by SUM. Our prior expectation is thus that ORIGEQ should be insignificant. Model 4 tests whether R&D has any effect on adoption. Model 5 is an almost linear version of model 1, with only FEMP appearing in log form.

The estimation of many models on the same data has been pejoratively termed 'data mining'. The effect of data mining is to create a bias in favour of finding significant coefficients. Thus, this experimentation should predispose us to find a significant distance effect.

Before examining the estimated coefficients, it is worthwhile discussing the behaviour of b_2. Recall that the estimation of the model is accomplished by calculating SUM for plausible values of b_2. A standard logit routine is then used to estimate the model for each value of b_2. The b_2 that provides the highest value of the log-likelihood function is taken as the estimated value, and it is the coefficients from this logit that are presented in the tables.

In the geographic distance estimations, b_2 was searched over values ranging from 0.01 to 100. When $b_2 = 100$, the zone of influence extends beyond 200 miles. Because we are looking for the effect of face-to-face contacts, 200 miles seems to be a reasonable limit.

In fact, it turns out that the estimates of b_2 are quite small. Figure 2 (see over) shows the graph of the value of the likelihood function versus b_2 for model 1, using ROBOT1B as the influential firm. Note that the likelihood function for this problem is always negative. In the remainder of the paper, we will refer to the negative of the value of the likelihood function as VOF. Thus to maximise the likelihood function, we minimise VOF.

The VOF shown in figure 2 reaches a relative minimum at $b_2 = 0.5$. It reaches a maximum at $b_2 = 4$, and then starts to decline. Table 1 (see over) shows the searched values of b_2 and the associated VOF and estimated value $\hat{b}_1$. This table shows that $\hat{b}_1$ is positive for small values of b_2, but declines as b_2 increases, becoming negative for values of b_2 greater than 5. We consider only values of b_2 that result in a positive $\hat{b}_1$, for two reasons. First, the robot

[6] We actually estimated many more models than the ones presented here; however, the results are essentially the same, and so we do not present these results.

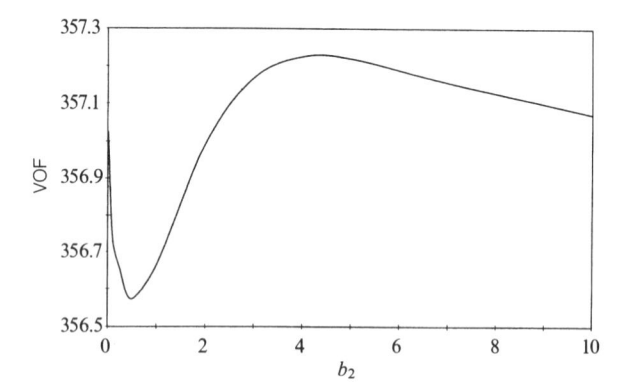

Figure 2. VOF versus b_2 for model 1, obtained by using geographic distance with ROBOT1B as the influential firm.

Table 1. Values of b_2 searched, with associated VOF and estimated values of $\hat{b}_1$, for model 1, obtained by using geographic distance and ROBOT1B as the influential firm.

b_2	VOF	$\hat{b}_1$	b_2	VOF	$\hat{b}_1$
0.01	178.51	0.2471	3	178.58	0.0538
0.1	178.37	0.3723	4	178.61	0.0087
0.25	178.33	0.3933	5	178.61	−0.0113
0.5	178.29	0.4024	7	178.58	−0.0239
1	178.33	0.3302	10	178.54	−0.0247

technology is not widely diffused in our data and, thus, increased information should increase the likelihood of adoption (that is, A_i should be positive). Second, there is no plausible value of b_2 at which VOF reaches a global minimum. We searched as high as 1000, and VOF continued to decrease. In light of this discussion $\hat{b}_2$ is taken to be 0.5.

Table 2 presents the estimation results obtained by using geographic distance and ROBOT1B as the influential firm (that is, any prior adopter is considered to be influential). The estimate of b_2 is chosen by taking the value that minimises VOF, subject to the constraint that $\hat{b}_1$ is positive. There are 294 potential adopters.

Table 2 shows that scaling the variables by taking logs appears to be beneficial. Model 1 has a lower VOF and more significant variables than model 5. The only firm characteristics which are consistently significant are NEWPROD and COMPLEX. Both variables have positive coefficients, as expected, indicating that robots are more profitable for firms with rapidly changing product lines and for firms which produce complex products. Firm employment is never significant, which is surprising given its importance in the literature. The trust variables, FAIR and SWITCH, are never significant. Neither is URBAN, indicating that being in an urban area has little effect on the probability of adoption. R&D expenditures also have no effect on adoption probability. When ORIGEQ

Table 2. Estimation results obtained by using geographic distance and ROBOT1B as the influential firm (t-statistics are given in parentheses).

Variables	Model				
	1	2	3	4	5
Constant	−3.0137	−3.3466	−4.7156	−3.2067	−2.2824
	(−3.47)	(−3.92)	(−4.05)	(−4.08)	(−3.18)
ln(FEMP)	0.1092	0.1122	0.1102	0.1147	0.0994
	(1.79)	(1.84)	(1.78)	(1.86)	(1.64)
ln(NEWPROD)	0.3550	0.3539	0.3663	0.3673	
	(2.60)	(2.59)	(2.65)	(2.65)	
ln(MKTSHR)	−0.0906	−0.0853	−0.1189	−0.0831	
	(0.81)	(−0.76)	(−1.04)	(−0.74)	
ln(ORIGEQ)			0.4594		
			(2.33)		
COMPLEX	0.2665	0.2708	0.2107	0.2731	0.2626
	(1.98)	(2.01)	(1.53)	(2.01)	(1.96)
FAIR	−0.0528		−0.0206		−0.0525
	(−0.50)		(−0.19)		(−0.50)
SWITCH		0.2209			
		(0.82)			
URBAN	−0.0015	−0.0117	−0.01164		−0.0181
	(−0.00)	(−0.04)	(−0.04)		(−0.06)
PCTRD				−0.0371	
				(−0.49)	
NEWPROD					0.0098
					(1.87)
MKTSHR					−0.0039
					(−0.88)
SUM	0.4024	0.4069	0.4039	0.3824	0.4329
b_2	0.5	0.5	0.5	0.5	0.5
N	294	294	294	294	294
VOF	178.29	178.08	175.21	178.29	180.02
Restricted VOF	178.61	178.41	174.54	178.59	180.40
Likelihood ratio	0.64	0.66	0.66	0.60	0.76

is in the model, it is significant, and COMPLEX becomes insignificant. We believe that model 3 is misspecified, however, and that ORIGEQ measures information, rather than the profitability of adoption.

With respect to this chapter, the most important result contained in table 2 is that SUM is never significant. This can be seen by examining the likelihood ratio row of table 2, which shows the values of the likelihood ratio test (this number is twice the difference between the VOF for the model and the restricted VOF).[7] Because this test has two degrees of freedom (for b_1

---

[7] The restricted VOF is the negative of the value of the likelihood function for the model excluding SUM (that is, b_1 and b_2 are restricted to be zero).

and b_2), the critical value at the 5% level is 5.99. All of the numbers shown in the likelihood ratio row are substantially below this value.

The insignificance of the information term implies that distance to prior adopters does *not* influence adoption, and that geographic distance does *not* represent a barrier to information transfer. Estimation results for models 1–5 for the remaining definitions of influential firm are presented in appendix B. The likelihood ratio row of these tables show that the information term is not significant in *any* of the models.

However, when similarity space, rather than geographic space, is used to measure proximity, a different picture emerges. Table 3 presents the results of model-3 estimations for all of the definitions of influential firm. Recall that we believe model 3 to be a misspecified version of model 1 (because it includes ORIGEQ as an independent variable).

Table 3. Results of model 3 (*t*-statistics are given in parentheses).

Variable	ROBOT1B	ROBOT2B	ROBOT3B	ROBOT4B
Constant	−4.74667	−4.88758	−5.07737	−5.73776
	(−4.15)	(−4.25)	(−4.38)	(−4.68)
ln(FEMP)	0.20675	0.20322	0.19809	0.20255
	(2.83)	(2.81)	(2.77)	(2.81)
ln(NEWPROD)	0.39659	0.39658	0.39752	0.39926
	(2.84)	(2.84)	(2.84)	(2.85)
ln(MKTSHR)	−0.12418	−0.12628	−0.12689	−0.12884
	(−1.08)	(−1.10)	(−1.10)	(−1.12)
ln(ORIGEQ)	−0.24082	−0.24133	−0.20286	−0.16578
	(−0.75)	(−0.74)	(−0.64)	(−0.54)
COMPLEX	0.16319	0.16743	0.16889	0.17133
	(1.17)	(1.20)	(1.21)	(1.23)
FAIR	−0.00385	−0.00378	−0.00580	−0.00565
	(−0.04)	(−0.04)	(−0.05)	(−0.05)
URBAN	0.10719	0.10596	0.10563	0.10865
	(0.35)	(0.34)	(0.34)	(0.35)
SUM	0.02116	0.03442	0.05762	0.13825
b_2	1.75	2	2	2.75
N	294	294	294	294
VOF	171.94	171.99	172.02	171.92
Restricted VOF	175.54	175.54	175.54	175.54
Likelihood ratio	7.20	7.10	7.04	7.24

The common findings in the distance and similarity analyses are that ln(MKTSHR), FAIR, and URBAN are all insignificant, and NEWPROD is significant and positive. COMPLEX is significant in all but the model-3 estimations, when geographic distance is used. COMPLEX is never significant when similarity distance is used (only the model-3 results are shown). An important difference is that, in the similarity specification, FEMP is significant and positive. This result increases our confidence in the similarity specification because the literature on technology adoption has found firm size to be an important factor.

The major difference between the geographic and similarity analyses is that SUM is significant (the likelihood ratio statistics are all greater than 5.99) and ORIGEQ is insignificant in the similarity analysis. We have argued that ORIGEQ measures information rather than profitability. Therefore, when a good measure of information is included in the logit, ORIGEQ should be insignificant. This has occurred and reinforces our belief that the similarity measure does, in fact, measure information exchange (or shared knowledge).

The likelihood function, with respect to b_2, is much better behaved when the similarity measure is used. Figure 3 shows VOF versus b_2 for ROBOT1B. This graph shows that VOF, as a function of b_2, has a unique minimum, and thus there is no need to constrain $\hat{b}_1$ in any fashion. This occurs because the 'spatial' effect is much stronger.

To aid in interpreting these results, it is helpful to graph ρ_{ij} versus separation distance, as shown in figure 4. Recall that ρ_{ij} represents the effect of

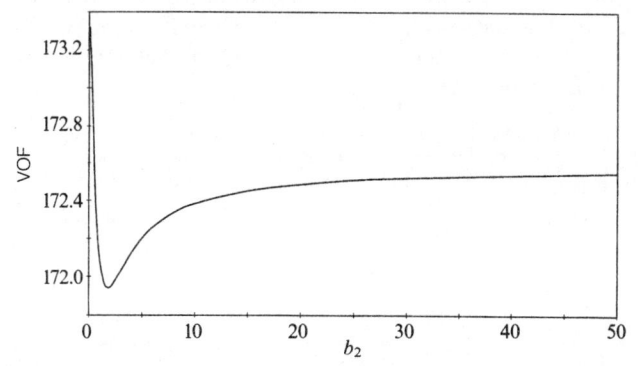

Figure 3. VOF versus b_2 for model 1, obtained by using similarity distance with ROBOT1B as the influential firm.

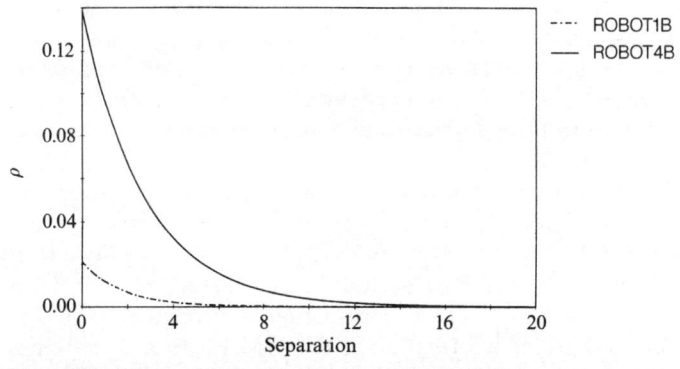

Figure 4. Estimated influence functions in similarity space.

prior adopter j on potential adopter i, and that this influence is stronger for similar firms. Let

$$\text{TERM}_i = X_{it}\beta + \sum_{j=1}^{N} \rho_{ij} y_{j,t-1} \,.$$

With the ROBOT4B results, an additional influential firm will increase TERM_i by 0.138, if the prior adopter is identical to firm i. The probability of adoption is $\exp(\text{TERM}_i)/[1+\exp(\text{TERM}_i)]$, and depends on firm i's characteristics as well as its similarity to prior adopters. With the means of the independent variables (including SUM) and the ROBOT4B results, the probability of adoption is 0.346. An additional, identical, influential firm will increase the adoption probability to 0.378, given the characteristics of the prior adopters.

If the additional influential firm is not identical to firm i, the effect on the adoption probability is smaller. Consider a potential adopter with the characteristics shown in table 4. This firm has an adoption probability of 0.342. If an additional firm becomes influential, and that firm has employment of 100 000 and sells only 10% of its output to auto producers, the adoption probability increases to 0.3516 (the similarity measure for these two firms is 3.31 and ρ is 0.0415). But if the additional firm is identical, the adoption probability increases to 0.374. Thus the effects of an additional influential firm diminish as the firms become less similar.

Table 4. Characteristics of hypothetical firm.

Variable	Hypothetical firm	Variable	Hypothetical firm
FEMP	20	COMPLEX	4
NEWPROD	70	FAIR	2
MKTSHR	10	URBAN	1
ORIGEQ	90	SUM	22

5 Conclusions

Our results indicate that, with respect to the adoption of robots, geographic space does not present a barrier to information transfer. Perhaps this is not surprising. The improvement in communications technology over the course of this century has been remarkable. We now have at our fingertips faxes, conference calls, and e-mail. Each improvement in communications technology reduces the importance of physical face-to-face contacts, and thus the impact of geographic space.

However, we do find that distance in similarity space is important. By this we mean that firms which are similar in size and industry appear to share information about the technology. Our similarity measure is significant at the 0.01 level, and has the correct sign without being constrained. Our belief that the measure does in fact measure information is strengthened by the fact that the other measure of information in the model (ORIGEQ) becomes insignificant in its presence. The model appears to be well specified in that firm size has a significant effect, unlike models when the similarity measure is omitted.

These results are preliminary in the sense that other explanations of our results may be found. For example, perhaps geographic space is important, but the firms in our sample are talking to nearby firms outside the auto supplier industry. Or perhaps the similarity measure is capturing profit rather than information. We intend to explore these issues further in our future research.

References
Brusco S, 1982, "The Emilian model: productive decentralization and social integration" *Cambridge Journal of Economics* **6** 167–184
Case A, 1992, "Neighborhood influence and technological change" *Regional Science and Urban Economics* **22** 491–508
Dubin R, 1995, "Estimating logit models with spatial dependence", in *New Directions in Spatial Econometrics* Eds L Anselin, R Florax (Springer, Berlin) pp 229–242
Grilliches Z, 1960, "Hybrid corn and the economics of innovation" *Science* **342** 275–280
Marshall A, 1898 *Principles of Economics* 4th edition (Macmillan, London)

continued over

APPENDIX A

Distance calculations

For most plants the following formula was used to calculate geographic distance:

$$D_{ij} = \sin y_i \sin y_j + \cos y_i \cos y_j \cos(x_i - x_j),$$

where the coordinates x and y are latitude and longitude, expressed in radians. However, if the two plants are separated by one of the Great Lakes, then a node was inserted into the trip, so that travel would occur over land. For example, if one plant is located in Toronto, Ontario and the other is located in Detroit, Michigan, the separation distance is calculated by adding the distance from Toronto to Sarnia to the distance from Sarnia to Detroit. Table A1 shows the nodes that we used for various plant locations. Although the first column is headed 'Origin' and the second 'Destination', the two columns are interchangeable because the distance matrix is symmetric.

Table A1. Nodes.

Origin	Destination	Node
Michigan	Eastern Ohio	Toledo
Michigan	Ontario	Sarnia
Michigan	Wisconsin	Chicago
Michigan	Northern Illinois	Chicago
Michigan	Pennsylvania	Toledo
Ontario	Ohio	Toledo
Ontario	New York	Buffalo
Ontario	Pennsylvania	Buffalo

APPENDIX B

Table B1. Additional results (t-statistics are given in parentheses).

Variables	Model				
	1	2	3	4	5
(a) Influential firm: ROBOT2B					
Constant	−3.03772	−3.32503	−4.68457	−3.23220	−2.28072
	(−3.48)	(−3.91)	(−4.04)	(−4.08)	(−3.17)
ln(FEMP)	0.11198	0.11456	0.11196	0.11728	0.10135
	(1.83)	(1.87)	(1.81)	(1.90)	(1.67)
ln(NEWPROD)	0.36593	0.36517	0.37643	0.37703	
	(2.69)	(2.68)	(2.74)	(2.73)	
ln(MKTSHR)	−0.09449	−0.08998	−0.12351	−0.08677	
	(−0.85)	(−0.81)	(−1.09)	(−0.78)	
ln(ORIGEQ)			0.45060		
			(2.29)		

Table B1 (continued)

Variables	Model				
	1	2	3	4	5
COMPLEX	0.25719	0.26104	0.20163	0.26350	0.25137
	(1.93)	(1.96)	(1.48)	(1.96)	(1.89)
FAIR	−0.04359		−0.01208		−0.04241
	(−0.42)		(−0.11)		(−0.41)
SWITCH		0.19949			
		(0.74)			
URBAN	−0.04012	−0.04757	−0.04207		−0.05521
	(−0.13)	(−0.16)	(−0.14)		(−0.18)
PCTRD				−0.03494	
				(−0.46)	
NEWPROD					0.01028
					(1.97)
MKTSHR					−0.00405
					(−0.90)
SUM	0.25656	0.25040	0.19735	0.3334	0.24365
b_2	3	3	3	3	3
N	294	294	294	294	294
VOF	178.33	178.14	175.37	178.32	180.15
Restricted VOF	178.61	178.41	175.54	178.59	180.40
Likelihood ratio	0.56	0.54	0.34	0.54	0.50
(b) Influential firm: ROBOT3B					
Constant	−3.08769	−3.35001	−4.78181	−3.26984	−2.32528
	(−3.51)	(−3.93)	(−4.10)	(−4.13)	(−3.22)
ln(FEMP)	0.11954	0.11974	0.11875	0.12266	0.10721
	(1.93)	(1.94)	(1.90)	(1.97)	(1.75)
ln(NEWPROD)	0.36050	0.35716	0.36797	0.36905	
	(2.65)	(2.62)	(2.67)	(2.67)	
ln(MKTSHR)	−0.08818	−0.08432	−0.11719	−0.08080	
	(−0.79)	(−0.76)	(−1.03)	(−0.72)	
ln(ORIGEQ)			0.45977		
			(2.34)		
COMPLEX	0.25233	0.26010	0.20073	0.26261	0.25068
	(1.89)	(1.95)	(1.47)	(1.95)	(1.89)
FAIR	−0.04001		−0.00667		−0.03881
	(−0.38)		(−0.06)		(−0.37)
SWITCH		0.19420			
		(0.72)			
URBAN	−0.06596	−0.05555	−0.06045		−0.06639
	(−0.21)	(−0.18)	(−0.20)		(−0.22)
PCTRD				−0.03506	
				(−0.46)	
NEWPROD					0.00990
					(1.89)

Table B1 (continued)

Variables	Model				
	1	2	3	4	5
MKTSHR					−0.00392
					(−0.87)
SUM	0.52341	0.60992	0.64095	0.60830	0.62669
b_2	4	3	3	3	3
N	294	294	294	294	294
VOF	178.03	177.85	174.93	178.01	179.80
Restricted VOF	178.61	178.41	175.54	178.59	180.40
Likelihood ratio	1.16	1.12	1.22	1.16	1.20
(c) Influential firm: ROBOT4B					
Constant	−3.00589	−3.30301	−4.70000	−3.19207	−2.24947
	(−3.46)	(−3.89)	(−4.05)	(−4.07)	(−3.15)
ln(FEMP)	0.11282	0.11521	0.11317	0.11701	0.10206
	(1.84)	(1.88)	(1.83)	(1.90)	(1.68)
ln(NEWPROD)	0.36678	0.36607	0.37811	0.37764	
	(2.70)	(2.69)	(2.75)	(2.74)	
ln(MKTSHR)	−0.09608	−0.09167	−0.12565	−0.08945	
	(−0.87)	(−0.83)	(−1.11)	(−0.80)	
ln(ORIGEQ)			0.45857		
			(2.34)		
COMPLEX	0.25048	0.25444	0.19506	0.25814	0.24515
	(1.87)	(1.91)	(1.43)	(1.92)	(1.85)
FAIR	−0.04818		−0.01484		−0.04702
	(−0.46)		(−0.14)		(−0.45)
SWITCH		0.19782			
		(0.73)			
URBAN	−0.03079	−0.03722	−0.03438		−0.04796
	(−0.10)	(−0.12)	(−0.11)		(−0.16)
PCTRD				−0.03613	
				(−0.48)	
NEWPROD					0.01034
					(1.98)
MKTSHR					−0.00419
					(−0.94)
SUM	0.47577	0.44412	0.57756	0.54015	0.47206
b_2	5	5	4	4	5
N	294	294	294	294	294
VOF	178.324	178.161	175.385	178.320	180.113
Restricted VOF	178.61	178.41	175.54	178.59	180.40
Likelihood ratio	0.58	0.50	0.30	0.54	0.58

Cluster Formation in the Framework of the Treuhand Approach: From Socialist to Market-oriented Clusters

P Friedrich, X Feng
Universität der Bundeswehr München, Neubiberg

1 The problem

One of the most promising strategies of business promotion to induce local economic growth seems to be cluster formation. Therefore, regional policy may support cluster formation. Whether this is a successful and promising policy is debated elsewhere (Steiner, 1997). We refer to the phenomenon that a transformation policy also has to be considered to offer the chance to form clusters. There is the need to change from a socialist centrally planned economy to a market or social market economy. If clusters play the role indicated above, a transformation policy needs to be developed which will lead to future clusters. However, transformation policy refers to the building up of a new private sector, a new public sector, and sometimes a non-profit-making third sector, as well as other institutions, laws, regulations, etc, and has to take place under many political and economic restrictions (Friedrich and Feng, 1997). A transformation policy is not primarily shaped for cluster formation. Especially with regard to those transformation policies which are to produce rapid transformations in the course of a big bang approach, it is debatable whether they prevent or hinder cluster formation. Such a policy is the so-called *Treuhand approach* to transformation in Germany.

The Treuhandanstalt (Treuhand) was a federal institution of public law which took part in almost all of the activities concerning transformation (Fischer et al, 1993; Treuhand, 1994). The Treuhand controlled a substantial share of property in the new federal states. Properties used by centrally commanded people-owned firms, state firms operating in agriculture and forestry, those properties used for military purposes, along with the former properties of the STASI (communist Secret Service), of the Communist Party and other parties, including those of socialist organisations, were all transferred to the Treuhand.

According to the Constitution, the Treaty for Social and Economic German Union, the Treaty of Unification, the Treuhand Law, the Treuhand Constitution, business guidelines, statutes and statutory requirements, specific laws concerning industrial and sector economies and restitutional matters, and economic and political needs, the following groups of aims (Friedrich, 1994) were demanded of the Treuhand: establishment of a privately dominated market economy; building up of a decentralised public sector; safeguarding of the transformation process; financing of the transformation process; support for the process of democratisation; and protection of the environment.

To achieve these important objectives many functions (Friedrich, 1994) had to be carried out, among them: partial or total privatisation of property, restructuring or liquidation of firms; decisions about which enterprises were to remain on the federal records; sales, leasing, and renting; the establishment of new firms; borrowing, transferring credits, and taking over losses; giving firms back to former private or public owners; transferring property to public bodies; the legal conversion of people-owned firms; formulation of statutes, programmes, staffing, etc; and measures to improve the infrastructure and the environment.

With respect to industrial, trade, and service firms, the policies chosen by the Treuhand were mainly for sales based on restructuring proposals formulated by the new owners (Ebbing, 1995; Jürgs, 1997; Treuhand, 1994). In agriculture and forestry the main activities concerned leasing. In cases involving public bodies, transfer of property was preferred. A sale to the buyer offering the best microeconomic isolated restructuring plan for a real asset was the rule. Sales conditions included obligations on the part of the buyer to restructure, to invest, to employ a specified number of workers, and to pay the asking price.

The Treuhand was not supposed to follow a structural regional policy, although it was involved in many actions to improve economic regional conditions in the new states. It was involved in making proposals for new laws, implementing European Recovery Programmes, EU programmes, employment and labour-market programmes, programmes for regions, sectors, and groups of enterprises, activities to promote small firms, and implementing some structural and environmental policies. Its emphasis was on microeconomic isolated spin-offs and restructuring of those firms sold to private or public owners or transferred to public bodies. The decisions were not based on mesoeconomic or macroeconomic insights, or knowledge of networks to be established in the framework of a regional cluster. Socialistic clusters were not transformed into market-oriented regional clusters. Only in exceptional cases, when industries were linked by technical processes or environmental effects, was cluster formation taken into account.

On the other hand, most decisions of the Treuhand showed structural effects on clusters in the new states. They were caused by the Treuhand's decision to sell, to restitute, to liquidate, to transfer, and to approve restructure plans for firms; investors' decisions to invest and employ, to restructure, etc; conditions fixed in sales contracts, and buyers' financial situations; economic circumstances in the region where investment was to take place; all-German and European economic conditions.

Therefore, we intend to analyse the chances for cluster formation within the framework of this approach. Cluster formation and decisions will be reflected in a sales and transfer model, showing the relationship to the economic conditions in the new states.[1]

[1] Contrary to expectations, the 'Complex Treuhandanstalt' or Treuhand Complex (Bundesministerium der Finanzen, 1994), still had many years from 1995 in which to

The following problems will be investigated:
(1) What are the possible policies for cluster formation within the Treuhand approach?
(2) How can the Treuhand's selling activities and regional conditions, as well as cluster requirements be shaped?
(3) To what extent can cluster formation be considered in terms of a model?
(4) What benefits and conditions, and in particular price levels, result from sales under conditions of cluster formation?
(5) Which form of sales and transfer policy seems the most promising?

Discussion of question 1 is based upon available analyses of forms of clusters and the Treuhand's possible successes (see Friedrich, 1995). For a discussion of question 2 it is necessary to consider the market systems relevent to cluster analysis of the Treuhand activities, and a simplified sales model. The discussion of question 3 includes the subsequent reactions of the contract partners within the model, in the context of different cluster-formation policies. Effects of these policies will then be deduced in the discussion of question 4. These findings will then help to ascertain which cluster-formation policy is the most promising, in question 5. In this chapter we focus on those Treuhand activities which affect the private sector and which form the main issues in cluster formation. The findings are based on empirical information as well as on model results.

2 The cluster policy of the Treuhand
2.1 Difficulties in the maintenance of socialist clusters
Three types of clusters can be distinguished (Sturn, 1997; Tichy, 1997):
(1) In a *leading firm cluster*, several (sometimes more) unconnected suppliers are focused around a leading firm.
(2) A *supplier cluster* shows hierarchical relations and vertical links between a final goods producer and component deliverers (sometimes a group of final or semifinal goods producers).
(3) A *network cluster* is characterised by multidirectional, mostly horizontal, links among firms or actors. The links may be caused by the exchange of goods, services, and information, as well as use of the same infrastructure.

In reality we find mixtures of these types of clusters. Sometimes clusters are defined with respect to goods or entrepreneurial functions, such as raw-material clusters, skill clusters, process clusters, distribution clusters,

---

Footnote (1) continued
carry out its tasks. In the course of the federal election campaign, the Treuhand was declared as having succeeded and was abandoned in 1995. A successor institution, the Bundesanstalt für vereinigungsbedingte Sonderaufgaben (BvS, or Federal Institution for Special Tasks relating to Reunification) was established. One of its tasks was and is the implementation of contract management (Balz, 1993; Küpper and Mayr, 1993). Confronted with another election campaign in 1997/98, it was again debated, whether the BvS should be closed down. But tasks will still remain, especially contract management (Brede et al, 1997).

product clusters, and R&D clusters (Tichy, 1997). One may add infrastructural clusters, security clusters, ethnic clusters, crime clusters, etc. Private enterprises, public firms, public offices (authorities), and research and academic institutions may be involved in clusters. We also have to consider *socialist clusters*, which were formed in Soviet-type, socialist, centrally planned economy.

In socialist economic planning, clusters were formed because people-owned firms were incorporated into *combines*. For the planners, the fewer institutions and people-owned firms, the easier the planning process. Therefore, resource concentration and the planning process were mutually reinforcing. Moreover, socialist clusters were based on technical dependencies among people-owned firms. People-owned firms were thus clusters. They were not firms in a Western sense but institutions of socialist life. Many socialist institutions were part of a people-owned firm, such as kindergartens, shops, restaurants, hospitals, housing, transportation units, holiday resorts, orchestras, singers' associations, sports clubs, convention halls and centres, hotels, communist combat units, medical centres, vocational schools, research units, and all kinds of repair, maintenance, and service institutions, as well as institutions run by the Communist Party and communist trade unions. Another driving force behind the formation of socialist clusters was the microeconomic behaviour of combines and people-owned firms which resulted from the planning process and which was affected by the obligation to fulfil plans. They had to produce planned outputs and had to consider binding constraints (with respect to inputs) and soft constraints (with respect to product demand and finance). Managers, in seeking to minimise risks and maximise the income of staff, pursued a policy of hoarding to ease resource constraints in the short run and a policy of expanding the people-owned firm or combine in the long run. This caused (Friedrich, 1992): self-enforced shortages by cumulative hoarding (short run); increased investment demand and increasing shortages (long run); vertical integration of production as far as possible; the upgrading and increased integration of social services and institutions; the tendency to avoid innovation.

Thus there was an inherent clustering policy, which caused an ageing and concentration process. The aim became more and more to utilise economies of scale and to save costs (inputs). It was tried to keep relations among people-owned firms stable, and there was a tendency to sustain institutions, planning procedures, and certain kinds of products, to avoid substitutions, and to defend hoarding. In small socialist countries some combines and clusters had as their regional basis the whole state. This was the case with some combines in the former GDR. It was often the case that aged products caused aged clusters (Tichy, 1997). Exceptions were found in the military–economic complex. Military and political competition were felt to be essential by communist leaders. Consequently new products and techniques were developed, and clusters of the military-economic complex and in the space industry were not as old as in other sectors.[2]

Such socialist clusters existed in the former GDR. Therefore, these clusters had to be abandoned or transformed into new clusters in the framework of the Treuhand approach. Socialist clusters were easily liquidated, but requirements for social-market-oriented clusters were not easy to meet. Market-oriented clusters presuppose: (a) communication of cluster partners through markets, other forms of exchange, or decrees of decentralised authorities, instead of a vertically integrated command system within big firms; (b) institutions within a formal or informal network which are able to transfer tacit knowledge; (c) regional proximity; (d) opportunities to use complementary components from cluster partners in order to achieve agglomeration economies and to use the existence of external effects to increase the utility of cluster members; (e) common goals, or the possibility to exchange and transfer utilities, and the opportunity to increase utilities for cluster members through cooperation; (f) cluster partners who are independent enough to fulfil commitments.

In the framework of the Treuhand approach the transformation of socialist clusters or the establishment of entirely new clusters caused considerable difficulties.

Socialist clusters could not survive for *economic and legal reasons*, including: (1) old techniques of production, old products, loss of clients, the breakdown of delivery relations and of COMECON, the sudden emergence of competition within Germany from the European Union and from worldwide suppliers; (2) legal requirements concerning the allocation of functions to the Federation (*Bund*), states (*Länder*), counties (*Kreise*), and towns (*Gemeinden*) as well as the establishment of institutions of social security, restitution, new property rights and the social market economy, European and German environmental regulations; (3) spin-offs to meet these requirements and new legal forms for firms and decision units; (4) high wage levels and an income policy to discourage migration, in order to fulfil the aspirations of the population; (5) regional proximity according to socialist central planning requirements and dependency on existing and old pipeline, railway, road traffic, and telecommunication systems.

Therefore, an adequate network, adequate decision units and cluster partners, legal conditions which enabled cooperation and coalition, and a means to exchange utility were lacking. Moreover, appropriate management knowledge to find markets, profit opportunities, and measures to express utilities were not available. Old socialist clusters, combines, and people-owned firms had no direct access to capital markets.

The *spin-off policy* of the Treuhand was primarily to dismantle socialist clusters. (a) Cluster members were transformed into new legal types of stock companies and limited liability companies according to old cluster structures.

---

[2] In agriculture some cluster structures survived because of EU agricultural policies, special restructural legislation for agriculture, leadership positions of former communists, and nonrestitution commitments fixed in the documents of unification.

Treuhand firms partly gained new scope for decisionmaking and began their own game of survival, but their business policies were partly controlled by the Treuhand and were sales oriented. Commercial activities favouring new clusters were extremely difficult. Treuhand firms did not know their future cluster partners, and their own efforts to find partners sometimes ended in conflict with the Treuhand. (b) Spin-offs were set up to transfer functions, people-owned firms, and property to local authorities, the states, the Federation, and to newly founded public and nonprofit institutions. These institutions which had to stabilise themselves with respect to execution of their tasks and democratic decisionmaking were oriented neither to old socialist nor to new clusters. (c) Spin-offs to private sectors took place more or less according to buyers' wishes.[3] Thus relations among the old cluster members were destroyed, new relations were not close and stable, and some component production was closed down because the buyers had new priorities. (d) With spin-offs the Treuhand had to consider restitution claims. Parts of clusters such as small people-owned firms had to be given back to former owners. New firms wanted to get rid of socialist cluster ties, and they looked for new markets and not for new cluster dependencies. (e) The sales policy of the Treuhand was not oriented to clusters. Many buyers who had firms in the western states did not establish ties to clusters in new states but to their activities located in the western states. (f) Restitution policy primarily followed legal requirements. The liquidation was also not cluster oriented. The Treuhand tried to find new owners to create individual economic activities. (g) Sometimes the Treuhand promoted the formation of small clusters by selling real estate to create industrial parks. But cluster building itself resulted from the efforts of other business-promotion agencies.

Treuhand activities were hindered by *ideological positions*, in favour of restructuring by means of middle-sized firms on an isolated individual market-oriented basis. Therefore, efforts to restructure socialist clusters or to build up new clusters were rare.[4] Moreover, there was an extreme time pressure to sell as quickly as possible and revenues of the Treuhand often had priority over chances to form clusters. Foreign investors, who were not very interested in clustering, often got priority in sales decisions. The Treuhand ended up with an unsaleable socialist cluster infrastructure.[5] Mostly this equipment was dismantled by so-called employment firms which were subsidised by the Treuhand and the labour administration. Treuhand activities were bound by EU policies on sales prices, subsidisation, and restructuring

---

[3] These include the establishment of a maintenance and service firm, use of property for storage and trade activities, and selling part of a plant or a production line.

[4] Political pressure resulted in a few attempts at cluster formation, in the chemical and steel industries, in coalmining, and in energy production.

[5] Technical networks such as underground systems, and combine-owned telephone systems did not meet new technical standards to serve as the basis of a future technical cluster.

measures. The political wishes of the Federation, states, and communes not directed to cluster building had to be considered.

The Treuhand had to overcome substantial *technical difficulties*. (1) The Treuhand had to staff itself, to staff Treuhand firms, to develop its procedures, to introduce legal accounting systems, to establish a contract management system, to control Treuhand firms, to finance, to consult, etc. The capacity for far-reaching cluster establishment (with the exception of financial help and provision of real estate) was not available. There was a lack of information about future market structures and developments. (2) The Treuhand approach was a quick muddling-through policy to reduce difficulties and to solve severe short-term problems of integration into an all-German and European economy and political system, into NATO, and into a development of the new states. Treuhand activities were not much concerned with visions of future competence clusters. (3) The Treuhand ran into sales difficulties if it tried to sell a firm on the basis of a socialist network or if the firm was related to counties, states, etc, which were in the course of establishing themselves.

2.2 From socialist clusters to market-oriented clusters

Few methods of forming clusters were left open to the Treuhand. One strategy would have been to establish a cluster of Treuhand firms by forming an industrial complex of public firms which would have done the restructuring with the assistance of the Treuhand.[6] But this alternative had no chance of realisation. Official ideological beliefs were in favour of privatisation.

It was not possible to sell a whole former socialist cluster to one buyer. Most buyers did not have the financial means to restructure clusters. Buyers were not willing to buy social institutions and infrastructure, and to do the job of the Treuhand (with respect to alteration of legal forms, coping with restitution claims, transferring property to public authorities, closing down production, etc). The purchaser refused to bear high risks caused by unclear real and financial conditions.[7] The buyer would have been confronted with extreme difficulties in firing the staff of former socialist clusters and with tremendous managerial problems. Managers with skills to manage and restructure a socialist cluster under market conditions were almost nonexistent. The Treuhand itself preferred to improve financial results through selling

[6] In the 1950s a totally different policy was followed in the western occupational zones and states. Remains of former clusters which survived war destruction, dismantling by occupational powers, reparations, and measures of the allied military government to split trusts, were reactivated mostly by Federal public firms. They served as a shield for small and medium sized firms of private founders, small private investors, and firms of the many refugees from former eastern provinces and German-speaking settlements in Eastern and Middle Europe.

[7] A new law was introduced which allowed the Treuhand to split firms and sell them without signing individual contracts with respect to individual assets. The related legal uncertainty referred to former debts, balances not available, evaluation of assets, size of equity capital, legal difficulties in splitting firms, and to difficulties which stemmed from employment laws, tax laws, accounting rules, stipulations to form commercially oriented balances, commercial laws, environmental laws, agricultural laws, and restitution laws.

individual assets and not clusters. The European Union was resistant to the selling of socialist clusters because of European overcapacity in the shipbuilding, steel, car, and textile industries, and partly in the chemical industry.

Instead of selling clusters, the Treuhand tried to sell stock companies and limited liability companies. But this policy turned out to be difficult too. Apart from the reasons already mentioned, additional causes played a role. Potential buyers were often former competitors who already had enough free capacity in the Western world. Foreign investors outside the European Union preferred to buy protected markets and not old plants to serve highly competitive markets within an integrated Europe. Greenfield investments in other European countries were judged as less risky and easier to manage. German and European investors liked to buy firms which had been partly restructured or nearly liquidated. They wanted to avoid the efforts already mentioned, or to build up a new industry with unknown cluster partners.

Therefore, the Treuhand concentrated on the sale of fields of activities in which investors were interested. Some of those activities were complementary to businesses the investor had in the western states. A demand existed for activities directed to serve future local markets such as construction, consumer goods production, retail, and services. Normally only small slices of socialist clusters could be sold. Above all, functions and equipment which connected and coordinated the cluster members and maintained the cluster network were lost. Surviving fields of business received an entirely new commercial orientation. Demand for some efficient services such as combined research units was lacking. Thus researchers were laid off, installations dismantled, and what remained became less and less attractive to buyers. As the most important industries were organised in socialist clusters, the closing down of nearly the whole industry had to be feared.

Against their ideological positions, politicians were forced to promise that so-called *industrial cores* should survive. This could be ensured only by the establishment of profitable commercially oriented clusters. Within the Treuhand approach an adequate sales policy had to be developed which would convince potential buyers. (1) The sales policy which supported clustering was to sell real estate, locations and old plants under the condition that the buyer was willing to deliver or to purchase goods and services from future buyers of Treuhand property or greenfield investors. (2) Sometimes buyers purchased from and negotiated with the Treuhand individually, but started to install linkages between each other, thereby establishing a cluster network. (3) The Treuhand tried to find investors who would buy real estate, former plants, etc, but who were willing to cooperate and restructure and to create new clusters. Some buyers jointly negotiated a common contract with the Treuhand. (4) Another policy was developed to sell property to a buyer, such as a municipality, a state, or a big firm, such as Jenooptics or Volkswagen, which in the course of business promotion attracted component suppliers to create a cluster. (5) In rare cases a buyer was found who took over several dismantled plants or firms after spin-offs and staff reductions to develop a new cluster.

To overcome the weaknesses of these policies, the Treuhand had to find buyers who were willing to cooperate, to accept obligations, to cooperate with unknown partners, to achieve agreement with the EU Commission on cluster creation, to organise the support of communes and state and federal governments, to develop special contract management, and to establish or operate an industrial infrastructure, such as energy and raw material provision, to serve the cluster members.

Implementation of policy 5 is unlikely. We have tackled policy 4 in other articles (Friedrich and Feng, 1993; Friedrich and Lindemann, 1993). Therefore, we concentrate on variants of policies 1 and 2 and alternative 3.

2.3 The example of the chemical industry

The validity of these remarks is demonstrated by the chemical industry, which was organised in socialist clusters. Table 1 shows 15 socialist clusters, comprising combines which were active in production, and one which traded in chemical products. The oil industry is also included. These 15 combines had 140 people-owned firms outside the main plants of the combines, at Bitterfeld (chemicals and photochemicals), Leuna, Buna (chemicals), Piesteritz (agrochemicals), Zeitz (petrochemicals and hydrocracking), and Böhlen (petrochemicals and refinery). One could speak of the chemistry triangle Bitterfeld–Halle–Merseburg–Leipzig. Chemical industries in the area employed approximately

Table 1. Socialist cluster in the chemical industry, 1989 (source: VCI, 1991).

Combine[a]	Production	
	million DM	%
Chemiefaserkombinat Schwarza (10)	5 475.3	6.98
Chemiekombinat Bitterfeld (7)	7 582.0	8.66
Agrochemisches Kombinat Piesteritz (7)	6 705.5	8.09
Petrochemisches Kombinat Schwedt (5)	18 357.9	20.94
Kombinat Plast-/Elastverarbeitung (14)	5 912.7	6.59
Pharmazeutisches Kombinat GERMED (14)	5 149.9	3.39
Kombinat Lacke/Farben (12)	3 598.2	4.38
Fotochemisches Kombinat Wolfen (7)	2 874.8	3.49
Kombinat Kosmetik Berlin (8)	3 115.6	2.65
Leuna-Werke (2)	10 421.7	12.63
Chemische Werke Buna (5)	9 422.8	11.13
Reifenkombinat Fürstenwalde (6)	3 516.3	3.49
Synthesewerk Schwarzheide (6)	3 651.5	4.34
Kombinat Chemieanlagenbau Leipzig-Grimma (12)	2 710.7	3.24

Total production (1989 prices): 88 500 million DM
Total number of employed: 308 008

[a] Number of legally and economically independent firms within the combine given in parentheses.
[b] In relation to the average of all combines of the Ministry of Chemical Industry: 287 283 Mark per employee.

110 000 (the whole of the chemical industry employed about 308 000). The chemical industry was an old industry which was founded at the end of last century, developed between 1916 and the 1930s, was destroyed, dismantled for reparations, and became Soviet stock companies until 1952 for Soviet use only. Afterwards, these big plants became combines and people-owned firms. The combines were vertically integrated, but also had horizontal supply relations to geographically neighbouring combines and people-owned firms. The most important supply relations which partly existed in the past and which are necessary in the future are demonstrated in figure 1. Within the combines at the main locations, a future network of technical input – output relations will exist which are shown with respect to materials used and deliveries and plants connected in figure 2. Clusters also existed among combines, especially at the main locations. The technology was outdated and based mainly on coal and potash. The coal-based industry had a high (30%) proportion of production (Breitenstein, 1996). One third of all equipment was over 50 years old. At Leuna and Buna, a considerable part of the equipment was invested before the Second World War. In the whole chemical industry of the GDR 50 000 employees were engaged in the repair of installations (Breitenstein, 1996, page 87) and about 20% of the employees did tasks which were atypical for the chemical industry. There were terrible emissions of dust and sulphur

Employees		Productivity		Product
no (E)	%	DM/E	%[b]	
29161	9.47	187 761	65.4	synthetic fibre
28800	9.35	263 264	91.6	chemistry
19236	6.24	348 591	121.3	agricultural chemistry
28562	9.27	642 739	223.7	petrochemistry
30663	9.95	192 828	67.1	plastic
16799	5.45	306 560	106.7	pharmaceutic
7763	2.52	463 506	161.3	lacquer
21491	6.98	133 768	46.6	photographic
8712	2.83	357 622	124.5	cosmetic
30173	9.79	345 398	120.2	chemical processing
27720	9.00	339 928	118.3	chemistry
11699	3.80	300 564	104.6	tyres
12197	3.96	299 377	104.2	synthesis
35090	11.39	77 250	26.9	chemical equipment

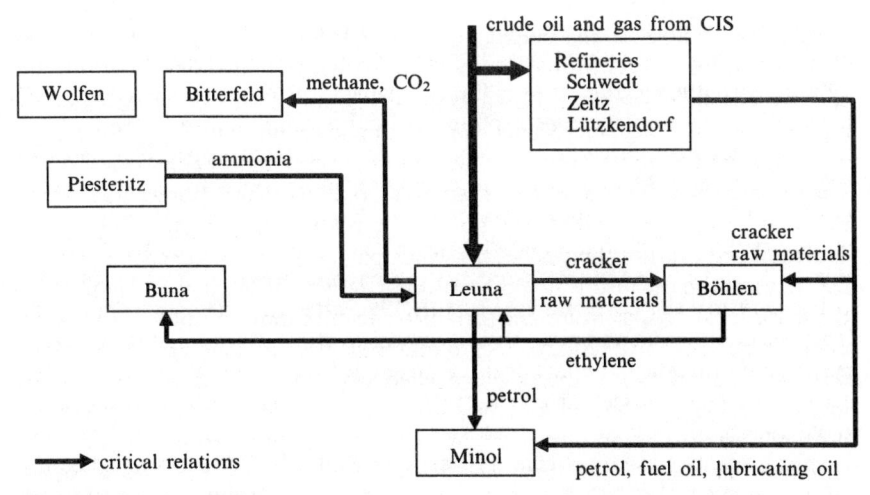

Figure 1. Supply relations.

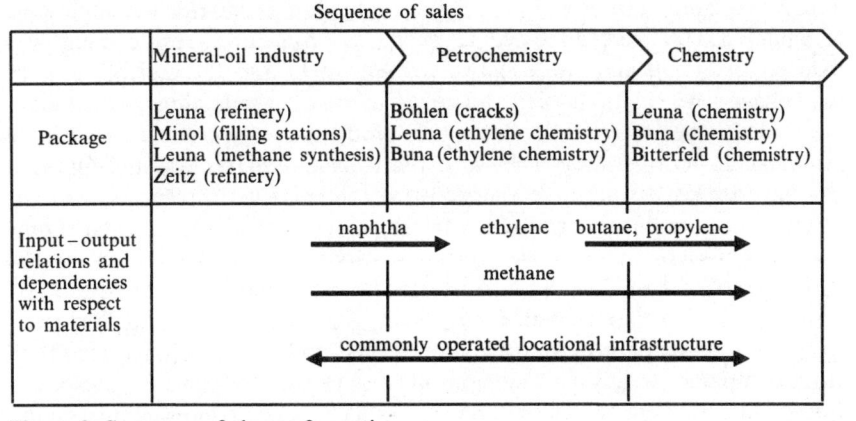

```
                                    crude oil and gas from CIS
                                              ↓
                                         ┌──────────┐
                                         │Refineries│
                         methane, CO₂    │ Schwedt  │
  ┌──────┐  ┌──────────┐ ─────────→      │  Zeitz   │
  │Wolfen│  │Bitterfeld│                 │Lützkendorf│
  └──────┘  └──────────┘                 └──────────┘
                                                                  cracker
              ammonia                                          raw materials
  ┌─────────┐ ─────────→                    cracker
  │Piesteritz│                   ┌─────┐ ─────────→ ┌──────┐
  └─────────┘                    │Leuna│raw materials│Böhlen│
  ┌────┐                         └─────┘            └──────┘
  │Buna│ ←─────────────────────
  └────┘                            │     ethylene
                                    │ petrol
                                    ↓
                                 ┌─────┐
                                 │Minol│ ← petrol, fuel oil, lubricating oil
                                 └─────┘

  ──→ critical relations
```

Figure 2. Sequence of cluster formation.

Sequence of sales

	Mineral-oil industry	Petrochemistry	Chemistry
Package	Leuna (refinery) Minol (filling stations) Leuna (methane synthesis) Zeitz (refinery)	Böhlen (cracks) Leuna (ethylene chemistry) Buna (ethylene chemistry)	Leuna (chemistry) Buna (chemistry) Bitterfeld (chemistry)
Input–output relations and dependencies with respect to materials	naphtha → ethylene butane, propylene → methane → ← commonly operated locational infrastructure →		

dioxide; two thirds of the water used in production was uncleaned. In 1990, because of the currency change, the abolition of subsidy of exports to Western countries, the breakdown of COMECON, and a worldwide recession in chemical products, it was estimated that 60% of people-owned chemical firms suffered heavy losses [1600 million DM, on a turnover of 4200 million DM (Breitenstein, 1996, page 88)]. Some experts argued that the whole chemical industry should close down and should not be reestablished (Fischer and Weissbach, 1993, page 65). The Treuhand was convinced that total restructuring was necessary and that new clusters must be created. The Treuhand started restructuring at an early stage and the decision of the federal government to maintain industrial cores, particularly the chemical industry in the region (Kohl, 1994) supported these efforts.

The activities of the Treuhand can be divided into five main phases: struggle for survival (1990); development of strategies to create new clusters (1991); restructuring and sales (1992); reorientation of strategies (1993 – 95); processes of investment, construction, and cooperation (since 1995).

These activities became embodied in the Treuhand approach. Between 1 April 1990 and 30 June 1990 the combines and people-owned firms were legally transferred to 17 stock companies and 205 limited liability companies.[8] In the first phase, Treuhand activities were characterised by a cluster-building policy (2) and immediate measures to prevent the breakdown of production and to improve the environment. In July 1990 after the currency change, the Treaty of Economic and Social Union came into effect. Because of the collapse in demand, liquidity support was arranged by the Treuhand in the GDR. Because of heavy losses and catastrophic environmental conditions related to old plants, the first closures occurred. Plants to process carbides (Piesteritz), chemical plants based on coal processing (Buna), tar processing (Leuna), ammonification (Leuna), aluminium production (Bitterfeld), and nitric acid production (Bitterfeld) were among the first to go. The first spin-offs of services took place, and staffing was reduced considerably. From 113 000 employees, of whom 83 000 were allocated to chemical production, employment was reduced to 35 600 employees between 1 July 1990 and the end of 1991. In that phase investigations were made as to which productions would have the best chances of survival and which productions were needed to keep a technical cluster operating. Therefore, the Treuhand guaranteed and financed new investments to improve energy provision, pipeline transports, and the environment (pollution, water treatment, and chemical treatments). Some investments were made to reinvest and modernise production lines. The main plants of combines were converted to stock companies. Chemie AG Bitterfeld-Wolfen, Leuna Werke AG, Buna AG, Böhlen AG, and Stickstoffwerke AG Wittenberg-Piesterritz were created and it was planned to sell all stock companies to buyers after spin-offs and closures of parts of enterprises. But the idea that the remainder and core of a cluster could be sold turned out to be an illusion. Buyers were not willing to restructure clusters with very high investments into networks, compound plants, and a multiproduct complex partly based on old technology.

Therefore, in the second phase, plans were developed to distinguish three categories of business. Core business activities were those which seemed to have the ability to survive as they could achieve profitability in the future, or serve as a basis for future activities. Service activities would operate until new investors started their plants, whereas the rest were those which showed very high losses and were candidates for closure if no buyer was found for them quickly. The last category comprised services such as fire brigades, research units, and restaurants, which had not yet undergone the spin-off process.

---

[8] The number of enterprises rose in the course of time through spin-offs and the number in the portfolio of Treuhand department for chemical industry rose to 281.

These plans were accompanied by the first sales of plants, property, and locations. To back the core business, to finance the service business, and to finance the rest of the businesses, expenditures of 8 billion DM were made for the redemption of former debts, to cover losses, and for investment. With this policy, a closure of the cluster was planned, but keeping open the possibility for new clusters. During this period, decisionmakers became convinced that several partners had to cooperate—the Treuhand and Treuhand firm, the investors (buyer or new firm), the Federation, state and municipalities, and the federal labour administration. In the earlier phases, planners overlooked the fact that the European Economic Union had to confirm investments, if—which was normally the case—subsidisation was involved. These institutions were differently engaged in a struggle of survival of Treuhand firms and reduction of losses, in strengthening competitiveness and rationalisation of Treuhand firms, selling business fields, in restructuring firms, in the establishment of cluster and public infrastructure, in performance of short-term ecological programmes, and in the mitigation of environmental damage.

At the end of the second phase, it became clear that only a business-field-oriented sale of activities had the chance of being successful within the Treuhand approach. It then turned out that businesses and activities must be sold which were related to each other. Therefore, ideas about future clusters and combinations of productions, goods, and services had to be elaborated. The structure of a whole cluster for the chemistry triangle was to be decided, and subclusters (for example, at the old main locations) had to be developed. The decision was made that the cluster should be based not on coal but on oil. Therefore, a chain of production activities and possible locations, as well as flows of goods had to be projected (figures 3 and 4). Refineries which served petrochemical complexes to deliver materials for chemical industries became very important. Figure 4 shows where the respective producers were to expect markets. These main cluster lines had to be accompanied by an industrial and public infrastructure to guarantee and stabilise links and a network of deliveries. Therefore, a range of productions had to be operated further on, and had to be improved. Other activities could be sold only jointly in such a way that mutual deliveries were possible, and new investors had to fit into this roughly elaborated network. Furthermore, new heavy investments had to be made in energy provision, public utilities, water clarification, pipeline systems, etc. For these purposes some special institutional arrangements were developed.

In the third phase, an adequate selling and restructuring strategy was also developed, and negotiations between the Treuhand and potential investors were opened. Cluster-building strategies required that investors must be found to develop new refineries. The investors should deliver the products to the cracking enterprises.

To find such investors, packages of activities had to be sold, as at Leuna. To find an investor for constructing a new refinery ready to take over the delivery of naphtha from Leuna, and to produce ethylene at the next stage of production,

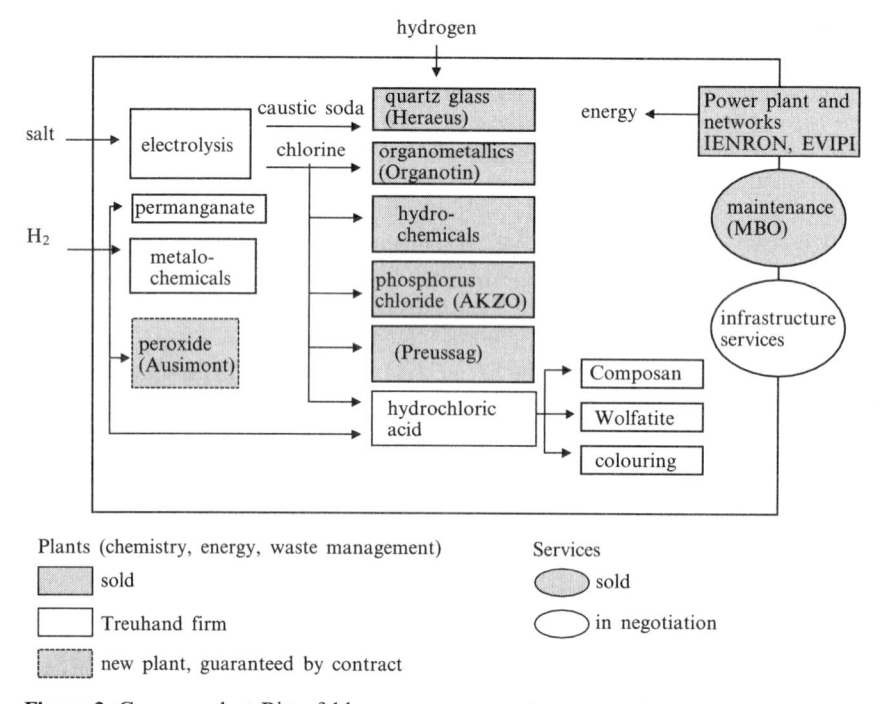

Figure 3. Compound at Bitterfeld.

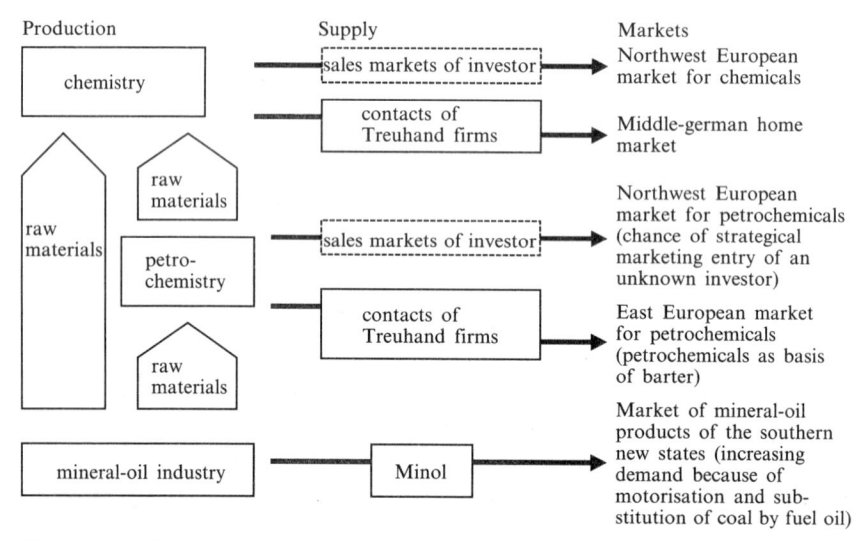

Figure 4. Markets and cluster partners based on refineries.

the Treuhand had to sell the petrol producer Minal AG at Leuna and the whole system of filling stations of the former GDR to Elf Aquitaine. Moreover, Treuhand had to guarantee to operate the old refineries at Leuna and Zeitz until the new one at Leuna was finished. Other sales of commercial activities took place and a special infrastructure company was formed under the participation of the Treuhand.

Another kind of package deal concerned the olefin industry at Buna and Böhlen, which was connected through chemical processes and deliveries of fluids and chemical materials. The Treuhand tried to find one buyer for the whole range of activities, including all the necessary infrastructural activities. All cracking activities, therefore, became concentrated at Buna, Böhlen, and Leuna, and the cracking cluster was bought and established by Dow Chemicals, one of the leading ethylene producers in the world. It turned out to be very costly, because the construction activities were heavily subsidised. Dow is going to attract further industries which will process their products, which in turn will enrich the cluster.

A different strategy was chosen at Bitterfeld–Wolfen. There, a precondition for survival was a cluster for the production of hydrogen and chlorine as a basis for the production of fibreglass, pharmaceutical products such as aspirin, and colouring materials. There, the Treuhand guaranteed and operated the basic productions and sold the processing firms. This policy was accompanied by spin-offs and the settlement of new productions based on the Treuhand cluster. But, as at Böhlen and Buna, buyers were less interested in purchasing the chemical and other infrastructure. Therefore, a development company was formed by private–public partnership which now delivers infrastructure services and develops locations related to the former Bitterfeld–Wolfen combine. It developed preconditions for an increase in the cluster through new investment and further sales.

At Wolfen sales of business activities were rare, so the policy there was to liquidate old production, to use employment firms to clear the locations, and to settle new firms promoted by the Treuhand and the development firm. A similar strategy was developed for Zeitz, because the refinery was to be closed. A chemical industrial park was founded which was to offer all kinds of services to support a network based on chemical production. The underlying idea was that in the future lean production will be introduced into the chemical industry too. This chemical park was sold to a private–public partnership company which must promote and settle new chemical producers and service firms. Another firm was founded to produce synthetics.

At Piesteritz the original policy to sell the core of the former people-owned firm was handled successfully. Therefore, one enterprise, SKV-Troisberg, took care of nitrogen production and will build up a kind of cluster which is also going to be supported by spin-offs. But public authorities had to erect a new clearing plant. This policy was possible because the losses from nitrogen production turned out to be relatively small, thus becoming attractive to profit-interested buyers.

In this third phase, much time was lost in complicated negotiations with investors who had to fit into a chemical production network and into the packages mentioned. Chains of contracts and many public–private joint actions had to be elaborated, and the approval of the EU Commission had to be gained. These activities concerned infrastructure, employment firms, social programmes, financial programmes, restitutions, physical and urban planning, and the development of new technical structures. In some cases BvS had to renegotiate contracts and had to find new buyers in the course of contract management in the fourth and fifth phases. As already described, the Treuhand was forced to follow a semiprogressive restructuring policy within the Treuhand approach. It had to maintain parts of the traditional cluster and material input–output network, but at the same time, to modernise the network and open the possibility for high-technology firms to fit into this network (Fischer and Weissbach, 1993, page 66). One of the consequences was that much of the research capacity which was previously available in combines was lost. Later, the Treuhand and BvS tried to concentrate research capacities at Leuna, Bitterfeld, and Zeitz, and had them concentrated on Leuna. In the course of restructuring in 1990, the main plants of the old combines became stock companies, and particularly after 1993, limited liability companies. After 1994 most of them were split into various small limited liability companies in order to restructure them, liquidate them, or sell them. The new clusters are already partly established, much of the infrastructure is installed, and some productions are already operating successfully, some at a small level, and other plants are in the course of investment. It was very difficult to find the optimal time for closure, restructuring measures, investments, spin-offs, sales, etc. Clusters have been physically erected to a considerable extent. These processes were supported by a helpful wage policy from the trade union active in the chemical industry. Nevertheless, employment fell to about 20 000 jobs, which remain or were newly established.

Cluster policies 1, 2, and 3 were chosen. Some aspects of policies 4 and 5 were also considered. The total expenditure was about 30 billion DM. Therefore, it has to be questioned, which of the three possible policies of cluster formation seems the most useful.

3 A sales model of the Treuhand
3.1 Basic cluster-formation policies
With the policies of the Treuhand, sales contracts and property transfers had to be negotiated and fixed. For purposes of analysis we shall try to model such a contract and its conditions.

The Treuhand follows an extremely complex goal system, which it has pursued with varying intensity during the phases of its development (Friedrich, 1995; Friedrich and Lindemann, 1993). As a result of the principal–agent relations (Meissner, 1994), in which the Treuhand takes on the role of principal and the buyer that of agent (Friedrich, 1995), an objective external evaluation of success must give way to an evaluation that is more action orientated. It is the attainment of levels of success, indicated in the contracts and terms of transferral, that serve as the gauge of success. Such yardsticks are determined by the Treuhand's goal system, which reduces sales to the following goals: increases in income, securing of jobs and investments, and the realisation of concepts of restructuring (Küpper and Mayr, 1993). The Treuhand was financed partly by credits and revenues from sales of property. There were limitations on these credits. But the Treuhand could expect those constraints to be flexible.[9]

The basis for determining success comprises four main goals and the utility value gained from their evaluation, including the level of utility value as understood in the completed contract.[10] As the sole fulfilment of some of these aims leads to a positive evaluation, an additive evaluation function should be brought into play. Measured goal fulfilment is dependent upon what the evaluation concentrates on.

For analytical purposes we consider a cluster in a very simplified way, concentrating mainly on private investors. A newly settled firm $i-1$ delivers output V_{i-1} to a newly settled firm i which uses V_{i-1} as input to produce V_i, which is delivered by i to consumer or firm $i+1$ as is demonstrated in figure 5.

Policy 1 is that of a sale to a buyer who intends to open commercial relations with other potential cluster members. This policy has two subcases. In policy 1(a) firm $i-1$ settles first and firm i settles later and has to adapt to firm $i-1$ (figure 6). In policy 1(b), however, purchaser i settles first but adapts to the later locating deliverer $i-1$ (figure 6).

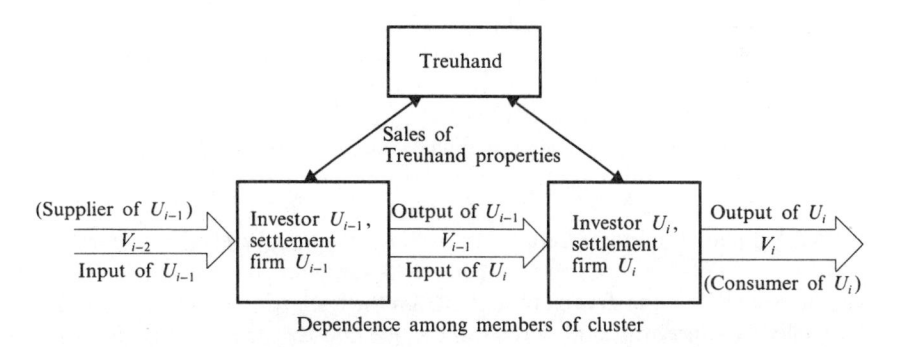

Dependence among members of cluster

Figure 5. Sales to cluster-dependent investors.

[9] For political reasons, measures to overcome transformation difficulties and the easing of unification process had priority. Therefore, the budget constraints of the Treuhand were partly soft. As a consequence, individual sale negotiations were not strictly financially linked with other sales. Financial needs were expressed mainly by the goal of increasing revenue within the goals system.
[10] The Treuhand's total achievement can thus be seen as comprising the utility value resulting from the first contract, including the addition or deduction of changes in utility value during the contract management phase (see Friedrich and Feng, 1997).

Policy 1: clusters within ordinary sales

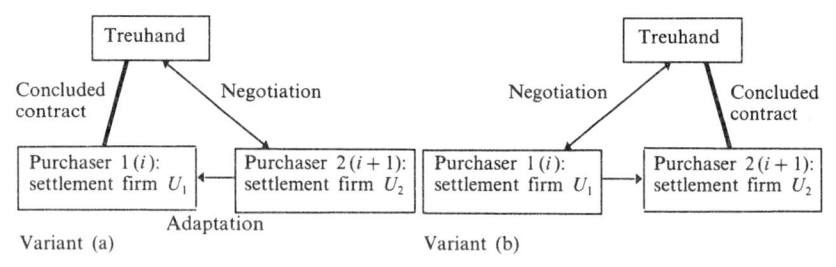

```
              Treuhand                                    Treuhand
Concluded  /          \ Negotiation      Negotiation  /            \ Concluded
contract  /            \                             /              \ contract
 ┌──────────────┐   ┌──────────────┐    ┌──────────────┐   ┌──────────────┐
 │ Purchaser 1(i):│  │Purchaser 2(i+1):│ │Purchaser 1(i): │  │Purchaser 2(i+1):│
 │ settlement firm U₁│◄─│settlement firm U₂│ │settlement firm U₁│─►│settlement firm U₂│
 └──────────────┘   └──────────────┘    └──────────────┘   └──────────────┘
             Adaptation
Variant (a)                              Variant (b)
```

Policy 2: cluster of mutually dependent investors

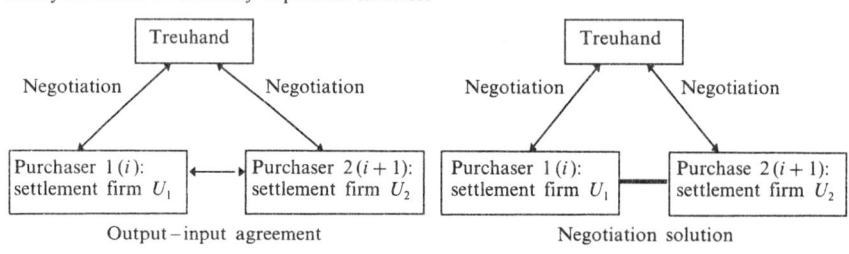

```
              Treuhand                                    Treuhand
Negotiation /        \ Negotiation       Negotiation /           \ Negotiation
 ┌──────────────┐   ┌──────────────┐    ┌──────────────┐   ┌──────────────┐
 │Purchaser 1(i): │  │Purchaser 2(i+1):│ │Purchaser 1(i):│  │Purchase 2(i+1): │
 │settlement firm U₁│◄─│settlement firm U₂│ │settlement firm U₁│──│settlement firm U₂│
 └──────────────┘   └──────────────┘    └──────────────┘   └──────────────┘
       Output–input agreement                    Negotiation solution
Variant (a): noncooperation among investors   Variant (b): cooperation among investors,
                                              separate negotiations with the Treuhand
```

Policy 3: cluster of mutually dependent investors

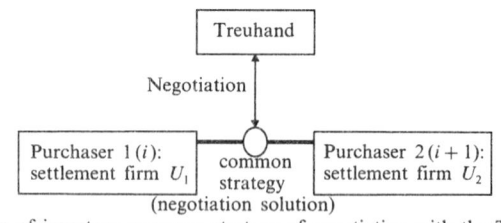

```
                         Treuhand
                            ▲
                       Negotiation
                            │
            ┌──────────────┬─┴─┬──────────────┐
            │Purchaser 1(i):│common│Purchaser 2(i+1):│
            │settlement firm U₁│strategy│settlement firm U₂│
            └──────────────┴──┴──────────────┘
                    (negotiation solution)
       coalition of investors, common strategy of negotiation with the Treuhand
```

Figure 6. Policies for sales of Treuhand properties with regard to cluster formation.

Policy 2 allows individual contracts between Treuhand and cluster members at the same time. This policy again has two subcases. In policy 2(a) cluster members coordinate with each other through the leadership of firm $i-1$ which fixes delivery conditions and purchaser i which chooses its best response when formulating an input–output agreement. Thus a Stackelberg case of cluster formation prevails (figure 6). This approach seems most likely because the deliverer may play a more active part in cluster formation. In policy 2(b), cluster member $i-1$ and cluster member i cooperate in mutual deliveries, but they sign contracts with the Treuhand individually (figure 6).

Policy 3 again incorporates buyer $i-1$ and buyer i, but they negotiate jointly with the Treuhand and come to one common contract with the Treuhand. They might establish for that deal a joint venture company which functions as the buyer. The buyers have to cooperate in their delivery network to formulate common offers to the Treuhand (figure 6).[11]

3.2 Features of the underlying model for sales to private investors

In contrast to Bös (1991), we refrain from offering a model which provides a basis for an analysis of privatisation and the different forms it can take. In his model of the decisions of privatisation, Bös includes the behaviour of the government, unions, and management, as well as the consequences of state expenses, tax revenue, and state deficit. Differing levels of privatisation (for example, mixed-economy enterprises) are possible. Here, however, we have concentrated on the relationship between the Treuhand and its goals within its sphere of activity. We will not, therefore, deal with models concerned with privatisation based on a general auction system (Sappington and Stiglitz, 1987). Equally, we have not provided a model articulating the relations between the national government and the Treuhand (see, alternatively, Meissner, 1994). Treuhand policy on the evaluation of agricultural land has been explored by Braun and Weikard (1994), and an econometric-based model of price fixing in relation to the confirmation of work places, has been developed by Lucke (1994). Demougin and Sinn's (1994) model includes some of the relevant economic goals and deals with the first contract only. It includes risks and risk aversion by decisionmakers and deals with the sale of participants, but does not investigate cluster policies. This model fails to serve our purposes adequately, as it refers to too few individual sales and economic goals. Furthermore, we will take the essential problems of cluster formation into consideration.[12]

Our model considers the specific economic goals of the Treuhand whilst exploring five general scenarios of clustering according to policies 1(a), 1(b), 2(a), 2(b), and 3. To stress the basic features of the model, we assume first that firm i buys property from Treuhand to erect a plant which fits into an already existing network. The relationship to the network is open insofar as firm i can buy as many intermediate products as it needs at prevailing prices. Firm i sells its products on a monopolistic market.

First, the influences of the contract-determining factors on the completion of the first contract will be discussed. The Treuhand (T) sells its property (B_i^0) at a price F_i^0 to private investors i. The Treuhand's goal function includes the confirmed number of jobs (A_i) and the investments of private investors K_i as well as Treuhand revenue (F_i), the costs of sale to the Treuhand (E_i), and subsidies (S_i) arising from the sale of property. In the framework of the concept for restructuring (V_i), subsidies (S_i) are granted for the elimination

---

[11] Policy 4, where property is sold to a municipality which sells to another buyer, thus forming a cluster of mutual delivery, was tackled within a similar model before. This is also true with respect to policy 5, where the buyer is obliged to build up its internal cluster. Therefore, we do not trace models for these policies here.

[12] Models which explain the extension of a network and price formation within networks do not fit the Treuhand case (see Gehrig, 1996).

of environmental damage. Furthermore, the goal function evaluates the restructuring concept (V_i). In accordance with the utility value analysis mentioned above, the utility value function of the Treuhand, shown in appendix A.2, forms the outcome (Friedrich and Feng, 1993).

The Treuhand's revenue (appendix A.2, A.3) results from the sale of property, which we have labelled (F_i^0). The subsidies (S_i) have been deducted from the sale price in order to ascertain the net income (F_i) which is reduced by the Treuhand costs of sale (E_i). The connection between the restructuring concept (V_i), which is represented by the output, and the subsidies, forms the relationship $\eta_i V_i$. The emphases of evaluation are positive.

The purchasing private investor's utility function corresponds with its profit function as a result of its aims of profit maximisation [appendix A.2 (2)].[13] Turnovers are determined by outputs, and delivery price P_{V_i}, which is itself determined according to a demand function. Demand depends on mill price, $P_{V_i}(V_i)$, and size of transportation costs, $P_{D_i}(D_i)$. This is a demand function of intermediate product supplier $i-1$ or consumption goods producer i. Costs (cost function) are based on assumed factor prices, transportation costs to the mill, $T_W(D_W)$, resulting tax payments, and a production function, which shows labour (A_i), capital (K_i), and municipal intermediate inputs (W_i), distance to municipality D_W, and intermediate supplies V_{i-1} as factors of production.[14] Capital inputs differ from the property bought from the Treuhand (B_i^0).[15] The company produces at minimal cost.

It is possible for decisionmakers to attain a minimum level of utility which includes alternative utilities, which, in turn, can then be taken up should the contract negotiations fall through. The minimum level of goal achievement of the Treuhand ($N_{T_i}^0$), when selling, is illustrated in its own rehabilitation of firms, the transferral of assets to municipalities, or in different alternatives of privatisation. The buyer's minimum utility ($N_{U_i}^0$) symbolises the opportunity utilities related to alternative financial investments made with its own means, and which have been designated for the purchase and restructuring of the firm. The Treuhand utility and the utility of the purchasing investor both

---

[13] The temporary tax exemption from business, capital, and property tax, at the time the first contract was signed, is taken into account in the profit function. These payments are otherwise included in r_B

[14] The intermediate municipal services (W) are mainly related to the location where a private enterprise is going to settle. The intermediate supplies from other firms V_{i-1} are also related to the location. Job and investment guarantees are related to labour and capital inputs.

[15] There are two kinds of real estate allocated to the production process: first, there is property which is bought and subsidised by the Treuhand (B_i^0); second, property which developed by the purchasing firms (B_i^1). The Treuhand supports the purchasing firm, by paying it compensation for the cost of improving the quality of the land and the environmental damage suffered by the real estate. The volume of supply and supply delivery price of the intermediate product equal in equilibrium the demanded volume and the demand price for the intermediate product which is purchased by the maker of the consumption good.

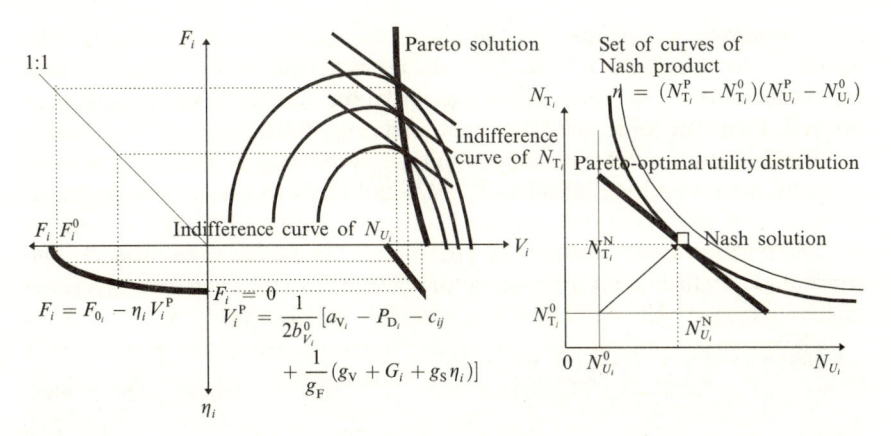

Figure 7. Indifference curves obtained when decisionmakers maximise their utilities.

depend on the intended output (restructuring concept) and the purchase price (F_i). Figure 7 shows the indifference curves that result when decisionmakers maximise their utilities. Both the purchase price and other conditions of purchase are negotiated. The point of tangency on the indifference curves marks a sequence of Pareto-optimal combinations of F_i^P and V_i^P.[16] The relationship between the net revenue F_i^P and the output V_i^P (restructuring concept) is also shown in figure 7. If the purchase price F_i^P increases, the output V_i^P decreases. The correspondence between the utilities of the Treuhand and the investor, depicted in figure 7, reveal a utility possibility curve which illustrates the utility distribution between the Treuhand the investor. The utility possibility curve shows the set of Pareto-optimal bargaining solutions related to the purchase price and restructuring concepts (V_i^P) indicated in figure 7 (Friedrich and Feng, 1993; Holler, 1992, page 25P; Krelle, 1976, page 773f; Rauhut et al, 1979, page 249f; Schneider, 1969, page 9f).[17]

Each bargaining solution on the Pareto limit is effective; however, not all utility distributions are acceptable to the decisionmakers, as the minimum utility levels lead to a reduction in the set of possible solutions. Furthermore, possible solutions are also restricted by the indifference curves corresponding

---

[16] The indifference curves related to the distribution of utility between the Treuhand and the private buyer, and the corresponding volume of output and sales prices can be deduced with the help of an expansion path, [appendix A.2(2)], from the utility functions. Conditions for the point of tangency are shown in figure 7. Restructuring, expressed in V_i, is supported in the framework of the Pareto-optimal contracts by subsidies. The Treuhand's increased financial support (S_i) thus increases its output and utility. Higher financial demands, however, also lead to a decrease in utility. This results in restrictions in the amount of subsidies and for the Pareto-optimal distribution of utility between the Treuhand (N_{T_i}) and the private buyer (N_{U_i}) (Friedrich and Feng, 1993, page 245f).

[17] The Nash solution results from the maximised Nash product, under the subcondition of Pareto-optimality, whereby the decisionmakers' points of conflict equal the minimum demand of the utility distribution, according to the minimax solution (Nash, 1953).

to the minimum utility levels. Application of the Nash solution[18] to figure 7 results in the utility distribution ($N_{T_i}^N$, $N_{U_i}^N$). The Treuhand utility is dependent on the minimum level of utility, the purchase price, and the output of restructuring. The sales conditions are fixed, and comprise the restructuring concept (output), purchase price, Treuhand subsidies for the elimination of environmental damage, as well as the jobs and investments guaranteed by the buyer.

The bargaining solution (see figure 7) and the related formulas, demonstrate that the formulation of the contract is influenced primarily by cluster factors.[19]

3.3 Cluster factors considered in the model

Many cluster factors are attached to the model and are listed in table 2. They express cluster factors which show special conditions prevailing at a specified site. The goals of the Treuhand normally refer to the whole area of the new

---

[18] One can calculate these results of negotiations by using the Nash concept. By subtracting the minimal utility levels $N_{T_i}^0$ and $N_{U_i}^0$ from the utilities realisable, a goal function results, which can be used to find the Nash solution. The points of conflict in the Nash solution, d_{T_i} and d_{U_i}, reflect the most disadvantageous position each party to the negotiations can be forced into by its opponent, in the case of conflict. As a decisionmaker is only able to insist on its minimum utility level, but can only reduce its conegotiator's utility to its own minimum level by using threats, the threat strategies converge with the minimum utilities, or the minimum utilities of both negotiating parties, according to the minimax solution: $d_{T_i} = N_{T_i}^0$, $d_{U_i} = N_{U_i}^0$ (Friedrich and Feng, 1993). The negotiation solution, illustrated in figure 7, results if this Nash product is maximised, according to the subconditions of the Pareto-optimal distribution of utility.

[19] The Treuhand's privatisation policy is reflected in the valuation weights g_V, g_A, g_K, g_F and g_S. Alternatives for property use and the Treuhand's initial position regarding privatisation, reflect the market form and the minimum levels of utility. Depending on the alternatives of communalisation, self-restructuring, etc, different levels of minimum utility result, in combination with evaluations affected by political, economic, financial, and ideological goals, respectively. The minimum utility levels reveal whether such alternatives should be considered. Past environmental damage to a Treuhand company's property and the evaluation methods applied to it, are reflected in B_i, r_B, and F_i. The Treuhand company's location, or its prospective location, are reflected in W, D_W, B_i^0, r_B, and T_W, t_B, t_G. Each prevailing tax policy shapes the parameters t_K and t_B as well as t_G. The size of demand is determined by the business cycle, which is expressed by the values of the parameters of the demand function [through a_{V_i}, b_{V_i} and average transportation costs $P_{D_i}(D_i)$] and by the size of the factor prices, r_A, r_K. Wage policy is taken into account, along with the wage rate, r_A. Money policy shapes the level of the interest rate, r_K. The prospective firm's technical development level of production is clarified by the parameter of the production function, $V_i^0(W, D_W)$, and exponents of production factors. Possible alternatives of investment for the prospective firm and its market chances are illustrated by its minimum utility level ($N_{U_i}^0$). This is because of its competitiveness with national and international rivals situated at other locations. Moreover, the market chances can be clarified by the parameters of the price functions $P_{V_i}(V_i)$. The latter also refer indirectly to different national and international business-cycle developments, economic policies, and differences in location, etc.

states, but also to the subregions where clusters are to be formed. The production technique of a settlement firm is primarily linked to a sector or to firm-typical processes. It depends on conditions of deliveries and locational characteristics as well as on cluster partners. The distance to component suppliers and the quality and volume of municipal intermediate products which influence the shape of the production function express cluster dependencies. The links between cluster firms become effective through delivery and price formation. Fees for municipal services, business tax rates, and real estate tax rates differ from town to town. Capital prices at different locations in the new states do not show substantial discrepancies. Actual labour prices differ regionally depending on unemployment rates, and existing sectors in a region, etc. However, labour prices fixed by trade unions and entrepreneurial associations for the new states are seen to be the same everywhere. We treat demand for the products of a buyer of a real estate as being dependent on the cluster partner. In the case of differing distances from the buyer to the client, we must consider varying demands. Transportation costs should also not be forgotten. As will be shown later, the Treuhand cluster-formation policy will itself become a cluster-determining factor.

The effects of varying cluster conditions can be deduced. Their impacts are demonstrated in table 2 (see over). In order to discuss the cluster-formation policy resulting from the Treuhand approach we turn to different cluster-formation policies and to the basic model presented above.

4 Cluster formation policies
4.1 Cluster policy 1
4.1.1 *Cluster policy 1(a)*
The mode of sales contract will always undergo changes when different cluster policies are taken into account. According to cluster policy 1(a), a property is going to be sold by the Treuhand to firm i which is now named firm 2. This firm buys intermediate products from a cluster partner $i-1$ which is now introduced as firm 1. Firm 1 has already settled and has formed a contract with the Treuhand. For supplies of intermediate goods firm 2 depends on firm 1. Therefore, firm 1 may close down if the Treuhand does not come to terms with firm 2. Firm 2 adapts to the delivery conditions of firm 1 and firm 1 fixes its price P_1 which is accepted by firm 2. From the point of view of firm 2, the price elasticity of the buyer P_1^s with respect to volume V_1^d turns out to be zero. But this is not the case when we consider firm 1. It expects a relative variation in demand if it changes its price. A negative elasticity results [appendix A.4(3)].

In policy 1(a) the minimum utilities vary to some extent. There is a minimum utility of firm 1 and of firm 2. The Treuhand shows a minimum condition relating to firm 1. Minimum requirements also exist for firm 2. They are reduced by disutilities caused by sunk costs and negotiation costs which were met when the contract was signed with firm 1. Therefore, the minimum utility is reduced by an amount which is bigger than $E_1/2g_F$ [appendix A.4(4)].

Table 2. Cluster factors.

		d (solution)/d (determining factor)		
		$N_{T_i}^N$	$N_{U_i}^N$	V_i^N
Weights of Treuhand's goal				
concept for restructuring	g_V	>0	>0	>0
guaranteed jobs	g_A	>0	>0	>0
investments of purchaser	g_K	>0	>0	>0
receipts from the sale	g_F	>0	<0	<0
grant for restructuring	g_S	>0	>0	>0
Speciality of Treuhand property	B_i^0	>0	>0	=0
Property of settlement firm				
technological parameter	V_i^0	>0	>0	>0
exponent of labour	α_i	>0	>0	>0
exponent of capital	β_i	>0	>0	>0
exponent of land	γ_i	>0	>0	>0
exponent of intermediate input	δ_i	>0	>0	>0
Threat points, minimum utility				
Treuhand	$N_{T_i}^0$ or d_{T_i}	>0	<0	=0
purchaser U_i	$N_{U_i}^0$ or d_{U_i}	<0	>0	=0
Economic condition				
business profit tax	t_G	<0	<0	=0
corporate profit tax	t_K	<0	<0	=0
real estate tax	t_B	<0	<0	<0
real price	r_B	<0	<0	<0
labour price	r_A	<0	<0	<0
capital price	r_K	<0	<0	<0
price of intermediate input	P_{i-1}	<0	<0	<0
municipal intermediate inputs	W	>0	>0	>0
distance between the production site and the supply site	$D_{i-1}(D_W)$	<0	<0	<0
fees of municipal intermediate inputs	T_W	<0	<0	=0
demand situation	a_{V_j}	>0	>0	>0
	b_{V_j}	<0	<0	<0
distance to sales area	D_i	<0	<0	<0
transportation costs	P_{D_i}	<0	<0	<0

Table 2. Cluster factors.

A_i^N	K_i^N	F_i^N	S_i^N	η_i^N
>0	>0	<0	>0	>0
>0	>0	<0	>0	>0
>0	>0	<0	>0	>0
<0	<0	>0	<0	<0
>0	>0	<0	>0	>0
$=0$	$=0$	>0	$=0$	$=0$
$< \frac{dV_{ij}^N}{dV_i^0}$	$< \frac{dV_{ij}^N}{dV_i^0}$	>0	<0	<0
$> \frac{dV_i^N}{d\alpha_i}$	$< \frac{dV_i^N}{d\alpha_i}$	>0	<0	<0
$< \frac{dV_i^N}{d\beta_i}$	$> \frac{dV_i^N}{d\beta_i}$	>0	<0	<0
$< \frac{dV_i^N}{d\gamma_i}$	$< \frac{dV_i^N}{d\gamma_i}$	>0	<0	<0
$< \frac{dV_i^N}{d\delta_i}$	$< \frac{dV_i^N}{d\delta_i}$	>0	<0	<0
$=0$	$=0$	>0	$=0$	$=0$
$=0$	$=0$	<0	$=0$	$=0$
$=0$	$=0$	<0	$=0$	$=0$
$=0$	$=0$	<0	$=0$	$=0$
$> \frac{dV_i^N}{dt_B}$	$> \frac{dV_i^N}{dt_B}$	<0	>0	>0
$> \frac{dV_i^N}{dr_B}$	$> \frac{dV_i^N}{dr_B}$	$> \frac{dF_i^N}{dt_B}$	>0	>0
$< \frac{dV_i^N}{dr_A}$	$> \frac{dV_i^N}{dr_A}$	<0	>0	>0
$> \frac{dV_i^N}{dr_K}$	$< \frac{dV_i^N}{dr_K}$	<0	>0	>0
$> \frac{dV_i^N}{dP_{i-1}}$	$> \frac{dV_i^N}{dP_{i-1}}$	<0	>0	>0
$< \frac{dV_i^N}{dW}$	$< \frac{dV_i^N}{dtW}$	>0	<0	<0
$> \frac{dV_i^N}{dD_{i-1}}$	$> \frac{dV_i^N}{dD_{i-1}}$	<0	>0	>0
$=0$	$=0$	<0	$=0$	$=0$
>0	>0	<0	<0	<0
<0	<0	<0	<0	$=0$
<0	<0	<0	>0	>0
<0	<0	<0	>0	>0

As demonstrated above, Pareto solutions are derived and solutions of the contract with firm 2 are found by application of the Nash concept. The utilities of the Treuhand ($N_{T_2}^{N1a}$) of firm 2 ($N_{U_2}^{N1a}$) as well as promised jobs (A_2^{N1a}) and investments (K_2^{N1a}), the Treuhand sales price (F_2^{N1a}), and the amount of subsidies are determined [appendix A.4(5)].

The contract already existing between firm 1 and the Treuhand was—in the context of policy 1(a)—based on estimations about a probable buyer, firm 2. In particular, a demand function for intermediate products by a potential buyer, firm 2, was assumed [appendix A.4(6)]. Therefore, the contract with firm 2 may have effects on firm 1. Firm 1 fixes an actual profit-maximising price for its sales, but cannot change the stipulation with respect to the property price with the Treuhand. Firm 1 is confronted with the actual demand curve for intermediate products (figure 8, second graph). As long as a new price for deliveries of firm 1 to firm 2 is situated northeast of the price which was assumed with respect to sales when the contract was signed between the Treuhand and firm 1, partner 1 is going to move to a profit-maximisation position (figure 8). New solutions are derived from the profit maximisation for the output (V_1^{s1a}), the price (P_1^{s1a}), etc. For firm 1, the utility ($N_{U_1}^{1a}$) from the new solution is higher than from the existing contract with the Treuhand. The Treuhand also gains in comparison with the results of the original contract. Output (V_1^{1a}) and factor inputs are higher with the actual solution. The Treuhand utility with firm 1 ($N_{T_1}^{1a}$) turns out higher than in the first situation. Firm 1 is overfulfilling its economic obligations. Renegotiation of

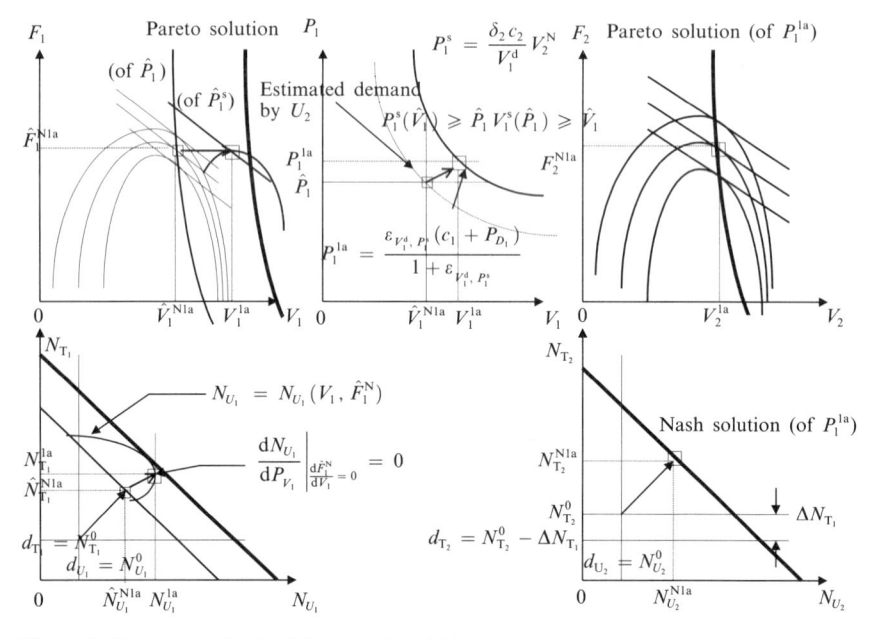

Figure 8. Solutions obtained from policy 1(a).

the contract with firm 1 would lead to increased output, a higher utility for firm 1, and to increased Treuhand utility. In the case where the demand curve for intermediate products does not shift northeast (figure 8), renegotiation must take place and policy 1(a) ends up as policy 2(a).

4.1.2 Cluster policy 1(b)

In policy 1(b), firm 2 is the first to sign a contract with the Treuhand. Firm 1 locates later. For the negotiation of the contract between the Treuhand and firm 2, a price for the intermediate product supplied by firm 1 was assumed (appendix A.5 (10)]. This assumption is made explicit in the price P_1^s. Moreover, the minimum utility conditions vary. The minimum utility of the Treuhand with respect to firm 1 subsequently decreases. That part of the Treuhand utility which relates to sunk costs and negotiation costs with firm 2 has been deducted [appendix A.5 (9) and (12)].

In the case of policy 1(b), firm 2 adapts to the price policy chosen by firm 1. If firm 1 fixes a lower price which is situated southeast of the supposed price, a higher output for firm 2 as well as higher labour input and investments result (figure 9). For the Treuhand and firm 1, we set out a relationship to Pareto solutions. By application of the Nash principle to the negotiations, a solution is found for utilities of the Treuhand and firm 1. Jobs, investments, and the Treuhand property price in connection with firm 1 [appendix A.5 (12)] are deduced.

The utility of firm 2 turns out to be higher than that resulting from the situation assumed when the contract is signed with the Treuhand. The utility

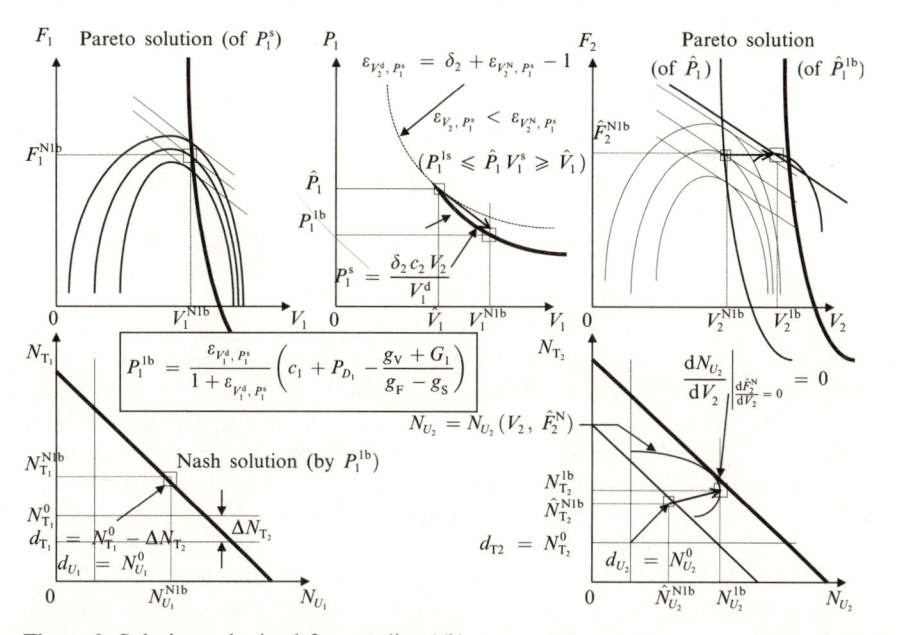

Figure 9. Solutions obtained from policy 1(b).

of the Treuhand with respect to firm 2 turns out to be higher compared with the situation assumed when a contract is signed with firm 2. The output of firm 2 increases too. A further output would stem from renegotiations with the Treuhand. The property price of the Treuhand and subsidies would change. The Treuhand and firm 2 could improve their utilities by renegotiations.

Policy 1(b) would end up as another policy [that is, policy 2(a)], if the resulting price for the intermediate product turned out to be higher combined with a smaller delivery.

4.2 Cluster policy 2—cluster of mutually dependent investors
4.2.1 *Cluster policy 2(a)—Stackelberg strategies among investors*
In policy 2(a) the Treuhand negotiates with two mutually dependent firms, 1 and 2, at the same time. There is no sequence of settlements. Again firm 1 takes a Stackelberg leader position with respect to deliveries to firm 2. Firm 1 fixes the price under the assumption of the best response by firm 2. Firm 2 behaves in analogy to policy 1(a). Minimum utilities of firms are given as $N_{U_1}^0$ and $N_{U_2}^0$, and the minimum utility of Treuhand as $N_{T_1}^0$ and $N_{T_2}^0$ [appendix A.6 (14)]. By elaborating the pareto-optimal solutions for negotiations between firm 2 and the Treuhand, and by applying the Nash concept under consideration of $N_{U_1}^0$ and $N_{U_2}^0$ to these solutions, the utilities of the Treuhand and of firm 2 are found together with output of firm 2, factor inputs, the price of Treuhand property, and the subsidisation (appendix A.6).

The solution between firm 1 and the Treuhand is evoked too, thus giving the utilities of the Treuhand and firm 1 as well as jobs, investments, output of firm 1, the price of Treuhand real estate, and subsidisation (appendix A.6). Moreover, the price for intermediate supplies and the amount of deliveries are indicated [appendix A.6 (15)]. The price is lower compared with policy 1(a), but higher compared with policy 1(b). Deliveries increase. The solution procedure is traced graphically in figure 10. An adapting process leads to a demand curve for intermediate products, which expresses Stackelberg delivery prices.

4.2.2 *Cluster policy 2(b)—Nash solutions among investors*
Policy 2(b) introduces more complications than the other policies. Cooperation is needed between firm 1 and firm 2, relating to the conditions of delivery from firm 1 to firm 2.

The minimum utility requirements vary [appendix A.7 (17)]. The minimum utilities of the firms undergo no changes. We assume that, if the firms and the Treuhand are not coordinated, they have to fall back on a sequential policy. Therefore, the minimum utility of the Treuhand with firm 1 comprises the minimum utility, less the difference in utility gained from a cooperative solution compared with the sequential one. The minimum utility with respect to firm 2 is defined in the same way. In bargaining between firm 1 and firm 2, the utilities which could be achieved without cooperation are considered as minimum utility requirements by the individual cluster partners [appendix A.7 (17)].

In the course of their cooperation, firm 1 and firm 2 have to find a solution for the delivery price P_1^c which shows a cost, an evaluation component, and an

Cluster formation and the Treuhand

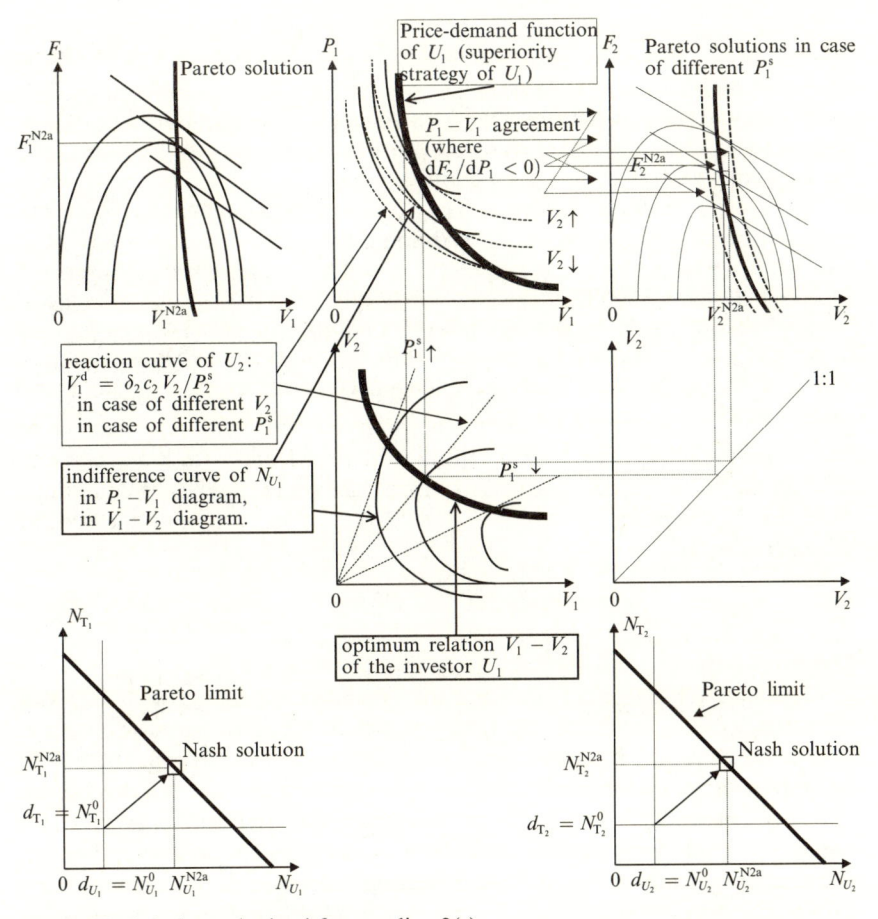

Figure 10. Solutions obtained from policy 2(a).

additional component P_1^{0c} [appendix A.7 (16)]. This additional component indicates how firm 1 and firm 2 shift additional utilities between the two settlement firms. The analysis starts by determining maximum utility of firm 2 if the utility of firm 1 and the Treuhand with respect to firm 1 are given. Cost-minimisation procedures are also applied. A cost function for firm 2 is found. This cost function serves as a basis to derive formulas for the solutions between the Treuhand and firm 2. They refer to the utility of the Treuhand and firm 2, the output of firm 2, jobs, investments, subsidies, and the price of Treuhand property. All formulas show the additional part of the delivery price for intermediate products.

In our next step we elaborate the solution for the Treuhand and firm 1. Because of the delivery dependencies [20] the cost function of firm 2 affects the solutions for firm 1. The respective formulas include the additional component in delivery price of intermediate products. Solution values for utilities of the Treuhand and firm 1, for jobs, investments, output, subsidisation, and the Treuhand property price are obtained.

The cluster partners, however, have to come to an agreement about the delivery price. They have to fix the additional component in such a way as to determine the delivery price [appendix A.7 (21)]. A set of Pareto-optimal utility combinations of firm 1 and firm 2 is specified. Again, the two cluster partners cooperate according to Nash. Therefore, under consideration of Pareto-optimal combinations of utilities of settlement firm 1 and settlement firm 2, the Nash product is maximised. The Nash product shows the multiplication of differences between utilities and minimum utility requirements of the cluster members. The relevant utilities of firm 1 and firm 2 are derived, and consequently the delivery price is ascertained and the cluster conditions are fixed. The solution is demonstrated in the second graph of the second row of figure 11.

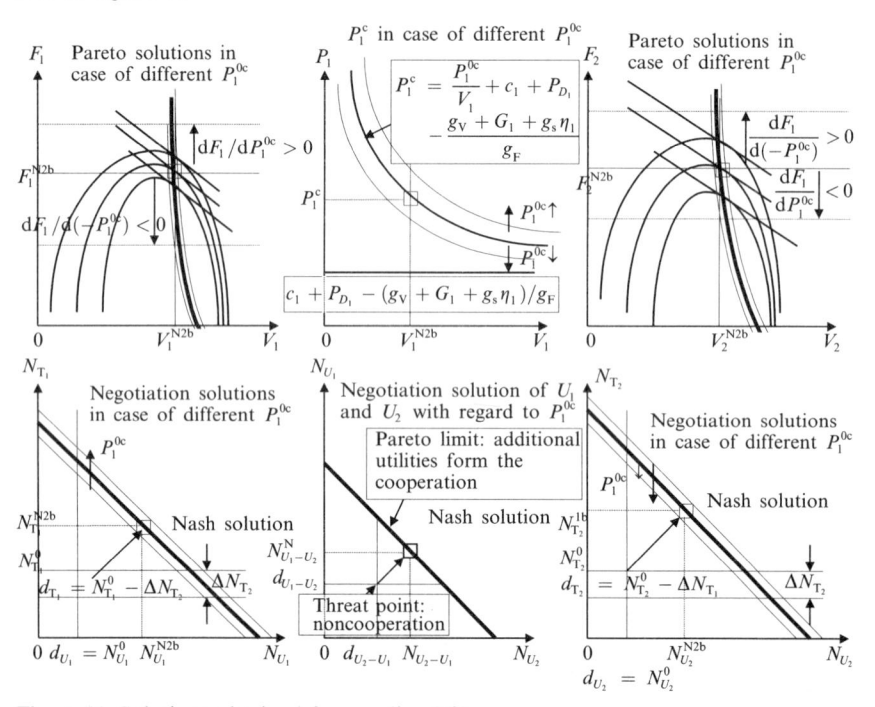

Figure 11. Solutions obtained from policy 2(b).

[20] These dependencies express c_2^c and P_1^{0c} in the cost function C_2 [appendix A.7 (16)].

4.3 Cluster policy 3—common negotiations with the Treuhand or joint ventures for cluster formation

By carrying out policy 3 the Treuhand signs one contract. The two firms form one negotiator. Two players are involved in the game.

The Treuhand utility results from an addition of goal fulfilment at the two plant locations. Jobs in the two firms as well as investments are added. One price will be paid for the two properties. Outputs of the two plants are evaluated equally. The same is true for subsidies [appendix A.8 (25)]. The utilities of the firms' decisionmaker (the profits of the plants) are also added [appendix A.8 (27)].

We gain minimum requirements of the Treuhand from the addition of minimum utilities related to the property and financial obligations at the plant locations. On the buyer's side the minimum utility requirements are summarised [appendix A.8 (26)].

The utility and profit function of the buyer are elaborated. The minimum costs function and factor output relations are determined by a cost-minimisation procedure [appendix A.8 (26)]. By respective substitutions, the utility function of the Treuhand is evolved. The determination of Pareto-optimal combinations of utility of the Treuhand and the buyer follows [appendix A.8 (28)].

Maximisation of the Nash product under the constraint of Pareto-optimal utility combinations delivers the solution formulas. The factors of the Nash product consist of the difference between the utility of the player achieved and the respective minimum utility requirement. The contract and cluster conditions result with respect to the utility of the Treuhand, the buyer, the two outputs, the two subsidies, and the purchase price [appendix A.8 (28)].

5 Best policies within the Treuhand approach to support cluster formation

Our model allows for conclusions on successful cluster structuring (Friedrich, 1995) by the Treuhand (BvS). A comparison of cluster policies is given in table 3. The total success of the Treuhand outlined in table 3 (see over) turns out to be highest with policy 3. Policies 1(a) and 1(b), which the Treuhand normally apply, show less favourable results. The Treuhand will achieve better results pursuing policy 2(a) rather than policies 1(a) or 1(b). Policy 3 is likely to be more fruitful than policies 2 or 1. With regard to the structure of clusters, one learns that the final output (V_2) as well as the intermediate output is again highest with policy 3. The marginal costs of the plants are at their lowest with policy 3, whereas sequential solutions 1(a) and 1(b) cause higher marginal costs. The revenues of the Treuhand are higher in the case of second policies as compared with sequential policies. Policy 3 can be related to higher or lower property prices. Similarly, it also depends on the evaluation weight for revenues. The producer of intermediate goods achieves higher sales prices in the case of sequential selling.

The model can also be used to analyse the successes anticipated when the Treuhand pursues a combination of different goals (Friedrich, 1995), when the

Table 3. Comparison of solutions from different cluster policies.

Policies	
Price of V_1 P_1	$P_1^{1a} \geq P_1^{2a} > P_1^{2b}, \quad P_1^{2a} \geq P_1^{1b}$
Production costs c_1	$c_1^{1a} = c_1^{1b} = c_1^{2a} = c_1^{2b} > c_1^{u3}$
c_2	$c_2^{1a} \geq c_2^{2a} > c_2^{c2b} > c_2^{u3}, \quad c_2^{1b} > c_2^{c2b}$
Threat points d_{T_1}	$d_{T_1}^{1a} = d_{T_1}^{2a} > d_{T_1}^{1b} > d_{T_1}^{2b}$
d_{T_2}	$d_{T_2}^{1b} = d_{T_2}^{2a} > d_{T_2}^{1a} > d_{T_2}^{2b}$
$d_T = d_{T_1} + d_{T_2}$	$d_T^3 = d_{T_1}^{2a} + d_{T_2}^{2a} > d_{T_1}^{1a} + d_{T_2}^{1a} > d_{T_1}^{2b} + d_{T_2}^{2b}$
d_{U_1}	$d_{U_1}^{1a} = d_{U_1}^{1b} = d_{U_1}^{2a} = d_{U_1}^{2b}$
d_{U_2}	$d_{U_2}^{1a} = d_{U_2}^{1b} = d_{U_2}^{2a} = d_{U_2}^{2b}$
$d_U = d_{U_1} + d_{U_2}$	$d_{U_1}^{1a} + d_{U_2}^{1a} = d_{U_1}^{1b} + d_{U_2}^{1b} = d_{U_1}^{2a} + d_{U_2}^{2a} = d_{U_1}^{2b} + d_{U_2}^{2b} = d_U^3$
Output of settlement firms V_1	$V_1^{N3} > V_1^{N2b} > V_1^{N2a} \geq V_1^{1a}, \quad V_1^{N2a} \geq V_1^{N1b}$
V_2	$V_2^{N3} > V_2^{N2b} > V_2^{N2a} \geq V_2^{N1a}, \quad V_2^{N2a} \geq V_2^{1b}$
Price of Treuhand properties $F = F_1 + F_2$	$F_1^{N2a} + F_2^{N2a} \geq F_1^{N1a} + F_2^{N1a}, \quad F_1^{N2a} + F_2^{N2a} \geq F_1^{N1b} + F_2^{N1b},$ $F_1^{N2a} + F_2^{N2a} > F_1^{N2b} + F_2^{N2b}$
Utilities of the Treuhand $N_T = N_{T_1} + N_{T_2}$	$N_T^{N3} > N_{T_1}^{N2a} + N_{T_2}^{N2a} > N_{T_1}^{N2b} + N_{T_2}^{N2b}, \quad N_{T_1}^{N2a} + N_{T_2}^{N2a} > N_{T_1}^{1a} + N_{T_2}^{1a},$ $N_{T_1}^{N2a} + N_{T_2}^{N2a} > N_{T_1}^{1b} + N_{T_2}^{1b}$

cluster conditions or locational factors vary, and different contract management policies are used. The success of the Treuhand differs considerably according to the type of cluster policy pursued. Policy 3 has proved particularly successful for the Treuhand. Moreover, the analytical framework allows us to investigate the effects on clustering if more than one Treuhand existed.

Furthermore, our empirically oriented analysis showed: that reclustering is necessary in the transformation process; the extreme difficulties in reshaping traditional clusters or forming new clusters; the necessity of building up public and private institutions when establishing a cluster; the limitations of cluster formation within a Treuhand approach; the tendency to build new clusters on the basis of traditional and socialist clusters; the high costs of clustering within a privatisation of clustering; the need to adapt Treuhand policies to the needs of clustering, and for public enterprises to discover clustering; and the possibility of successful cluster transformation of the chemical industry in the new states of Germany.

According to our results the Federal government should: assign cluster building and contract management to the Treuhand; concentrate on cluster policy 3 and not on policies 1(a), 1(b), 2(a), or 2(b); vary the goal function of the Treuhand (BvS) in such a way that restructuring, and the establishment of new Treuhand firms is permitted to strengthen cluster formation; protect Treuhand cluster formations against EU disapproval and competition, as well as against the policy of nonsubsidy.

Acknowledgements. For information about chemical industry we thank Prof. Dr. Hilmar Schmidt associated with BvS, Berlin.

References
Balz M, 1993, "Entwurf einer Richtlinie für das Vertragsmanagement", unpublished paper, Treuhandanstalt, c/o BvS, Alexanderplatz 6, 10178 Berlin
Bös D, 1991 *Privatisation: A Theoretical Treatment* (Clarendon Press, Oxford)
Braun J, Weikard H-P, 1994, "Auction models of privatisation of agricultural land in Eastern Germany", DP 9506, Institut für Agrarökonomie, Universität Göttingen
Brede H, Eichhorn P, Engelhardt W W, Freidrich P, Gottschalk W, Hirsch H, Oettle K, Püttner G, 1997 *Thesen zur Weiterführung der Bundesanstalt für vereinigungsbedingte Sonderaufgaben (BvS) vom Mai 1997* (Universität der Bundeswehr München, München)
Breitenstein P, 1996, "Erfolgreiche Umstrukturierung der Chemieregion", in *TLG, Dokumentation* c/o BvS, Alexanderplatz 6, 10178 Berlin, pp 86–96
Bundesministerium der Finanzen, 1994, "Finalisation of the fulfilment of remaining tasks of the Treuhand Law", 9 August *BGBl* part 1, pages 2062–2065
Demougin D, Sinn H-W, 1994, "Privatisation, risk-taking, and the communist firm" *Journal of Public Economics* **55** 203–231
Ebbing F, 1995 *Die Verkaufspraxis der Treuhandanstalt* (RWS Verlag Kommunikationsforum, Köln)
Fischer J, Weissach H-J, 1993, "Forschung und Entwicklung in Transformationsprozess der grosschemischen Industrie Sachsen-Anhalts", in *Der Transformationsprozess der gross-chemischen Industrie Sachsen-Anhalts, Forschungsbeiträge zum Arbeitsmarkt in Sachsen-Anhalt, Band 4* (Ministerium für Arbeit, Soziales und Gesundheit des Landes Sachsen-Anhalt, Magdeburg) pp 60–87
Fischer W, Hax H, Schneider H-K (Eds), 1993 *Treuhandanstalt, das Unmögliche wagen, Forschungsberichte* (Akademie Verlag, Berlin)
Friedrich P, 1992, "From centrally planned to market oriented economies: problems of transition in East Germany and Eastern Europe", in *Towards a New South African Economy: Comparative Perspectives* Eds H Weiland, M Braham (ABI, Freiburg) pp 62–73
Friedrich P, 1994, "Zukünftige Ziele und Aufgaben der Treuhandanstalt" *Zeitschrift für öffentliche und gemeinschaftwirtschaftliche Unternehmen* **17** 1–23
Friedrich P, 1995, "Die Politik der Treuhandanstalt – Erfolge und unbewältigte Probleme", in *Die Wettbewerbsfähigkeit der ostdeutschen Wirtschaft, Jahrestagung des Vereins für Socialpolitik Gesellschaft für Wirtschafts-und Sozialwissenschaften in Jena 1994* Eds G Gutmann, K W Rothschild, Schriften des Vereins für Socialpolitik, Band 239 (Duncker und Humblot, Berlin), pp 95–165
Friedrich P, Feng X, 1993, "Ansätze einer Theorie des Verkaufs von Treuhandvermögen an Kommunen" *Jahrbuch für Sozialwissenschaft* **44**(2) 233–277
Friedrich P, Feng X, 1997, "Regional aspects of the Treuhand contract management", Summer Institute, Norway, forthcoming; copy available from the author

Friedrich P, Lindemann S, 1993, "Die Treuhandanstalt ein Instrument zum Aufbau des Föderalismus?", in *Finanzierungsprobleme der deutschen Einheit I: Staatsverschuldung, EG-Regionalfonds, Treuhandanstalt* Ed. K-H Hansmeyer, Schriften des Vereins für Socialpolitik, Band 229/I, (Duncker und Humblot, Berlin) pp 77–173

Gehrig T, 1996, "Natural oligopoly and customer networks in intermediate markets" *International Journal of Industrial Organization* **14** 101–118

Holler M J, 1992 *Ökonomische Theorie der Verhandlungen* (R Oldenbourg, München)

Jürgs M, 1997 *Die Treuhändler* (Droemer Knaur, München)

Kohl H, 1994, "Rede zur Regionalkonferenz 'Industriestandort Sachsen Anhalt— Zukunft der Chemieregion vom 1.12 1992'", in *Treuhandanstalt, Dokumentation 1990–1994* Treuhandanstalt, Berlin, pp 290–296

Krelle W, 1976 *Preistheorie II* (J C B Mohr, Tübingen)

Küpper H-U, Mayr R, 1993, "Vertragsgestaltung and Vertragsmanagement der Treuhand", in *Treuhandanstalt, Das Unmögliche wagen, Forschungsberichte* Eds W Fischer, H Hax, H-K Schneider (Duncker und Humblot, Berlin) pp 315–353

Lucke B, 1994, "Die Privatisierungspolitik der Treuhandanstalt—Eine ökonomische Analyse", DP14/94, Institut für Statistik und Ökonometrie, Freie Universität, Berlin

Meissner T, 1994, "Die Wirkung der institutionellen Ausgestaltung der Treuhandanstalt, Berlin, auf die Privatisierung in den neuen Bundesländern", in *Schriften der Universität Trier Fachbereich IV* Eds D Dickertmann, M Lehrmann (Universität Trier, Trier)

Nash J, 1953, "Two-person cooperative games" *Econometrica* **21** 128–140

Rauhut B, Schmitz N, Zachow E-W, 1979 *Spieltheorie* (B G Teubner, Stuttgart)

Sappington D E M, Stiglitz J E, 1987, "Privatisation, information and incentives" *Journal of Policy Analysis and Management* **6** 567–582

Schneider H, 1969 *Das allgemeine Gleichgewicht in der Marktwirtschaft* (J C B Mohr, Tübingen)

Steiner M (Ed.), 1997 *Competence Clusters* (Leykam, Graz)

Sturn D, 1997, "Motors of cooperation and main actors in clusters", in *Competence Clusters* Ed. M Steiner (Leykam, Graz)

Tichy G, 1997, "Are today's clusters the problem areas of tomorrow?", in *Competence Clusters* Ed. M Steiner (Leykam, Graz) pp 94–100

Treuhand, 1994 *Treuhandanstalt, Dokumentation 1990–1994* Treuhandanstalt, c/o BvS, Alexanderplatz 6, 10178 Berlin

VCI, 1991, "Landesverband Ost, Strukturwandel der ostdeutschen Chemie", Verband der Chemischen Industrie e.V., Harz 51, 06108 Halle

APPENDIX
A.1 Glossary

A_i	guaranteed jobs of purchaser i	P	superscript indicating Pareto solution
a_{V_i}	parameter of demand function of firm i	r_A	labour price
a	superscript indicating model variant (a)	r_B	real price
		r_K	capital price
b_{V_i}	parameter of demand function of firm i	s	superscript indicating supply
		S_i	price reduction to support the firm's restructuring of purchaser i
b	superscript indicating model variant (b)	t_B	real estate tax
B_i	real estate of settlement firm i	t_G	business profit tax
B_i^0	property of Treuhand	t_K	corporate profit tax
B_i^1	property developed by purchasing firm i	T_W	fees for municipal intermediate inputs
c_i	marginal costs of settlement firm i	T	subscript indicating Treuhand
C_i	cost function of firm i	U_i	purchaser i, settlement firms i
d_X	threat point of purchaser or Treuhand	V_i	output of firm i
		V_i^0	technological parameter of firm i
d_{X-Y}	threat point of purchaser X by negotiation with another member, Y, of cluster	W	municipal intermediate inputs
		Z_j^{xy}	solution Z for decisionmaker j of cluster policy x, variant y
d	superscript indicating demand	Z_{X-Y}	negotiation solution members X and Y of cluster
D_W	distance to municipality		
E_i	cost of sales to Treuhand	Z^c	result Z of cooperation among members of cluster
F_i	sales price of Treuhand property		
F_i^0	results from sale of property	Z^N	Nash solution from negotiation between the Treuhand and the purchaser
g_Y	weight of goal Y of Treuhand		
G_i	a variable standing for $(g_A/m_{A_i}) + (g_K/m_{K_i})$	Z^P	Pareto solution between the Treuhand and the purchaser
I_i	investments of purchaser i	Z^u	result Z of common negotiations of cluster members with the Treuhand
K_i	capital stock of firm i		
L	Lagrange function		
m_Z	marginal productivity of factor Z	α_j	exponent of labour
n	Nash product	β_j	exponent of capital
N_X	utility for purchaser or Treuhand	γ_j	exponent of land
N_X^0	minimum standard of utility of purchaser or Treuhand	δ_j	exponent of intermediate input
		ε_{P^s,V^d}	price elasticity
N	superscript indicating Nash solution	ε_{V^d,P^s}	demand elasticity
		η_j	price-reduction coefficient for firm j
P_{D_i}	transportation costs		
P_{i-1}	price of intermediate input of firm i	λ	multiplier of Lagrange function
$P_{V_{i-1}}$	mill price of intermediate input V_{i-1}	$\hat{}$	indicates an estimate
		$-$	indicates an indifference curve
P^0	additional component of price to take account of firms shifting additional utilities		

A.2 Description of the decisionmakers
(1) Treuhand
Utility from the contract with the purchaser $U_i (i = 1, 2, \ldots, n)$:

$$N_{T_i} = g_V V_i + g_A A_i + g_K K_i + g_F (F_i - E_i) + g_S S_i \geq N_{T_i}^0,$$

where $g_V, g_A, g_K, g_F, g_S > 0$, $\quad g_S < g_F < 2g_S$, $\quad K_i = K_i^0 + I_i = 0 + I_i = I_i$;
Sales price of property: $F_i = F_i^0 - S_i$;
Price reduction to support the firm's restructuring: $S_i = \eta_i V_i$, where $\eta_i \geq 0$;
Truehand cost of sales: $E_i \geq 0$;

(2) Private investor, settlement firm i $(i = 1, 2, \ldots, n)$
Utility from the contract with the Treuhand:

$$N_{U_i} = (1 - t_K - t_G)(P_i V_i - C_i) \geq N_{U_i}^0;$$

Production function: $V_i = V_i^0(W, D_W) A_i^{\alpha_i} K_i^{\beta_i} B_i^{\gamma_i} V_{i-1}^{\delta_i}$,
where $\alpha_i, \beta_i, \gamma_i, \delta_i > 0$, $\quad \alpha_i + \beta_i + \gamma_i + \delta_i = 1$, $\quad V_i^0 > 0$, $\quad dV_i^0/dW > 0$,
$dV_i^0/dD_W < 0$, $\quad B_i = B_i^0 + B_i^1$, $\quad B_i^0 > 0$, $\quad B_i^1 \geq 0$;

Production costs, cost function, least-cost combination of inputs:

$$C_i = T_W(D_W) + r_A A_i + r_K K_i + (r_B + t_B) B_i^1 + P_{i-1} V_{i-1} + t_B B_i^0 + F_i$$
$$= T_W(D_W) + r_A A_i + r_K K_i + (r_B + t_B) B_i + P_{i-1} V_{i-1} + F_i - r_B B_i^0,$$

where $dT_W/dD_W > 0$;

$$\text{minimise}\{C_i | V_i = \bar{V}_i\}, \qquad L_{C_i} = C_i + \lambda_{V_i}(\bar{V}_i - V_i),$$

$$\frac{\partial L_{C_i}}{\partial A_i} = 0, \quad \frac{\partial L_{C_i}}{\partial K_i} = 0, \quad \frac{\partial L_{C_i}}{\partial B_i} = 0, \quad \frac{\partial L_{C_i}}{\partial V_{i-1}} = 0,$$

$$\frac{r_A A_i}{\alpha_i} = \frac{r_K K_i}{\beta_i} = \frac{(r_B + t_B) B_i}{\gamma_i} = \frac{P_{i-1}(1 + \varepsilon_{P_{i-1}^s, V_{i-1}^d}) V_{i-1}}{\delta_i} \quad \text{(expansion path)},$$

where

$$\varepsilon_{P_{i-1}^s, V_{i-1}^d} = \frac{V_{i-1}}{P_{i-1}} = \frac{dP_{i-1}^s}{dV_{i-1}^d},$$

$$C_i = c_i V_i + T_W(D_W) + F_i - r_B B_i^0,$$

where

$$c_i = [V_i^0(W, D_W)]^{-1} \left(\frac{r_A}{\alpha_i}\right)^{\alpha_i} \left(\frac{r_K}{\beta_i}\right)^{\beta_i} \left(\frac{r_B + t_B}{\gamma_i}\right)^{\gamma_i} \left[\frac{P_{i-1}(1 + \varepsilon_{P_{i-1}^s, V_{i-1}^d})}{\delta_i}\right]^{\delta_i} \varepsilon_\delta,$$

where

$$\varepsilon_\delta = 1 - \delta_i \frac{\varepsilon_{P_{i-1}^s, V_{i-1}^d}}{1 + \varepsilon_{P_{i-1}^s, V_{i-1}^d}},$$

$$\frac{V_i}{A_i} = \frac{\varepsilon_\delta r_A}{\alpha_i c_i} = m_{A_i}, \qquad \frac{V_i}{K_i} = \frac{\varepsilon_\delta r_K}{\beta_i c_i} = m_{K_i},$$

$$\frac{V_i}{B_i} = \frac{\varepsilon_\delta (r_B + t_B)}{\gamma_i c_i}, \qquad \frac{V_i}{V_{i-1}} = \frac{P_{i-1}}{\delta_i c_i}\left[1 + \varepsilon_{P_{i-1}^s, V_{i-1}^d}(1 - \delta_i)\right];$$

Price-demand function

$$P_i = P_{V_i}(V_i) + P_{D_i}(D_i) \quad (\text{price} = \text{ex-factory price} + \text{transportation cost}),$$

where
$$P_{V_i} = P_{V_i^d}(V_i^s), \quad \frac{dP_{V_i^d}}{dV_i^s} < 0, \quad \frac{dP_{D_i}}{dD_i} > 0, \quad V_i^s = V_i^d = V_i, \quad P_i^s = P_i^d = P_i,$$
(equilibrium condition).

A.3 Negotiation solution of sales of Treuhand property

Sales of Treuhand property
Threat strategies by negotiations
Threat points of investor: $d_{U_i} = N_{U_i}^0$; threat points of the Treuhand: $d_{T_i} = N_{T_i}^0$;

Negotiation solution
Pareto solution: maximise $\{N_i | N_{U_i} = \bar{N}_{U_i}\}$,

$$N_{T_i} = (g_V + G_i + g_S \eta_i)V_i + g_F(F_i - E_i), \quad G_i = \frac{g_A}{m_{A_i}} + \frac{g_K}{m_{K_i}},$$

$$\bar{N}_{U_i} = (1 - t_K - t_G)[P_{V_i} V_i - T_W(D_W) - c_i V_i - F_i + r_B B_i^0],$$

where $P_i = a_{V_i} - b_{V_i} V$, $P_{V_i} = P_i - P_{D_i} = a_{V_i} - P_{D_i} - b_{V_i} V$, $a_{V_i}, b_{V_i} > 0$, $a_{V_i} - P_{D_i} > c_i$, $\varepsilon_{P_0^s, V_0^d} = 0$,

$$L_{N_{T_i}} = N_{T_i} + \lambda_{N_{U_i}}(\bar{N}_{U_i} - N_{U_i}), \quad \frac{\partial L_{N_{T_i}}}{\partial V_i} = \frac{\partial N_{T_i}}{\partial V_i} - \lambda_{N_{U_i}} \frac{\partial N_{U_i}}{\partial V_i} = 0,$$

$$\frac{\partial L_{N_{T_i}}}{\partial F_i} = \frac{\partial N_{T_i}}{\partial F_i} - \lambda_{N_{U_i}} \frac{\partial N_{U_i}}{\partial F_i} = 0,$$

$$V_i^P = \frac{1}{2b_{V_i}}\left[a_{V_i} - (c_i + P_{D_i}) + \frac{1}{g_F}(g_V + G_i + g_S \eta_i)\right],$$

where $\dfrac{dV_i^P}{d\eta_i} = \dfrac{g_S}{2b_{V_i} g_F} > 0$,

$$\eta_i \Big|_{\frac{dN_{T_i}(V_i^P)}{d\eta_i} = 0} = \frac{g_V + G_i}{g_F - g_S} - 2b_{V_i} \frac{g_F}{g_S} V_i^P \geq 0,$$

where

$$\frac{d^2 N_{T_i}(V_i^P)}{d\eta_i^2} = -\frac{(g_F - g_S)g_S}{b_{V_i} g_F} < 0, \quad \frac{g_F}{2g_S} < 1, \quad 2c_i > \frac{g_V + G_i}{g_F g_S} > a_{V_i},$$

$$\frac{N_{T_i}^P}{g_F} + \frac{N_{U_i}^P}{1 - t_K - t_G} = \frac{1}{16 b_{V_i}}\left(a_{V_i} - c_i + \frac{g_V + G_i}{g_F - g_S}\right)^2 - T_W(D_W) + r_B B_i^0 - E_i,$$

where $N_{T_i}^P \geq d_{T_i}$, $N_{U_i}^P \geq d_{U_i}$;

Nash solution : maximise $\{n_i\}$, $n_i = (N_{T_i}^P - d_{T_i})(N_{U_i}^P - d_{U_i})$,

$$L_{n_i} = n_i + \lambda_{n_i}\left[\frac{N_{T_i}^P}{g_F} + \frac{N_{U_i}^P}{1 - t_K - t_G} - \frac{1}{16 b_{V_i}}\left(a_{V_i} - c_i - P_{D_i} + \frac{g_V + G_i}{g_F - g_S}\right)^2 + T_W(D_W) - r_B B_i^0 + E_i\right],$$

$$\frac{\partial L_{n_i}}{\partial N_{T_i}} = N_{U_i}^P - d_{U_i} + \frac{\lambda_{n_i}}{g_F} = 0, \quad \frac{\partial L_{n_i}}{\partial N_{U_i}} = N_{T_i}^P - d_{T_i} + \frac{\lambda_{n_i}}{1 - t_K - t_G} = 0,$$

and

$$N_{T_i}^N = \frac{g_F}{2}\left[\frac{d_{T_i}}{g_F} - \frac{d_{U_i}}{1 - t_K - t_G} - T_W(D_W) + r_B B_i^0 - E_i\right.$$
$$\left. + \frac{1}{16 b_{V_i}}\left(a_{V_i} - c_i - P_{D_i} + \frac{g_V + G_i}{g_F - g_S}\right)^2\right],$$

$$N_{U_i}^N = \frac{1 - t_K - t_G}{2}\left[\frac{d_{U_i}}{1 - t_K - t_G} - \frac{d_{T_i}}{g_F} - T_W(D_W) + r_B B_i^0 - E_i\right.$$
$$\left. + \frac{1}{16 b_{V_i}}\left(a_{V_i} - c_i - P_{D_i} + \frac{g_V + G_i}{g_F g_S}\right)^2\right],$$

$$V_i^N = \frac{1}{4 b_{V_i}}\left(a_{V_i} - c_i - P_{D_i} + \frac{g_V + G_i}{g_F - g_S}\right), \quad A_i^N = \frac{V_i^N}{m_{A_i}}, \quad K_i^N = \frac{V_i^N}{m_{K_i}},$$

$$\eta_i^N = \left(1 - \frac{g_F}{2 g_S}\right)\frac{g_V + G_i}{g_F - g_S} - \frac{g_F}{2 g_S}(a_{V_i} - c_i - P_{D_i}),$$

$$F_i^N = \frac{1}{2}\left\langle\frac{d_{T_i}}{g_F} - \frac{d_{U_i}}{1 - t_K - t_G} - T_W(D_W) + r_B B_i^0 + E_i\right.$$
$$\left. + \frac{1}{16 b_{V_i}}\left\{\left(a_{V_i} - c_i - P_{D_i} + \frac{g_V + G_i}{g_F - g_S}\right)^2 + 4\left[(a_{V_i} - c_i - P_{D_i})^2 - \frac{(g_V + G_i)^2}{g_F - g_S}\right]\right\}\right\rangle.$$

A.4 Solutions of policy 1(a)

(3) Behaviour of investors under output–input adaptation, compare formula (2)
Demand behaviour of U_2 with regard to input V_1:

$$V_1^d = \frac{\delta_2 c_2}{P_1^s} V_2, \quad \frac{dV_1^d}{dV_2} = \frac{\delta_2 c_2}{P_1^s}, \quad \varepsilon_{P_1^s, V_1^d} = 0,$$

Supply behaviour of U_1 with regard to price of output V_1:

$$P_1^s = \frac{\delta_2 c_2}{V_i^d} V_2, \quad \frac{dV_1^d}{dP_i^s} = \frac{\partial V_1^d}{\partial P_1^s} + \frac{\partial V_1^d}{\partial V_2}\frac{dV_2}{dP_i^s}, \quad \varepsilon_{V_1^d, P_1^s} = \delta_2 + \varepsilon_{V_2, P_1^s} - 1 \leqslant 0;$$

(4) Threat strategies by negotiations
Threat points of investors: $d_{U_1} = d_{U_1}^{la} = N_{U_1}^0$, $d_{U_2} = d_{U_2}^{la} = N_{U_2}^0$,
Threat point of the Treuhand: $d_{T_1} = d_{T_1}^{la} = N_{T_1}^0$, $d_{T_2} = d_{T_2}^{la} = N_{T_2}^0 - \Delta N_{T_1} \leqslant N_{T_2}^0$,
$\Delta N_{T_1} \geqslant E_1/2 g_F$;

(5) Negotiation solution between the Treuhand and purchaser 2
Pareto solution: maximise $\{N_{T_2}|N_{U_2} = \bar{N}_{U_2}\}$,

$$N_{T_2} = (g_V + G_2 + g_S \eta_2)V_2 + g_F(F_2 - E_2),$$
$$\bar{N}_{U_2} = (1 - t_K - t_G)[P_{V_2} V_2 - T_W(D_W) - c_2 V_2 - F_2 + r_B B_2^0],$$

where $\partial V_2/\partial P_1 < 0$, $\partial N_{T_2}/\partial P_1 < 0$, $g_V > G_2$, $P_{V_2} = a_{V_2} - P_{D_2} - b_{V_2} V_2$, $a_{V_2}, b_{V_2} > 0$,
$a_{V_2} > c_2 + P_{D_2}$

$$L_{N_{T_i}} = N_{T_2} + \lambda_{N_{U_2}}(\bar{N}_{U_2} - N_{U_2}),$$

Cluster formation and the Treuhand

and

$$\frac{\partial L_{N_{T_2}}}{\partial V_2} = \frac{\partial N_{T_2}}{\partial V_2} - \lambda_{N_{U_2}} \frac{\partial N_{U_2}}{\partial V_2} = 0, \quad \frac{\partial L_{N_{T_2}}}{\partial F_2} = \frac{\partial N_{T_2}}{\partial F_2} - \lambda_{N_{U_2}} \frac{\partial N_{U_2}}{\partial F_2} = 0,$$

$$V_2^P = \frac{1}{2b_{V_2}} \left(a_{V_2} - c_2 - P_{D_2} + \frac{g_V + G_2 + g_S \eta_2}{g_F} \right), \text{ where } \frac{dV_2^P}{d\eta_2} = \frac{g_S}{2b_{V_2} g_F} > 0,$$

$$\eta_2 \Big|_{\frac{dN_{T_2}(V_2^P)}{d\eta_2} = 0} = \frac{g_V + G_2}{g_F - g_S} - 2b_{V_2} \frac{g_F}{g_S} V_2^P \geq 0,$$

where

$$\frac{d^2 N_{T_2}(V_2^P)}{d^2 \eta_2} = -\frac{(g_F - g_S) g_S}{b_{V_2} g_F} < 0, \quad \frac{g_F}{2g_S} < 1, \quad 2c_2 > \frac{g_V + G_2}{g_F - g_S} > a_{V_2} - P_{D_2},$$

$$\frac{N_{T_2}^P}{g_F} + \frac{N_{U_2}^P}{1 - t_K - t_G} = \frac{1}{16 b_{V_2}} \left(a_{V_2} - c_2 - P_{D_2} + \frac{g_V + G_2}{g_F - g_S} \right)^2 - T_W(D_W) + r_B B_2^0 - E_2,$$

where $N_{T_2}^P \geq d_{T_2}$, $N_{U_2}^P \geq d_{U_2}$;

Nash solution: maximise $\{n_2\}$, $n_2 = (N_{T_2}^P - d_{T_2})(N_{U_2}^P - d_{U_2})$,

$$L_{n_2} = n_2 + \lambda_{n_2} \left[\frac{N_{T_2}^P}{g_F} + \frac{N_{U_2}^P}{1 - t_K - t_G} - \frac{1}{16 b_{V_2}} \left(a_{V_2} - c_2 - P_{D_2} + \frac{g_V G_2}{g_F - g_S} \right)^2 \right.$$

$$\left. + T_W(D_W) - r_B B_2^0 + E_2 \right],$$

$$\frac{\partial L_{n_2}}{\partial N_{T_2}} = N_{U_2}^P - d_{U_2} + \frac{\lambda_{n_2}}{g_F} = 0, \quad \frac{\partial L_{n_2}}{\partial N_{U_2}} = N_{T_2}^P - d_{T_2} + \frac{\lambda_{n_2}}{1 - t_K - t_G} = 0,$$

$$N_{T_2}^{N1a} = \frac{g_F}{2} \left[\frac{d_{T_2}}{g_F} - \frac{d_{U_2}}{1 - t_K - t_G} - T_W(D_W) + r_B B_2^0 - E_2 \right.$$

$$\left. + \frac{1}{16 b_{V_2}} \left(a_{V_2} - c_2 - P_{D_2} + \frac{g_V + G_2}{g_F - g_S} \right)^2 \right],$$

where

$$\frac{dN_{T_2}^{N1a}}{dP_1} = -\delta_2 \frac{g_F}{4P_1} \left(c_2 - \frac{G_2}{g_F - g_S} \right) V_2^N < 0, \quad c_2 - \frac{G_2}{g_F - g_S} > 0,$$

$$N_{U_2}^{N1a} = \frac{1 - t_K - t_G}{2} \left[\frac{d_{U_2}}{1 - t_K - t_G} - \frac{d_{T_2}}{g_F} - T_W(D_W) + r_B B_2^0 - E_2 \right.$$

$$\left. + \frac{1}{16 b_{V_2}} \left(a_{V_2} - c_2 - P_{D_2} + \frac{g_V + G_2}{g_F - g_S} \right)^2 \right],$$

where

$$\frac{dN_{U_2}^{N1a}}{dP_1} = -\delta_2 \frac{1 - t_K - t_G}{4P_1} \left(c_2 - \frac{G_2}{g_F - g_S} V_2^N \right) < 0,$$

$$V_2^{N1a} = \frac{1}{4bV_2} \left(a_{V_2} - c_2 - P_{D_2} + \frac{g_V + G_2}{g_F - g_S} \right), \quad A_2^{N1a} = \frac{V_2^{N1a}}{m_{A_2}}, \quad K_2^{N1a} = \frac{V_2^{N1a}}{m_{K_2}},$$

where
$$\frac{dV_2^{\text{N1a}}}{dP_1} = -\frac{\delta_2}{4b_{V_2} P_1}\left(c_2 - \frac{G_2}{g_F - g_S}\right) < 0,$$
where
$$\varepsilon_{V_2^N, P_1^s} = -\delta_2\left(c_2 - \frac{G_2}{g_F - g_S}\right)\left(a_{V_2} - c_2 - P_{D_2} + \frac{g_V + G_2}{g_F - g_S}\right)^{-1} < -\delta_2,$$

$$\frac{d\varepsilon_{V_2^N, P_1^s}}{dP_1^s} = \frac{\delta_2}{P_1}\left(a_{V_2} + \frac{g_V}{g_F + g_S}\right)\varepsilon_{V_2^N, P_1^s}\left(a_{V_2} - c_2 - P_{D_2} + \frac{g_V + G_2}{g_F - g_S}\right)^{-1} < 0,$$

$$\eta_2^{\text{N1a}} = \left(1 - \frac{g_F}{2g_S}\right)\frac{g_V + G_2}{g_F - g_S} - \frac{g_F(a_{V_2} - c_2 - P_{D_2})}{2g_S}, \quad \text{where } \frac{d\eta_2^{\text{N1a}}}{dP_1} > 0,$$

$$F_2^{\text{N1a}} = \frac{1}{2}\Bigg\langle\frac{d_{T_2}}{g_F} - \frac{d_{U_2}}{1 - t_K - t_G} - T_W(D_W) + r_B B_2^0 + E_2$$

$$+ \frac{1}{16b_{V_2}}\Bigg\{\left(a_{V_2} - c_2 - P_{D_2} + \frac{g_V + G_2}{g_F - g_S}\right) + 4\left[(a_{V_2} - c_2 - P_{D_2})^2 - \frac{(g_V + G_2)^3}{g_F - g_S}\right]\Bigg\}\Bigg\rangle,$$

where $dF_2^{\text{N1a}}/dP_1 < 0$;

(6) Concluded contract between the Treuhand and purchaser 1, compare formula (12)
Estimated sales curve: $\hat{P}_{V_1} = \hat{P}_1 - P_{D_1}$, $d\hat{P}_{V_1}/d\hat{V}_1 = d\hat{P}_{V_1^D}/d\hat{V}_1^s < 0$,
Negotiation solution: $\hat{N}_{T_1}^{\text{N1a}} = \hat{N}_{T_1}^{\text{N1a}}(\hat{P}_{V_1})$, $\hat{N}_{U_1}^{\text{N1a}} = \hat{N}_{U_1}^{\text{N1a}}(\hat{P}_{V_1})$, $\hat{V}_1^{\text{N1a}} = \hat{V}_1^{\text{N1a}}(\hat{P}_{V_1})$,
$\hat{A}_1^{\text{N1a}} = \hat{A}_1^{\text{N1a}}(\hat{P}_{V_1})$, $\hat{K}_1^{\text{N1a}} = \hat{K}_1^{\text{N1a}}(\hat{P}_{V_1})$, $\hat{\eta}_1^{\text{N1a}} = \hat{\eta}_1^{\text{N1a}}(\hat{P}_{V_1})$, $\hat{F}_1^{\text{N1a}} = \hat{F}_1^{\text{N1a}}(\hat{P}_{V_1})$,
where

$$\frac{d\hat{N}_{T_1}^{\text{N1a}}}{d\hat{P}_{V_1}} > 0, \quad \frac{d\hat{N}_{U_1}^{\text{N1a}}}{d\hat{P}_{V_1}} > 0, \quad \frac{d\hat{V}_1^{\text{N1a}}}{d\hat{P}_{V_1}} > 0, \quad \frac{d\hat{\eta}_1^{\text{N1a}}}{d\hat{P}_{V_1}} > 0, \quad \frac{d\hat{F}^{\text{N1a}}}{d\hat{P}_{V_1}} > 0;$$

(7) Input–output adaptation between the settlement firms, new profit maximisation of firm 1: $N_{U_1} = (1 - t_K - t_G)[P_{V_1} V_1 - C_1(V_1, F_1^N)]$,
where $dF_1^N/dV_1 = 0$, $V_1^d(\hat{P}_1^s) = V_1^d(\hat{P}_{V_1} + P_{D_1}) \geq \hat{V}_1^{\text{N1a}}$, $P_{V_1}(\hat{V}_1^N) \geq \hat{P}_{V_1}(\hat{V}_1^N)$,

$$\frac{dN_{U_1}}{dP_{V_1}} = \frac{\partial N_{U_1}}{\partial P_{V_1}} + \frac{\partial N_{U_1}}{\partial V_1}\frac{dV_1^d}{dP_1^s}\frac{dP_1^s}{dP_{V_1}} = (1 - t_K - t_G)\left[V_1 + (P_{V_1} - c_1)\frac{dV_1^d}{dP_1^s}\right] = 0,$$

where $V_1 = V_1^d = \delta_2 c_2 V_2^N/P_1^S$, $P_{V_1} = P_1^S - P_{D_1}$,

$$P_1^{\text{la}} = P_1^{\text{sla}} = \left(1 - \frac{1}{\delta_2 + \varepsilon_{V_2^N, P_1^s}}\right)(c_1 + P_{D_1}) = \frac{\varepsilon_{V_1^d, P_1^s}}{1 + \varepsilon_{V_1^d, P_1^s}}(c_1 + P_{D_1}) > c_1 + P_{D_1},$$

$$V_1^{\text{la}} = V_1^{\text{sla}} = V_1^{\text{dla}} = \delta_2 c_2 V_2^{\text{N1a}}\left[\left(1 - \frac{1}{\delta_2 + \varepsilon_{V_2^N, P_1^s}}\right)(c_1 + P_{D_1})\right]^{-1}$$

$$= \frac{1 + \varepsilon_{V_1^d, P_1^s}}{\varepsilon_{V_1^d, P_1^s}}\frac{\delta_2 c_2 V_2^{\text{N1a}}}{c_1 + P_{D_1}};$$

Results with purchaser 1: $N_{U_1}^{\text{la}}(P_{V_1}^{\text{la}}) \geq N_{U_1}^{\text{N1a}}(P_{V_1}) \geq \hat{N}_{U_1}^{\text{N1a}}(\hat{P}_{V_1})$, $N_{T_1}^{\text{la}}(P_{V_1}^{\text{la}}) \geq \hat{N}_{T_1}^{\text{N1a}}(\hat{P}_{V_1})$,
$N_{T_1}^{\text{la}}(P_{V_1}^{\text{la}}) \leq N_{T_1}^{\text{N1a}}(P_{V_1})$, $V_1^{\text{la}}(P_{V_1}^{\text{la}}) \geq \hat{V}_1^{\text{N1a}}(\hat{P}_{V_1})$, $V_1^{\text{la}}(P_{V_1}^{\text{la}}) \leq V_1^{\text{N1a}}(P_{V_1})$.

A.5 Solutions of policy 1(b)

Behaviour of investors by the output–input adaptation, compare formula (3)
(9) Threat strategies
Threat point of investors: $d_{U_1} = d_{U_1}^{1b} = N_{U_1}^0$, $d_{U_2} = d_{U_2}^{1b} = N_{U_2}^0$,

Threat point of the Treuhand: $d_{T_1} = d_{T_1}^{1b} = N_{T_1}^0 - \Delta N_{T_1} \leq N_{T_1}^0$, $\Delta N_{T_1} \geq E_2/2g_F$,
$d_{T_2} = d_{T_2}^{1b} = N_{T_2}^0$;

(10) Concluded contract between the Treuhand and purchaser 2, compare formula (5)

$$\hat{N}_{T_2}^{N1b} = \frac{g_F}{2}\left[\frac{d_{T_2}}{g_F} - \frac{d_{U_2}}{1 - t_K - t_G} - T_W(D_W) + r_B B_2^0 - E_2 \right.$$
$$\left. + \frac{1}{16 b_{V_2}}\left(a_{V_2} - \hat{c}_2 - P_{D_2} + \frac{g_V + \hat{G}_2}{g_F - g_S}\right)^2\right],$$

where $\hat{G}_2 = \frac{g_A}{\hat{m}_{A_2}} + \frac{g_K}{\hat{m}_{K_2}}$,

$$\hat{N}_{U_2}^{N1b} = \frac{1 - t_K - t_G}{2}\left[\frac{d_{U_2}}{1 - t_K - t_G} - \frac{d_{T_2}}{g_F} - T_W(D_W) + r_B B_2^0 - E_2 \right.$$
$$\left. + \frac{1}{16 b_{V_2}}\left(a_{V_2} - \hat{c}_2 - P_{D_2} + \frac{g_V + \hat{G}_2}{g_F - g_S}\right)^2\right],$$

$$\hat{V}_2^{N1b} = \frac{1}{4 b_{V_2}}\left(a_{V_2} - \hat{c}_2 - P_{D_2} + \frac{g_V + \hat{G}_2}{g_F - g_S}\right), \quad \hat{A}_2^{N1b} = \frac{\hat{V}_2^N}{\hat{m}_{A_2}}, \quad \hat{K}_2^{N1b} = \frac{\hat{V}_2^N}{\hat{m}_{K_2}},$$

$$\varepsilon_{\hat{V}_2^N, \hat{P}_1^s} = -\delta_2\left(\hat{c}_2 - \frac{\hat{G}_2}{g_F - g_S}\right)\left(a_{V_2} - \hat{c}_2 - P_{D_2} + \frac{g_V + \hat{G}_2}{g_F - g_S}\right)^{-1} < -\delta_2,$$

$$\hat{\eta}_2^{N1b} = \left(1 - \frac{g_F}{2g_S}\right)\frac{g_V + \hat{G}_2}{g_F - g_S} - \frac{g_F(a_{V_2} - \hat{c}_2 - P_{D_2})}{2g_S},$$

$$\hat{F}_2^{N1b} = \frac{1}{2}\left\langle \frac{d_{T_2}}{g_F} - \frac{d_{U2}}{1 - t_K - T_G} - T_W(D_W) + r_B B_2^0 + E_2\right.$$
$$\left. + \frac{1}{16 b_{V_2}}\left\{\left(a_{V_2} - \hat{c}_2 - P_{D_2} + \frac{g_V + \hat{G}_2}{g_F - g_S}\right)^2 + 4\left[(a_{V_2} - \hat{c}_2 - P_{D_2})^2 - \left(\frac{g_V + \hat{G}_2}{g_F - g_S}\right)^2\right]\right\}\right\rangle,$$

where $P_{V_2} = P_2 - P_{D_2} = a_{V_2} - P_{D_2} - b_{V_2} V_2$, $\hat{c}_2 = \hat{c}_2(\hat{P}_1^s)$, $d\hat{c}_2/d\hat{P}_1^s > 0$,
$\hat{m}_{A_2} = \hat{m}_{A_2}(\hat{c}_2)$, $\hat{m}_{K_2} = \hat{m}_{K_2}(\hat{c}_2)$;

(11) Input-output adaptation between the settlement firms, new profit maximisation of firm 2: $N_{U_2} = (1 - t_K - t_G)[P_{V_2} V_2 - c_2(V_2, F_2^N)]$, $dN_{U_2}/dV_2 = 0$,
where

$$\frac{dF_2^N}{dV_2} = 0, \quad V_1 = V_1^d = \frac{\delta_2 c_2 V_2}{P_1^s} = \frac{\delta_2 c_2 V_2}{P_{V_1} + P_{D_1}} = V_1^s, \quad P_1 = P_1^s \leq \hat{P}_1^s,$$

$$V_2^{1b} = \frac{a_{V_2} - c_2 - P_{D_2}}{2 b_{V_2}} \geq \hat{V}_2^{N1b}(\hat{P}_1^s) = \frac{1}{4 b_{V_2}}\left(a_{V_2} - \hat{c}_2 - P_{D_2} + \frac{g_V + \hat{G}_2}{g_F - g_S}\right),$$

$$\varepsilon_{V_2, P_1^s} = -\delta_2 \frac{c_2}{a_{V_2} c_2} < \varepsilon_{\hat{V}_2^N, \hat{P}_1^s} < -\delta_2,$$

where

$$\frac{d\varepsilon_{V_2,P_1^s}}{dP_1^s} = \frac{\delta_2}{P_1}\frac{a_{V_2}}{a_{V_2}-c_2}\varepsilon_{V_2,P_1^s} < \frac{d\varepsilon_{\hat{V}_2^N,P_1^s}}{d\hat{P}_1^s} < 0;$$

(12) Negotiation solution between the Treuhand and purchaser 1
Pareto solution: maximise $\{N_{U_1}|N_{T_1} = \bar{N}_{T_1}\}$, $L_{N_{U_1}} = N_{U_1} + \lambda_{N_{T_1}}(\bar{N}_{T_1} - N_{T_1})$,

$$\frac{\partial L_{N_{U_1}}}{\partial V_1} = \frac{\partial N_{U_1}}{\partial V_1} + \frac{\partial N_{U_1}}{\partial P_{V_1}}\frac{dP_{V_1}}{dP_1}\frac{dP_1^s}{dV_1^d} - \lambda_{N_{T_1}}\frac{\partial N_{T_1}}{\partial V_1} = 0,$$

$$\frac{\partial L_{N_{U_1}}}{\partial F_1} = \partial\frac{N_{U_1}}{\partial F_1} - \lambda_{N_{U_1}}\frac{\partial N_{T_1}}{\partial F_1} = 0,$$

$$P_1 = \left(c_1 + P_{D_1} - \frac{g_V + G_1 + g_S\eta_1}{g_F}\right)\left(1 - \frac{1}{\delta_2 + \varepsilon_{V_2,P_1^s}}\right)$$

$$= \left(c_1 + P_{D_1} - \frac{g_V + G_1 + g_S\eta_1}{g_F}\right)\frac{\varepsilon_{V_1^d,P_1^s}}{1+\varepsilon_{V_1^d,P_1^s}},$$

$$V_1^P = \delta_2\left(c_1 + P_{D_1} - \frac{g_V + G_1 + g_S\eta_1}{g_F}\right)^{-1} c_2 V_2 \left(1 - \frac{1}{\delta_2 + \varepsilon_{V_2,P_1^s}}\right)^{-1}$$

$$= \delta_2 c_2 V_2 \left(c_1 + P_{D_1} - \frac{g_V + G_1 + g_S\eta_1}{g_F}\right)^{-1}\frac{1+\varepsilon_{V_1^d,P_1^s}}{\varepsilon_{V_1^d,P_1^s}},$$

$$\frac{dV_1^d}{dP_1^s} = -\frac{(1-\delta_2-\varepsilon_{V_2^N,P_1^s})\delta_2 c_2}{P_1^2}V_2^N = \frac{\varepsilon_{V_1^d,P_1^s}}{P_1} < 0,$$

where $\varepsilon_{V_1^d,P_1^s} < -1+\delta_2 + \varepsilon_{V_2^N,\hat{P}_1^s} < -1$,

$$\frac{dP_1^s}{d\eta_1} = \frac{\partial P_1^s}{\partial \eta_1} + \frac{\partial P_1^s}{\partial \varepsilon_{V_1^d,P_1^s}}\frac{d\varepsilon_{V_1^d,P_1^s}}{dP_1^s}\frac{dP_1^s}{d\eta_1}, \quad \frac{dP_1^s}{d\eta_1} = \frac{g_S}{g_F(1-\delta_2)(1+\varepsilon_{V_1^d,P_1^s})} < 0,$$

$$\frac{dV_1^P}{d\eta_1} = \frac{dV_1^P}{dP_1^s}\frac{dP_1^s}{d\eta_1} = \frac{g_S}{g_F}\frac{\varepsilon_{V_1^d,P_1^s}}{(1-\delta_2)(1+\varepsilon_{V_1^d,P_1^s})}\frac{V_1^P}{P_1} > 0,$$

$$\eta_1\bigg|_{\frac{dN_{T_1}(V_1^P)}{d\eta_1}=0} = \frac{g_V+G_1}{g_F-g_S} - \frac{dV_1^P}{d\eta_1}V_1^P \geq 0,$$

$$\frac{N_{T_1}^P}{g_F} + \frac{N_{U_1}^P}{1-t_K-t_G} = -\frac{\delta_2 c_2}{2b_{V_2}\varepsilon_{V_1^d,P_1^s}}(a_{V_2}-c_2-P_{D_2})-T_W(D_W)+r_B B_1^0 - E_1,$$

where $N_{T_1}^P \geq d_{T_1}$, $N_{U_1}^P \geq d_{U_1}$;

Nash solution: maximise $\{n_1\}$, $n_1 = (N_{T_1}^P - d_{T_1})(N_{U_1}^P - d_{U_1})$,

$$L_{n_1} = n_1 + \lambda_{n_1}\left[\frac{N_{T_1}^P}{g_F} + \frac{N_{U_1}^P}{1-t_k-t_G} + \frac{\delta_2 c_2}{2b_{V_2}\varepsilon_{V_1^d}^d P_1^s}(a_{V_2}-c_2-P_{D_2})+T_W(D_W)\right.$$
$$\left. - r_B B_1^0 + +E_1\right],$$

$$\frac{\partial L_{n_1}}{\partial N_{T_1}} = N_{U_1}^P - d_{U_1} + \frac{\lambda_{n_1}}{g_F} = 0, \quad \frac{\partial L_{n_1}}{\partial N_{U_1}} = N_{T_1}^P - d_{T_1} + \frac{\lambda_{n_1}}{1-t_K-t_G} = 0,$$

Cluster formation and the Treuhand

and
$$N_{T_1}^{N1b} = \frac{g_F}{2}\left[\frac{d_{T_1}}{g_F} - \frac{d_{U_1}}{1-t_K-t_G} - \frac{\delta_2 c_2(a_{V_2}-c_2-P_{D_2})}{2b_{V_2}\varepsilon_{V_1^d,P_1^s}} - T_W(D_W) + r_B B_1^0 - E_1\right],$$

where $dN_{T_1}^{N1b}/d\varepsilon_{V_1^d,P_1^s} > 0$,

$$N_{U_1}^{N1b} = \frac{1-t_K-t_G}{2}\left[\frac{d_{U_1}}{1-t_K-t_G} - \frac{d_{T_1}}{g_F} - \frac{\delta_2 c_2(a_{V_2}-c_2-P_{D_2})}{2b_{V_2}\varepsilon_{V_1^d,P_1^s}}\right.$$
$$\left. - T_W(D_W) + r_B B_1^0 - E_1\right], \quad \text{where } \frac{dN_{U_1}^{N1b}}{d\varepsilon_{V_1^d,P_1^s}} > 0,$$

$$P_1^{1b} = \frac{\varepsilon_{V_1^d,P_1^s}}{1+\varepsilon_{V_1^d,P_1^s}}\left(c_1 + P_{D_1} - \frac{g_V + G_1}{g_F - g_S}\right) \leq \hat{P}_1^s, \quad \text{where } \frac{dP_1^{1b}}{d\varepsilon_{V_1^d,P_1^s}} > 0,$$

$$\eta_1^{N1b} = \frac{g_V + G_1}{g_F - g_S} - \frac{g_F}{g_S}(1-\delta_2)\frac{1+\varepsilon_{V_1^d,P_1^s}}{\varepsilon_{V_1^d,P_1^s}}P_1 \geq 0, \quad \text{where } \frac{d\eta_1^{N1b}}{d\varepsilon_{V_1^d,P_1^s}} < 0,$$

$$V_1^{N1b} = \frac{1+\varepsilon_{V_1^d,P_1^s}}{\varepsilon_{V_1^d,P_1^s}}\frac{\delta_2 c_2}{2b_{V_2}}(a_{V_2}-c_2-P_{D_2})\left(c_1 + P_{D_1} - \frac{g_V+G_1}{g_F-g_S}\right)^{-1},$$

where $dV_1^{N1b}/d\varepsilon_{V_1^d,P_1^s} > 0$,

$$A_1^{N1b} = \frac{V_1^{N1b}}{m_{A_1}}, \quad K_1^{N1b} = \frac{V_1^{N1b}}{m_{K_1}},$$

$$F_1^{N1b} = \tfrac{1}{2}\left\{\frac{d_{T_1}}{g_F} - \frac{d_{U_1}}{1-t_K-t_G} - T_W(D_W) + r_B B_1^0 - E_1 + \frac{1+\varepsilon_{V_1^d,P_1^s}}{\varepsilon_{V_1^d,P_1^s}}\frac{\delta_2 c_2}{4b_{V_2}}\right.$$
$$\left. \times (a_{V_2}-c_2-P_{D_2})\left[\frac{2+3\varepsilon_{V_1^d,P_1^s}}{1+\varepsilon_{V_1^d,P_1^s}} - 2\delta_2(c_1+P_{D_1})\left(c_1+P_{D_1}-\frac{g_V+G_1}{g_F-g_S}\right)^{-1}\right]\right\},$$

where $dF_1^{N1b}/d\varepsilon_{V_1^d,P_1^s} > 0$;

(13) Results with purchaser 2: $N_{U_2}^{1b}(P_1^{1b}) \geq N_{U_2}^{N1b}(P_1^s) \geq \hat{N}_{U_2}^{N1b}(\hat{P}_1^s)$,
$N_{T_2}^{1b}(P_1^{1b}) \geq \hat{N}_{T_2}^{N1b}(\hat{P}_1^s)$,
$N_{T_2}^{1b}(P_1^{1b}) \leq N_{T_2}^{N1b}(P_1^s), V_2^{1b}(P_1^{1b}) \geq \hat{V}_2^{N1b}(\hat{P}_1^{1b})$,
$V_2^{1b}(P_1^{1b}) \leq V_2^{N1b}(P_1^s)$.

A.5 Solutions of policy 2(a)
Behaviour of investors under output–input agreement, noncooperative behaviour, compare formula (3)
(14) Threat strategies
Threat points of investors: $d_{U_1} = d_{U_1}^{2a} = N_{U_1}^0$, $d_{U_2} = d_{U_2}^{2a} = N_{U_2}^0$,
Threat points of the Treuhand: $d_{T_1} = d_{T_1}^{2a} = N_{T_1}^0$, $d_{T_2} = d_{T_2}^{2a} = N_{T_2}^0$;
Negotiation solution between the Treuhand and purchaser 2, compare formula (5)

$$N_{T_2}^{N2a} = \frac{g_F}{2}\left[\frac{d_{T_2}}{g_F} - \frac{d_{U_2}}{1-t_K-t_G} - T_W(D_W) + r_B B_1^0 - E_2\right.$$
$$\left. + \frac{1}{16b_{V_2}}\left(a_{V_2}-c_2-P_{D_2}+\frac{g_V+G_2}{g_F-g_S}\right)^2\right],$$

and
$$N_{U_2}^{N2a} = \frac{1-t_K-t_G}{2}\left[\frac{d_{U_2}}{1-t_K-t_G} - \frac{d_{T_2}}{g_F} - T_W(D_W) + r_B B_1^0 - E_2\right.$$
$$\left. + \frac{1}{16b_{V_2}}\left(a_{V_2} - c_2 - P_{D_2} + \frac{g_V+G_2}{g_F-g_S}\right)^2\right],$$

$$V_2^{N2a} = \frac{1}{4b_{V_2}}\left(a_{V_2} - c_2 - P_{D_2} + \frac{g_V+G_2}{g_F-g_S}\right), \quad A_2^{N2a} = \frac{V_2^{N2a}}{m_{A_2}}, \quad K_2^{N2a} = \frac{V_2^{N2a}}{m_{K_2}},$$

$$\varepsilon_{V_2^N,P_1^s} = -\delta_2\left(c_2 - \frac{G_2}{g_F-g_S}\right)\left(a_{V_2} - c_2 - P_{D_2} + \frac{g_V+G_2}{g_F-g_S}\right)^{-1},$$

$$\eta_2^{N2a} = \left(1 - \frac{g_F}{2g_S}\right)\frac{g_V+G_2}{g_F-g_S} - \frac{g_F}{2g_S}(a_{V_2} - c_2 - P_{D_2}),$$

$$F_2^{N2a} = \frac{1}{2}\left\langle \frac{d_{T_2}}{g_F} - \frac{d_{U_2}}{1-t_K-t_G} - T_W(D_W) + r_B B_2^0 - E_2\right.$$
$$\left. + \frac{1}{16b_{V_2}}\left\{\left(a_{V_2} - c_2 - P_{D_2} + \frac{g_V+G_2}{g_F-g_S}\right)^2 + 4\left[(a_{V_2} - c_2 - P_{D_2})^2 - \frac{(g_V+G_2)^2}{g_F-g_S}\right]\right\}\right\rangle;$$

Negotiation solution between the Treuhand and purchaser 1, compare formula (12)

$$N_{T_1}^{N2a} = \frac{g_F}{2}\left[\frac{d_{T_1}}{g_F} - \frac{d_{U_1}}{1-t_K-t_G} - \delta_2 c_2\left(a_{V_2} - c_2 - P_{D_2} + \frac{g_V+G_2}{g_F-g_S}\right)\left(4b_{V_2}\varepsilon_{V_1^d,P_1^s}\right)^{-1}\right.$$
$$\left. - T_W(D_W) + r_B B_1^0 - E_2\right],$$

$$N_{U_1}^{N2a} = \frac{1-t_K-t_G}{2}\left[\frac{d_{U_1}}{1-t_K-t_G} - \frac{d_{T_1}}{g_F} - \delta_2 c_2\left(a_{V_2} - c_2 - P_{D_2} + \frac{g_V+G_2}{g_F-g_S}\right)\right.$$
$$\left. \times \left(4b_{V_2}\varepsilon_{V_1^d,P_1^s}\right)^{-1} - T_W(D_W) + r_B B_1^0 - E_2\right],$$

$$\eta_1^{N2a} = \frac{g_V+G_1}{g_F-g_S} - \frac{g_F}{g_S}(1-\delta_2)\frac{1+\varepsilon_{V_1^d,P_1^s}}{\varepsilon_{V_1^d,P_1^s}}P_1 \geq 0,$$

$$V_1^{N2a} = \frac{1+\varepsilon_{V_1^d,P_1^s}}{\varepsilon_{V_1^d,P_1^s}}\frac{\delta_2 c_2}{4b_{V_2}}\left(a_{V_2} - c_2 - P_{D_2} + \frac{g_V+G_2}{g_F-g_S}\right)\left(c_1 + P_{D_1} - \frac{g_V+G_1}{g_F-g_S}\right)^{-1},$$

$$A_1^{N2a} = \frac{V_1^{N2a}}{m_{A_1}}, \quad K_1^{N2a} = \frac{V_1^{N2a}}{m_{K_1}},$$

$$F_1^{N2a} = \frac{1}{2}\left\{\frac{d_{T_1}}{g_F} - \frac{d_{U_1}}{1-t_K-t_G} - T_W(D_W) + r_B B_1^0 + E_1 + \frac{1+\varepsilon_{V_1^d,P_1^s}}{\varepsilon_{V_1^d,P_1^s}}\frac{\delta_2 c_2}{4b_{V_2}}\right.$$
$$\times \left(a_{V_2} - c_2 - P_{D_2} + \frac{g_V+G_2}{g_F-g_S}\right)$$
$$\left. \times \left[\frac{2+3\varepsilon_{V_1^d,P_1^s}}{1+\varepsilon_{V_1^d,P_1^s}} - 2\delta_2(c_1 + P_{D_1})\left(c_1 + P_{D_1} - \frac{g_V+G_1}{g_F-g_S}\right)^{-1}\right]\right\};$$

Cluster formation and the Treuhand

(15) Input-output agreement between the settlement firms

$$P_1^{2a} = \frac{\varepsilon_{V_1^d, P_1^s}}{1 + \varepsilon_{V_1^d, P_1^s}} \left(c_1 + P_{D_1} - \frac{g_V + G_1}{g_F - g_S} \right) \geqslant P_1^{1b},$$

$$P_1^{2a} < \frac{\varepsilon_{V_1^d, P_1^s} (c_1 + P_{D_1})}{1 + \varepsilon_{V_1^d, P_1^s}} \leqslant P_1^{1a},$$

$$V_1^{N2a} \geqslant \frac{1 + \varepsilon_{V_1^d, P_1^s}}{\varepsilon_{V_1^d, P_1^s}} \frac{\delta_2 c_2}{4 b_{V_2}} \left(a_{V_2} - c_2 - P_{D_2} + \frac{g_V + G_2}{g_F - g_S} \right) (c_1 + P_{D_1})^{-1} \geqslant V_1^{1a},$$

$$V_1^{N2a} \geqslant \frac{1 + \varepsilon_{V_1^d, P_1^s}}{\varepsilon_{V_1^d, P_1^s}} \frac{\delta_2 c_2}{2 b_{V_2}} (a_{V_2} - c_2 - P_{D_2}) \left(c_1 + P_{D_1} - \frac{g_V + G_1}{g_F - g_S} \right)^{-1} \geqslant V_1^{N1b},$$

where

$$\varepsilon_{V_1^d, P_1^s} = -\delta_2 \left[c_2 - \frac{G_2}{g_F - g_S} + (1 - \delta_2) \left(a_{V_2} - c_2 - P_{D_2} + \frac{g_V + G_2}{g_F - g_S} \right) \right]$$

$$\times \left(a_{V_2} - c_2 - P_{D_2} + \frac{g_V + G_2}{g_F - g_S} \right)^{-1} \geqslant \varepsilon_{V_1^d, P_1^s}^{1b},$$

$$c_2^{2a} \leqslant c_2^{1a}, \quad \varepsilon_{V_1^d, P_1^s} \geqslant \varepsilon_{V_1^d, P_1^s}^{1a}.$$

A.7 Solutions of policy 2(b)
(16) Cooperative demand and supply behaviour of investors
minimise $\{ C_2 | V_2 = \bar{V}_2, N_{U_1} = \bar{N}_{U_1}, N_{T_1} = \bar{N}_{T_1} \}$ (Pareto optimum),
where

$$P_1^c = P_1^s = P_1^d, \quad V_1 = V_1^s = V_1^d, \quad \frac{dV_1^d}{dP_1^s} = \frac{dV_1}{dP_1^c} = \left(\frac{dP_1^c}{dV_1} \right)^{-1} = \left(\frac{dP_1^s}{dV_1^d} \right)^{-1}$$

(cooperative solutions),

$$L_{C_2} = C_2 + \lambda_{V_2}(\bar{V}_2 - V_2) + \lambda_{N_{U_1}}(\bar{N}_{U_1} - N_{U_1}) + \lambda_{N_{T_1}}(\bar{N}_{T_1} - N_{T_1}),$$

$$\frac{\partial L_{C_2}}{\partial A_2} = 0, \quad \frac{\partial L_{C_2}}{\partial K_2} = 0, \quad \frac{\partial L_{C_2}}{\partial B_2} = 0, \quad \frac{\partial L_{C_2}}{\partial V_1} = 0, \quad \frac{\partial L_{C_2}}{\partial P_1} = 0, \quad \frac{\partial L_{C_2}}{\partial F_1} = 0,$$

$$\frac{r_A A_2}{\alpha_2} = \frac{r_K K_2}{\beta_2} = \frac{(r_B + t_B) B_2}{\gamma_2} = \left(c_1 + P_{D_1} - \frac{g_V + G_1 + g_S \eta_1}{g_F} \right) \frac{V_1}{\delta_2},$$

$$P_1^c = \left(c_1 + P_{D_1} - \frac{g_V + G_1 + g_S \eta_1}{g_F} \right) \frac{\varepsilon_{V_1, P_1^c}}{1 + \varepsilon_{V_1, P_1^c}},$$

$$\frac{dV_1}{V_1} = dP_1^c \left(c_1 + P_{D_1} - \frac{g_V + G_1 + g_S \eta_1}{g_F} - P_1^c \right)^{-1},$$

$$\int \frac{1}{V_1} dV_1 = \int \left(c_1 + P_{D_1} - \frac{g_V + G_1 + g_S \eta_1}{g_F} - P_1^c \right)^{-1} dP_1^c,$$

$$P_1^c = c_1 + P_{D_1} - \frac{g_V + G_1 + g_S \eta_1}{g_F} + \frac{P_1^0}{V_1}, \quad \text{where } dP_1^{0c}/dV_1 = 0,$$

and
$$\frac{V_2}{V_1} = \frac{P_1^c}{\delta_2 c_2}\left[1+\varepsilon_{P_1^c,V_1}(1-\delta_2)\right]$$

$$= V_2^0(W,D_W)\left(\frac{\alpha_2}{r_A}\right)^{\alpha_2}\left(\frac{\beta_2}{r_K}\right)^{\beta_2}\left(\frac{\gamma_2}{r_B+t_B}\right)^{\gamma_2}\left[\delta_2\left(c_1+P_{D_1}-\frac{g_V+G_1+g_S\eta_1}{g_F}\right)^{-1}\right]^{\delta_2-1},$$

$$\frac{dV_1}{dP_1^c} = \frac{V_1}{V_2}\frac{dV_2}{dP_1^c}, \quad \varepsilon_{V_1,P_1^c} = \varepsilon_{V_2,P_1^c},$$

$$C_2 = c_2^c V_2 + P_1^{0c} + T_W(D_W) + F_2 - r_B B_2^0 =$$

$$c_2 V_2 + T_W(D_W) + F_2 - r_B B_2^0,$$

$$c_2^c = \frac{1}{V_2^0(W,D_W)}\left(\frac{r_A}{\alpha_2}\right)^{\alpha_2}\left(\frac{r_K}{\beta_2}\right)^{\beta_2}\left(\frac{r_B+t_B}{\gamma_2}\right)^{\gamma_2}\left[\frac{1}{\delta_2}\left(c_1+P_{D_1}-\frac{g_V+G_1+g_S\eta_1}{g_F}\right)\right]^{\delta_2}$$

$$\leq c_2^c\left[\delta_2 P_1^c + (1-\delta_2)\left(c_1+P_{D_1}-\frac{g_V+G_1+g_S\eta_1}{g_F}\right)\right]$$

$$\times\left(c_1+P_{D_1}-\frac{g_V+G_1+g_S\eta_1}{g_F}\right)^{-1} = c_2,$$

$$\frac{V_2}{A_2} = \frac{r_A}{\alpha_2 c_2^c} = m_{A_2^c}, \quad \frac{V_2}{K_2} = \frac{r_K}{\beta_2 c_2^c} = m_{K_2^c},$$

$$\frac{V_2}{B_2} = \frac{r_B+t_B}{\gamma_2 c_2^c}, \quad \frac{V_2}{V_1} = \frac{1}{\delta_2 c_2^c}\left(c_1+P_{D_1}-\frac{g_V+G_1+g_S\eta_1}{g_F}\right) \geq \frac{P_1^c(1+\varepsilon_{V_1,P_1^c})}{\delta_2 c_2 \varepsilon_{V_1,P_1^c}};$$

(17) Threat strategies
Threat points of investors in negotiations with the Treuhand: $d_{U_1} = d_{U_1}^{2b} = N_{U_1}^O$, $d_{U_2} = d_{U_2}^{2b} = N_{U_2}^O$,
Threat points of the Treuhand: $d_{T_1} = d_{T_1}^{2b} = N_{T_1}^0 - \Delta N_{T_1}$, $\Delta N_{T_2} = N_{T_2}^{N2b} - N_{T_2}^{1b}$, $d_{T_2} = d_{T_2}^{2b} = N_{T_2}^0 - \Delta N_{T_2}$, $\Delta N_{T_1} = N_{T_1}^{N2b} - N_{T_1}^{1a}$,
Threat points of investors in negotiations with each other: $d_{U_1-U_2} = N_{U_1}^{N2a}$, $d_{U_2-U_1} = N_{U_2}^{N2a}$.

(18) Negotiation solution between the Treuhand and purchaser 2, compare formula (5)

$$N_{T_2}^{N2b} = \frac{g_F}{2}\left\{\frac{d_{T_2}}{g_F} - \frac{d_{U_2}}{1-t_K-t_G} - P_1^{0c} - T_W(D_W) + r_B B_2^0 - E_2\right.$$

$$\left. + \frac{1}{16b_{V_2}}\left[a_{V_2} - c_2^c - P_{D_2} + \frac{(g_V+G_2^c)}{g_F-g_S}\right]^2\right\}, \quad \text{where } G_2^c = \frac{g_A}{m_{A_2^c}} + \frac{g_K}{m_{K_2^c}},$$

$$N_{U_2}^{N2b} = \frac{1-t_K-t_G}{2}\left\{\frac{d_{U_2}}{1-t_K-t_G} - \frac{d_{T_2}}{g_F} - P_1^{0c} - T_W(D_W) + r_B B_2^0 - E_2\right.$$

$$\left. + \frac{1}{16b_{V_2}}\left[a_{V_2} - c_2^c - P_{D_2} + \frac{(g_V+G_2^c)}{g_F-g_S}\right]^2\right\},$$

and
$$V_2^{N2b} = \frac{1}{4b_{V_2}}\left(a_{V_2} - c_2^c - P_{D_2} + \frac{g_V + G_2^c}{g_F - g_S}\right), \quad \text{where} \quad \frac{dV_2^{N2b}}{dc_2^c} < 0,$$

$$A_2^{N2b} = \frac{V_2^{N2b}}{m_{A_2^c}}, \quad K_2^{N2b} = \frac{V_2^{N2b}}{m_{K_2^c}},$$

$$\eta_2^{N2b} = \left(1 - \frac{g_F}{2g_S}\right)\frac{g_V + G_2^c}{g_F - g_S} - \frac{g_F}{2g_S}(a_{V_2} - c_2^c - P_{D_2}), \quad \text{where} \quad \frac{d\eta_2^{N2b}}{dc_2^c} > 0,$$

$$F_2^{N2b} = \tfrac{1}{2}\left\langle \frac{d_{T_2}}{g_F} - \frac{d_{U_2}}{1 - t_K - t_G} - P_1^{0c} - T_W(D_W) + r_B B_2^0 + E_2 \right.$$

$$\left. + \frac{1}{16b_{V_2}}\left\{\left(a_{V_2} - c_2^c - P_{D_2} + \frac{g_V + G_2^c}{g_F - g_S}\right)^2 + 4\left[(a_{V_2} - c_2^c - P_{D_2})^2 - \left(\frac{g_V + G_2^c}{g_F - g_S}\right)^2\right]\right\}\right\rangle,$$

where
$$\frac{dF_2^{N2b}}{dc_2^c} < 0;$$

(19) Negotiation solution between the Treuhand and purchaser 1
Pareto solution, compare formulas (12) and (16)

$$V_1^P = \frac{\delta_2 c_2^c}{4b_{V_1}}\left(a_{V_2} - c_2^c - P_{D_2} + \frac{g_V + G_2^c}{g_F - g_S}\right)\left(c_1 + P_{D_1} - \frac{g_V + G_1 + g_S\eta_1}{g_F}\right)^{-1},$$

$$\frac{dV_1^P}{d\eta_1} = \frac{g_S}{g_F}V_1^P\left(c_1 + P_{D_1} - \frac{g_V + G_1 + g_S\eta_1}{g_F}\right)^{-1}$$

$$\times \left[1 - \delta_2 + \delta_2\left(c_2^c - \frac{G_2^c}{g_F - g_S}\right)\left(a_{V_2} - c_2^c - P_{D_2} + \frac{g_V + G_2^c}{g_F - g_S}\right)^{-1}\right] > 0,$$

$$\eta_1\bigg|_{\frac{dN_{T_1}(V_1^P)}{d\eta_1} = 0} = \frac{g_V + G_1}{g_F - g_S} - \left(\frac{dV_1^P}{d\eta_1}\right)^{-1} V_1^P \geq 0,$$

$$\frac{N_{T_1}^P}{g_F} + \frac{N_{U_1}^P}{1 - t_K - t_G} = P_1^{0c} - T_W(D_W) + r_B B_1^0 - E_1,$$

where $N_{T_1}^P \geq d_{T_1}$, $N_{U_1}^P \geq d_{U_1}$;

Nash solution: maximise $\{n_1\}$, $n_1 = (N_{T_1}^P - d_{T_1})(N_{U_1}^P - d_{U_1})$,

$$L_{n_1} = n_1 + \lambda_{n_1}\left(\frac{N_{T_1}}{g_F} + \frac{N_{U_1}}{1 - t_K - t_G} - P_1^{0c} + T_W(D_W) - r_B B_1^0 + E_1\right),$$

$$\frac{\partial L_{n_1}}{\partial N_{T_1}} = 0, \quad \frac{\partial L_{n_1}}{\partial N_{U_1}} = 0,$$

$$N_{T_1}^{N2b} = \frac{g_F}{2}\left[\frac{d_{T_1}}{g_F} - \frac{d_{U_1}}{1 - t_K - t_G} + P_1^{0c} - T_W(D_W) + r_B B_2^0 - E_1\right],$$

and
$$N_{U_1}^{\text{N2b}} = \frac{1-t_\text{K}-t_\text{G}}{2}\left[\frac{d_{U_1}}{1-t_\text{K}-t_\text{G}} - \frac{d_{T_1}}{g_\text{F}} + P_1^{0c} - T_\text{W}(D_\text{W}) + r_\text{B} B_2^0 - E_1\right],$$

$$\eta_1^{\text{N2b}} = \frac{g_\text{V}+G_1}{g_\text{F}-g_\text{S}} - \frac{g_\text{F}}{g_\text{S}}\left(c_1 + P_{D_1} - \frac{g_\text{V}+G_1}{g_\text{F}-g_\text{S}}\right)\left(a_{V_2} + c_2^c - P_{D_2} - \frac{g_\text{V}+G_2^c}{g_\text{F}-g_\text{S}}\right)$$
$$\times \left\{\delta_2\left[\left(c_2^c - \frac{G_2^c}{g_\text{F}-g_\text{S}}\right) - \left(a_{V_2} - c_2^c - P_{D_2} + \frac{g_\text{V}+G_2^c}{g_\text{F}-g_\text{S}}\right)\right]\right\}^{-1} \geq 0,$$

$$V_1^{\text{N2b}} = \frac{\delta_2 c_2^c}{4 b_{V_2}}\left[\delta_2\left(c_2^c - \frac{G_2^c}{g_\text{F}-g_\text{S}}\right) - \delta_2\left(a_{V_2} - c_2^c - P_{D_2} + \frac{g_\text{V}+G_2^c}{g_\text{F}-g_\text{S}}\right)\right]$$
$$\times \left[\delta_2\left(c_2^c - \frac{G_2^c}{g_\text{F}-g_\text{S}}\right) + (1-\delta_2)\left(a_{V_2} - c_2^c - P_{D_2} + \frac{g_\text{V}+G_2^c}{g_\text{F}-g_\text{S}}\right)\right]^{-1}$$
$$\times \left(a_{V_2} - c_2^c - P_{D_2} + \frac{g_\text{V}+G_2^c}{g_\text{F}-g_\text{S}}\right)\left(c_1 + P_{D_1} - \frac{g_\text{V}+G_1}{g_\text{F}-g_\text{S}}\right)^{-1},$$

$$P_1^c = \left(c_1 + P_{D_1} - \frac{g_\text{V}+G_1}{g_\text{F}-g_\text{S}}\right)\left[\delta_2\left(c_2^c - \frac{G_2^c}{g_\text{F}-g_\text{F}}\right) + (1-\delta_2)\right.$$
$$\left. \times \left(a_{V_2} - c_2^c - P_{D_2} + \frac{g_\text{V}+G_2^c}{g_\text{F}-g_\text{S}}\right)\right]$$
$$\times \left[\delta_2\left(c_2^c - \frac{G_2^c}{g_\text{F}-g_\text{S}}\right) - \delta_2\left(a_{V_2} - c_2^c - P_{D_2} + \frac{g_\text{V}+G_2^c}{g_\text{F}-g_\text{S}}\right)\right]^{-1} + \frac{P_1^{0c}}{V_1^\text{N}},$$

$$\varepsilon_{V_1, P_1^c} = -\frac{\delta_2 C_2^c}{4 b_{V_2} P_1^{0c}}\left(a_{V_2} - c_2^c - P_{D_2} + \frac{g_\text{V}+G_2^c}{g_\text{F}-g_\text{S}}\right) - 1 < -\frac{1}{\delta_2} < -1,$$
where $d\varepsilon_{V_1, P_1^c}/dc_2^c > 0,$

$$F_1^{\text{N2b}} = \frac{1}{2}\left\{\frac{d_{T_1}}{g_\text{F}} - \frac{d_{U_1}}{1-t_\text{K}-t_\text{G}} + P_1^{0c} - T_\text{W}(D_\text{W}) + r_\text{B} B_1^0 + E_1 \right.$$
$$- \frac{c_2^c}{4 b_{V_2}}\left(a_{V_2} - c_2^c - P_{D_2} + \frac{g_\text{V}+G_2^c}{g_\text{F}-g_\text{S}}\right)\left[\delta_2 \frac{g_\text{V}+G_1}{g_\text{F}-g_\text{S}}\left(c_1 + P_{D_1} - \frac{g_\text{V}+G_1}{g_\text{F}-g_\text{S}}\right)^{-1}\right.$$
$$\left. + \left(a_{V_2} - c_2^c - P_{D_2} + \frac{g_\text{V}+G_2^c}{g_\text{F}-g_\text{S}}\right)\left(a_{V_2} - 2c_2^c - P_{D_2} + \frac{g_\text{V}+2G_2^c}{g_\text{F}-g_\text{S}}\right)^{-1}\right]$$
$$\times \left[\delta_2\left(c_2^c - \frac{G_2^c}{g_\text{F}-g_\text{S}}\right) - \delta_2\left(a_{V_2} - c_2^c - P_{D_2} + \frac{g_\text{V}+G_2^c}{g_\text{F}-g_\text{S}}\right)\right]$$
$$\left. \times \left[\delta_2\left(c_2^c - \frac{G_2^c}{g_\text{F}-g_\text{S}}\right) + (1-\delta_2)\left(a_{V_2} - c_2^c - P_{D_2} + \frac{g_\text{V}+G_2^c}{g_\text{F}-g_\text{S}}\right)\right]^{-1}\right\};$$

Cluster formation and the Treuhand

(20) Total negotiation solutions

$$N_{T_1}^{N2b} + N_{T_2}^{N2b} = \frac{g_F}{2}\left[\frac{N_{T_1}^0 + N_{T_2}^0 - N_{T_1}^{N2b} - N_{T_2}^{N2b} + N_{T_1}^{1a} + N_{T_2}^{1b}}{g_F} - \frac{N_{U_1}^0 + N_{U_2}^0}{1 - t_K - t_G}\right.$$

$$\left. - 2T_W(D_W) + r_B(B_1^0 + B_2^0) - E_1 - E_2 + \frac{1}{16b_{V_2}}\left(a_{V_2} - c_2^c - P_{D_2} + \frac{g_V + G_2^c}{g_F - g_S}\right)^2\right]$$

$$< N_{T_1}^{N2a} + N_{T_2}^{N2c} - \tfrac{1}{2}\left(N_{T_1}^{N2b} + N_{T_2}^{N2b} - N_{T_1}^{1a} + N_{T_2}^{1b}\right) < N_{T_1}^{N2a} + N_{T_2}^{N2a},$$

$$N_{T_1}^{N2b} + N_{T_2}^{N2b} > N_{T_1}^{1a} + N_{T_2}^{1b},$$

where

$$2c_2^c = 2\frac{1 + \varepsilon_{V_1, P_1^c}}{1 + (1 - \delta_2)\varepsilon_{V_1, P_1^c}}\, c_2 \geqslant \left(1 + \frac{1}{1 - 2\delta_2 \varepsilon_{V_1, P_1^c}}\right) c_2 > c_2,$$

$$N_{U_1}^{N2b} + N_{U_2}^{N2b} = \frac{1 - t_K - t_G}{2}\left[\frac{N_{U_1}^0 + N_{U_2}^0}{1 - t_K - t_G}\right.$$

$$- \frac{N_{T_1}^0 + N_{T_2}^0 - N_{T_1}^{N2b} - N_{T_1}^{N2b} + N_{T_1}^{1a} + N_{T_2}^{1b}}{g_F} - 2T_W(D_W) + r_B(B_1^0 + B_2^0)$$

$$\left. - E_1 - E_2 + \frac{1}{16b_{V_2}}\left(a_{V_2} - c_2^c - P_{D_2} + \frac{g_V + G_2^c}{g_F - g_S}\right)^2\right] > N_{U_1}^{N2a} + N_{U_2}^{N2a},$$

$$F_{T_1}^{N2b} + F_{T_2}^{N2b} < F_{T_1}^{N2a} + F_{T_2}^{N2a}, \quad \eta_1^{N2b} > \eta_1^{N2a}, \quad \eta_2^{N2b} < \eta_2^{N2a},$$

$$V_1^{N2b} > V_1^{N2a}, \quad V_2^{N2b} > V_2^{N2a}$$

(21) Negotiation solution between investors with regard to P_1^{0c}

Pareto solution:

$$N_{U_1-U_2}^P + N_{U_2-U_1}^P - \frac{1 - t_K - t_G}{3}\left[\frac{N_{U_1}^0 + N_{U_2}^0}{1 - t_K - t_G} - \frac{N_{T_1}^0 + N_{T_2}^0 + N_{T_1}^{1a} + N_{T_2}^{1b}}{g_F}\right.$$

$$- 4T_W(D_W) + 2r_B(B_1^0 + B_2^0) - 2(E_1 - E_2)$$

$$\left. + \frac{1}{8b_{V_2}}\left(a_{V_2} - c_2^c - P_{D_2} + \frac{g_V + G_2^c}{g_F - g_S}\right)^2\right] = 0,$$

where $N_{U_1-U_2}^P = N_{U_1}^{N2b} \geqslant d_{U_1-U_2} = N_{U_1}^{N2a}$, $N_{U_2-U_1}^P = N_{U_2}^{N2b} \geqslant d_{U_2-U_1} = N_{U_2}^{N2a}$;

Nash solution: maximise $\{n_U\}$, $n_U = (N_{U_1-U_2}^P - d_{U_1-U_2})(N_{U_2-U_1}^P - d_{U_2-U_1})$,

$$L_{n_U} = n_U + \lambda_{N_{U_1-U_2}}\left\{N_{U_1-U_2}^P + N_{U_2-U_1}^P - \frac{1 - t_K - t_G}{3}\left[\frac{N_{U_1}^0 + N_{U_2}^0}{1 - t_K - t_G}\right.\right.$$

$$- \frac{N_{T_1}^0 + N_{T_2}^0 + N_{T_1}^{1a} + N_{T_2}^{1b}}{g_F} - 4T_W(D_W) + 2r_B(B_1^0 + B_2^0) - 2(E_1 - E_2)$$

$$\left.\left. + \frac{1}{8b_{V_2}}\left(a_{V_2} - c_2^c - P_{D_2} + \frac{g_V + G_2^c}{g_F - g_S}\right)^2\right]\right\},$$

$$\frac{\partial L_{n_U}}{\partial N_{U_1-U_2}^P} = 0, \quad \frac{\partial L_{n_U}}{\partial N_{U_2-U_1}^P} = 0,$$

and
$$P_1^{0c} = \tfrac{1}{2}\left(N_{U_1}^{N2b} + N_{U_2}^{N2b} - d_{U_1-U_2} - d_{U_2-U_1}\right) = \tfrac{1}{2}\left(N_{U_1}^{N2b} + N_{U_2}^{N2b} - N_{U_1}^{N2a} - N_{U_2}^{N2a}\right)$$

$$= \frac{1-t_K-t_G}{6}\left\{\frac{N_{U_1}^0 + N_{U_2}^0 - 3(N_{U_1}^{N2a} + N_{U_2}^{N2a})}{1-t_K-t_G} - \frac{N_{T_1}^0 + N_{T_2}^0 + N_{T_1}^{1a} + N_{T_2}^{1b}}{g_F}\right.$$

$$-4T_W(D_W) + 2\left[r_B(B_1^0 + B_2^0) - E_1 - E_2\right]$$

$$\left. + \frac{1}{8b_{V_2}}\left(a_{V_2} - c_2^c - P_{D_2} + \frac{g_V + G_2^c}{g_F - g_S}\right)^2\right\},$$

where
$$N_{U_1-U_2}^N = N_{U_1}^{N2b} = \tfrac{1}{2}\left(d_{U_1-U_2} - d_{U_2-U_1} + N_{U_1}^{N2b} + N_{U_2}^{N2b}\right)$$

$$= \frac{1-t_K-t_G}{6}\left\{\frac{3(N_{U_1}^{N2a} - N_{U_2}^{N2a}) + N_{U_1}^0 + N_{U_2}^0}{1-t_K-t_G} - \frac{N_{T_1}^0 + N_{T_2}^0 + N_{T_1}^{1a} + N_{T_2}^{1b}}{g_F}\right.$$

$$-4T_W(D_W) + 2\left[r_B(B_1^0 + B_2^0) - E_1 - E_2\right] + \frac{1}{8b_{V_2}}\left(a_{V_2} - c_2^c - P_{D_2} + \frac{g_V + G_2^c}{g_F - g_S}\right)^2\right\},$$

$$N_{U_2-U_1}^N = N_{U_2}^{N2b} = \tfrac{1}{2}\left(d_{U_2-U_1} - d_{U_1-U_2} + N_{U_1}^{N2b} + N_{U_2}^{N2b}\right)$$

$$= \frac{1-t_K-t_G}{6}\left\{\frac{3(N_{U_2}^{N2a} - N_{U_1}^{N2a}) + N_{U_1}^0 + N_{U_2}^0}{1-t_K-t_G} - \frac{N_{T_1}^0 + N_{T_2}^0 + N_{T_1}^{1a} + N_{T_2}^{1b}}{g_F}\right.$$

$$-4T_W(D_W) + 2\left[r_B(B_1^0 + B_2^0) - E_1 - E_2\right] + \frac{1}{8b_{V_2}}\left(a_{V_2} - c_2^c - P_{D_2} + \frac{g_V + G_2^c}{g_F - g_S}\right)^2\right\}.$$

A.8 Solutions of policy 3

(25) Treuhand: utility from the contract with the purchaser $U_1 + U_2$, compare formula (1)
$$N_T = N_{T_1} + N_{T_2} = g_V(V_1 + V_2) + g_A A + g_K K + g_F(F-E) + g_S(S_1+S_2)$$
$$\geq N_{T_1}^0 + N_{T_2}^0 = N_T^0,$$

where $A = A_1 + A_2$, $K = K_1 + K_2$, $F = F_1 + F_2$, $E = E_1 + E_2$;

(26) Cluster, settlement firms $U_1 + U_2$, compare formula (2)
Utility from the contract with the Treuhand:
$$N_U = N_{U_1} + N_{U_2} = (1-t_K-t_G)(P_2 V_2 - C_1 - C_2) \geq N_{U_1}^0 + N_{U_2}^0 = N_U^0;$$

Production functions:
$$V_1 = V_1^0(W, D_W) A_1^{\alpha_1} K_1^{\beta_1} B_1^{\gamma_1}(V^0)^{\delta_1}, \quad V_2 = V_2^0(W, D_W) A_2^{\alpha_2} K_2^{\beta_2} V_1^{\delta_2};$$

Production costs, cost function, least-cost combination of inputs:
$$C = C_1 + C_2 = 2T_W(D_W) + r_A A + r_K K + (r_B + t_B)B + P^0 V^0 + P_{D_1} V_1 + F - r_B B^0$$
where $B = B_1 + B_2$, $B^0 = B_1^0 + B_2^0$, $\varepsilon_{P^{0s}, V^{0d}} = 0$,

minimise $\{C|V_2 = \bar{V}_2, N_T = \bar{N}_T\}$,

$$L_C = C + \lambda_{V_2}(\bar{V}_2 - V_2) + \lambda_{N_T}(\bar{N}_T - N_T) + \lambda_{V_1}\left[V_1 - V_1^0(W, D_W) A_1^{\alpha_1} K_1^{\beta_1} B_1^{\gamma_1}(V^0)^{\delta_1}\right],$$

and
$$\frac{\partial L_C}{\partial A_1} = \frac{\partial L_C}{\partial A_2} = \frac{\partial L_C}{\partial K_1} = \frac{\partial L_C}{\partial K_2} = \frac{\partial L_C}{\partial B_1} = \frac{\partial L_C}{\partial B_2} = \frac{\partial L_C}{\partial V_1} = \frac{\partial L_C}{\partial V^0} = \frac{\partial L_C}{\partial F} = 0,$$

$$\left(r_A - \frac{g_A}{g_F}\right)\frac{A_1}{\alpha_1} = \left(r_K - \frac{g_K}{g_F}\right)\frac{K_1}{\beta_1} = (r_B + t_B)\frac{B_1}{\gamma_1} = \frac{P^0 V^0}{\delta_1},$$

$$\left(r_A - \frac{g_A}{g_F}\right)\frac{A_2}{\alpha_2} = \left(r_K - \frac{g_K}{g_F}\right)\frac{K_2}{\beta_2} = (r_B + t_B)\frac{B_2}{\gamma_2} = \left(c_1 + P_{D_1} - \frac{g_V}{g_F}\right)\frac{V_1}{\delta_2},$$

$$c_1^u = \left[1 + \alpha_1 r_A \left(r_A - \frac{g_A}{g_F}\right)^{-1} + \beta_1 r_K \left(r_K - \frac{g_K}{g_F}\right)^{-1} + \gamma_1\right]\frac{1}{V_1^0(W, D_W)}$$
$$\times \left[\frac{1}{\alpha_1}\left(r_A - \frac{g_A}{g_F}\right)\right]^{\alpha_1}\left[\frac{1}{\beta_1}\left(r_K - \frac{g_K}{g_F}\right)\right]^{\beta_1}\left(\frac{r_B + t_B}{\gamma_1}\right)^{\gamma_1}\left(\frac{P^0}{\delta_1}\right)^{\delta_1} < c_1,$$

$$c_2^u = \left[\delta_2(c_1 + P_{D_1})\left(c_1 + P_{D_1} - \frac{g_V}{g_F}\right)^{-1} + \alpha_1 r_A \left(r_A - \frac{g_A}{g_F}\right)^{-1} + \beta_1 r_K \left(r_K - \frac{g_K}{g_F}\right)^{-1} + \gamma_1\right]$$
$$\times \frac{1}{V_2^0(W, D_W)}\left[\frac{1}{\alpha_2}\left(r_A - \frac{g_A}{g_F}\right)\right]^{\alpha_2}\left[\frac{1}{\beta_2}\left(r_K - \frac{g_K}{g_F}\right)\right]^{\beta_2}\left(\frac{r_B + t_B}{\gamma_2}\right)^{\gamma_2}$$
$$\times \left[\frac{1}{\delta_2}\left(c_1 + P_{D_1} - \frac{g_V}{g_F}\right)\right]^{\delta_2} \leq c_2^c < c_2,$$

where $C_1 = c_1^u V_1 + T_W(D_W) + F_1 - r_B B_1^0$, $\quad C = c_2^u V_2 + 2T_W(D_W) + F - r_B B^0$,

$$\frac{V_1}{A_1} = m_{A_1}^u, \quad \frac{V_1}{K_1} = m_{K_1}^u, \quad \frac{V_2}{A_2} = m_{A_2}^u, \quad \frac{V_2}{K_2} = m_{K_2}^u, \quad \frac{V_2}{V_1} = \frac{c_1^u + P_{D_1}}{\delta_2 c_2^u};$$

Price-demand function: $P_2 = P_{V_2}(V_2) + P_{D_2}(D_2)$, $P_2 = a_{V_2} - b_{V_2} V_2$,
where $a_{V_2}, b_{V_2} > 0$, $\quad a_{V_2} > c_2^u + P_{D_2}$;
(27) Threat point of investor: $d_U = d_U^3 = N_{U_1}^0 + N_{U_2}^0 = N_U^0$,
Threat point of the Treuhand: $d_T = d_T^3 = N_{T_1}^0 + N_{T_2}^0 = N_T^0$;

(28) Negotiation solution between the Treuhand and the cluster
Pareto solution: maximise $\{N_T | N_U = \bar{N}_U\}$,

$$L_{N_T} = N_T + \lambda_{N_U}(\bar{N}_U - N_U), \quad \frac{\partial L_{N_T}}{\partial V_2} = 0, \quad \frac{\partial N_T}{\partial F} = 0,$$

where
$$N_T = \left[g_V + G_2^u + g_S \eta_2 + (g_V + G_1^u + g_S \eta_1)\frac{\delta_2 c_2^u}{c_1 P_{D_1}}\right] V_2 + g_F(F - E),$$

where
$$G_i^u = \frac{g_A}{m_{A_i}^u} + \frac{g_K}{m_{K_i}^u},$$
$$N_U = (1 - t_K - t_G[(a_{V_2} - b_{V_2} V_2 - c_2^u)V_2 + 2T_W(D_W) - F + r_B B^0],$$

$$V_2^P = \frac{1}{2b_{V_2}}\left\{a_{V_2} - c_2^u - P_{D_2} + \frac{1}{g_F}\left[g_V + G_2^u + g_S \eta_2 + (g_V + G_1^u + g_S \eta_1)\frac{\delta_2 C_2^u}{c_1 + P_{D_1}}\right]\right\},$$

$$\frac{dV_2^P}{d\eta_1} = \frac{g_S}{2b_{V_2} g_F}\frac{\delta_2 c_2^u}{c_1 + P_{D_1}} > 0, \quad \frac{dV_2^P}{d\eta_2} = \frac{g_S}{2b_{V_2} g_F} > 0,$$

and

$$\frac{N_T^P}{g_F} + \frac{N_U^P}{1 - t_K - t_G} = \frac{1}{16b_{V_2}} \left\{ a_{V_2} - c_2^u - P_{D_2} + \frac{1}{g_F - g_S} \left[g_V + G_2^u \right. \right.$$
$$\left. \left. + (g_V + G_1^u) \frac{\delta_2 c_2^u}{c_1 + P_{D_1}} \right] \right\}^2 - 2T_W(D_W) + r_B B_0 - E,$$

where $N_T^P \geq d_T$, $N_U^P \geq d_U$;

Nash solution: maximise $\{n\}$, $n = (N_T^P - d_T)(N_U^P - d_U)$,

$$L_n = n + \lambda_n \left\langle \frac{N_{T_2}^P}{g_F} + \frac{N_{U_2}^P}{1 - t_K - t_G} + 2T_W(D_W) - r_B B^0 + E \right.$$
$$\left. - \frac{1}{16b_{V_2}} \left\{ a_{V_2} - c_2^u - P_{D_2} + \frac{1}{g_F - g_S} \left[g_V + G_2^u + (g_V + G_1^u) \frac{\delta_2 c_2^u}{c_1 + P_{D_1}} \right] \right\}^2 \right\rangle,$$

$$\frac{\partial L_n}{\partial N_T} = 0, \quad \frac{\partial L_n}{\partial N_U} = 0,$$

$$N_T^{N3} = \frac{g_F}{2} \left\langle \frac{d_T}{g_F} - \frac{d_U}{1 - t_K - t_G} - 2T_W(D_W) + r_B B^0 - E \right.$$
$$\left. + \frac{1}{16b_{V_2}} \left\{ a_{V_2} - c_2^u - P_{D_2} + \frac{1}{g_F - g_S} \left[g_V + G_2^u + (g_V + G_1^u) \frac{\delta_2 c_2^u}{c_1 + P_{D_1}} \right] \right\}^2 \right\rangle,$$

$$N_U^{N3} = \frac{1 - t_K - t_G}{2} \left\langle \frac{d_U}{1 - t_K - t_G} - \frac{d_T}{g_F} - 2T_W(D_W) + r_B B^0 - E \right.$$
$$\left. + \frac{1}{16b_{V_2}} \left\{ a_{V_2} - c_2^u - P_{D_2} + \frac{1}{g_F - g_S} \left[g_V + G_2^u + (g_V + G_1^u) \frac{\delta_2 c_2^u}{c_1 + P_{D_1}} \right] \right\}^2 \right\rangle,$$

$$V_2^{N3} = \frac{1}{4b_{V_2}} \left\{ a_{V_2} - c_2^u - P_{D_2} + \frac{1}{g_F - g_S} \left[g_V + G_2^u + (g_V + G_1^u) \frac{\delta_2 c_2^u}{c_1 + P_{D_1}} \right] \right\},$$

$$V_1^{N3} = \frac{\delta_2 c_2^u}{c_1 + P_{D_1}} V_2^{N3}, \quad A_2^{N3} = \frac{V_2^{N3}}{m_{A_2}^u}, \quad K_2^{N3} = \frac{V_2^{N3}}{m_{K_2}^u}, \quad A_1^{N3} = \frac{V_1^{N3}}{m_{A_1}^u}, \quad K_1^{N3} = \frac{V_1^{N3}}{m_{A_1}^u},$$

$$\eta_2^{N3} = \left(1 - \frac{g_F}{2g_S}\right) \frac{1}{g_F - g_S} (g_V + G_2^u) - \frac{g_F}{2g_S} (a_{V_2} - c_2^u - P_{D_2}),$$

$$\eta_1^{N3} = \left[\left(1 - \frac{g_F}{2g_S}\right) \frac{1}{g_F - g_S} (g_V + G_1^u) - \frac{g_F}{2g_S} (a_{V_2} - c_2^u - P_{D_2})\right] \frac{\delta_2 c_2^u}{c_1 + P_{D_1}},$$

$$F^{N3} = \frac{1}{2} \left\langle \frac{d_T}{g_F} - \frac{d_U}{1 - t_K - t_G} - 2T_W(D_W) + r_B B^0 + E \right.$$
$$+ \frac{1}{16b_{V_2}} \left\{ a_{V_2} - c_2^u - P_{D_2} + \frac{1}{g_F - g_S} \left[g_V + G_2^u + (g_V + G_1^U) \frac{\delta_2 c_2^u}{c_1 + P_{D_1}} \right] \right\}^2$$
$$\left. + \frac{1}{4b_{V_2}} \left\{ (a_{V_2} - c_2^u - P_{D_2})^2 - \frac{1}{g_F - g_S} \left[g_V + G_2^u + (g_V + G_1^u) \frac{\delta_2 c_2^u}{c_1 + P_{D_1}} \right]^2 \right\} \right\rangle.$$

A Risk-oriented Analysis of Regional Clusters

O M Fritz, H Mahringer
Joanneum Research, Vienna
M T Valderrama
Institute for Advanced Studies, Vienna

1 Introduction

Recently the promotion of industrial clusters has been regarded by regional policymakers and development economists alike as one possible alternative to the traditional regional economic policy which concentrates on supporting individual sectors or firms ('industrial or strategic targeting'). Cooperation within geographically concentrated networks of firms, research institutions, as well as governmental institutions is thought to make lagging regions more competitive, to accomplish regional economic sustainability in the long run, and thereby to protect regions from the dangers of globalisation. Prominent examples of well-functioning clusters in fast-growing regions are Silicon Valley, Little Italy, and, from an Austrian perspective, the automotive cluster in the Austrian province of Styria.

These policies, however, ignore the fact that there are some dangers and instabilities for the region introduced by the formation of certain types of clusters. In this chapter we intend to point out and discuss some of these potential negative effects, which are expressed in terms of risk. Two types of risk are distinguished: structural risk concerns the long-run development of the region, whereas cyclical risk is associated with regional growth fluctuations in the short and medium terms. Regional portfolio analysis, invented and first applied by Conroy in the early 1970s (Conroy, 1972; 1974; 1975) is proposed and illustrated as an appropriate framework to analyse the potential tradeoffs between the positive returns for the region from cluster formation and the (cyclical) risk that may be involved.

The chapter is organised as follows: in part 2 we summarise the theoretical basis for the formation of regional industrial clusters as well as the concept itself. In part 3 we discuss risk consideration in relation to regional clusters. Part 4 contains a short review of the regional portfolio literature. In part 5 we attempt to use this methodology in the analysis of regional clusters, and in the final section of the chapter we present a summary and conclusions.

2 Theoretical basis and characterisation of clusters

The notion and economic rationale of spatially concentrated industrial clusters have been around for quite some time. For detailed reviews the reader can refer to Hutschenreiter and Peneder (1994), Meijboom and Rongen (1995), Penttinen (1994), Tichy (1995).

Marshall (1920) is commonly cited as the first to mention the occurrence of spatially concentrated industries (for example, see Hutschenreiter and

Peneder, 1994; Tichy, 1995). The basic cause of agglomeration according to Marshall is the presence of increasing returns to scale which are external to the firm. Based on this concept there are three principal factors for the formation of clusters: labour-market effects (that is, by agglomerating, industries create supply and demand for skilled labour), input – output interdependency (that is, clusters are formed by specialised firms supplying intermediate products or services to each other), and knowledge spillovers. Marshall's arguments for the presence of industrial clustering have been further used and developed in the new growth and trade theories, where human capital and therefore knowledge are an important part of the production function.

A new wave of popularity for clusters was certainly created by Porter (1990), who takes up many of the earlier arguments and includes clusters as one of the components of his famous diamond. He suggests that economic growth is based on Schumpeterian dynamics where interrelated firms in a geographically limited area become more competitive by constant pressure to innovate and by the benefit of being located close to each other. Clusters facilitate innovation and therefore facilitate the creation of the conditions necessary for the competitive advantages of an industry to persist. The presence of internationally competitive industries ensures, on the one hand, cost-effective and speedy delivery of components ('vertical support'). On the other hand, horizontal support is given when industries coordinate and share activities and thereby stimulate competition. Regional concentration promotes the flow of information and technological spillovers. Economies external to individual firms can thus be internalised within a cluster.

Given the various theoretical approaches and the different cluster definitions they implicitly or explicitly contain, the fuzziness of the cluster concept is quite obvious. In order to find a cluster definition on which our risk analysis can be based, we attempt to derive some typical cluster characteristics.

2.1 Interrelatedness

As Tichy (1997, page 94) puts it, clusters consist of "nodes linked by a network". These nodes are firms, industries, and other public and private institutions (for example, public regional development agencies, universities, etc) which are linked through supplier – buyer (input – output) relations and various other forms of cooperation (common research projects, informal contacts to exchange information, participation in regional associations, etc). Different types of networks or clusters such as hub-and-spoke or vertically disintegrated structures can be identified, depending on the underlying hierarchical structure.

2.2 Geographic proximity

Geographic proximity facilitates cooperation between the members of the cluster or may even be a necessary condition to take advantage of externalities or spillovers. However, because the cost of information transfer is decreasing rapidly, the importance of geographic proximity as a cluster characteristic may diminish over time. Consequently, besides regional clusters, which by definition rely on geographic proximity, interregional, national, and international clusters have been formed. Some cluster benefits, on the other hand, are directly linked to spatial concentration, which applies especially to immobile resources such as human capital.

2.3 Synergies and spillovers
Synergies and spillovers are the benefits associated with the linkages between the various actors or 'networkers' and the benefits from geographic proximity (agglomeration economies). These synergies and spillovers, which are external to the firm but internal to the cluster, are the incentives to join an existing cluster or take part in its initiation. Examples of externalities are skilled labour markets or specialised business services.

3 Regional clusters and economic risk
Regional policymakers follow cluster strategies in order to make their region more competitive and thereby accomplish sustained economic growth. But when pursuing such policy, they should also be aware of the potential drawbacks that are associated with regional clusters. In this chapter we discuss two of these potential negative effects. Using an analogy to the regional portfolio concept, which is the methodological tool that will be applied, we speak of the risk associated with pursuing cluster policies.

At this stage of our research work, it is not our goal to carry out a very stringent and systematic analysis but rather to point out the problem, discuss some related aspects, and in particular suggest a methodology which appears useful in conducting further research on this matter.

The first type of risk, which we call *structural risk*, has been discussed to some extent by Austrian regional economists (foremost by Tichy, 1992; 1995; 1997). It is related to the long-run consequences for a region dominated by an industrial cluster. The second type of risk, often named *cyclical risk*, refers to economic stability which is, besides sustained growth, a common goal of regional policy.

3.1 Structural risk
History shows that the regional concentration of resources in one industry or product group, as may be implied by the formation of a cluster, bears the risk that a permanent decline in that industry or product group may bring down the whole region. Detroit is probably the best known example of a metropolitan region that became a so-called 'old industrial area' because of the crisis in the international automobile industry, which caused a crisis in the US automobile cluster, whose activities were heavily concentrated in and around Detroit. A similar crisis occurred in the Austrian region of upper Styria. Its steel cluster, composed mainly of nationalised firms which were quite successful in the 1950s and 1960s because of process innovations, began declining in the 1970s before it finally broke down completely in the 1980s. These clusters, once dynamic, successful, and innovative, could not stay at the competitive edge of their respective industries. They both failed to adjust to changing production regimes.

Tichy (1997) examines some factors responsible for what he calls ageing or petrifaction of clusters. At the core of his analysis is the theory of regional product cycles. This theory states that goods are created and initially produced in agglomerations because of their informational advantages in the face of high uncertainty, the concentrated specialised demand, and the availability of skilled labour. Later, when production processes become more standardised, production moves to the periphery, where labour is less skilled but cheaper and where economies of scale are utilised. It is at this stage, Tichy claims, that the region specialises and networks or clusters are formed. But once the products reach the final phase of their life cycle and their competitive advantage is shrinking, concentration of the number of firms and activities sets in.[1] Thus the networks become smaller and the information flow is reduced. Smaller clusters are less likely to stimulate innovations, so the region finally degenerates into an old industrialised area without any endogenous potential to regain its competitiveness.

A key issue in this respect, which has also been mentioned by Markusen (1996), is the link between the success of a cluster, its increasing specialisation, and the subsequent tendency to become a 'closed system'. As a result, the information necessary for rapid adaptation to market changes may not reach the cluster firms. Furthermore, success and specialisation in one industry may impede the development of other sectors in the region or even drive out firms, preventing the diversification of the regional economy. This may occur when public funds (for example, for education, research, or infrastructure) are allocated according to the importance of regional industries, so that clusters receive a larger share. Industries outside the cluster will be put at a disadvantage particularly if these publicly produced goods are indivisible. A cluster-oriented policymaker may also allocate more direct and indirect subsidies to cluster industries. Additionally, successful industries will attract workers by offering higher wages which may cause deterioration in the human resource basis of other, less productive, industries. As a result, the risk of becoming a problem area once the cluster declines increases, together with the exposure to cyclical instability (see below).

This conclusion holds no matter if the theory of regional product and/or industry cycles applies. Whenever a cluster impedes the development of firms or industries outside the cluster, the region's dependency on the cluster and thus its exposure to structural risk increases. It is therefore important to promote constant adaptation of the cluster, to keep it open to new entrants, and to avoid negative externalities transmitted by the cluster which block structural change.

---

[1] In their study on new manufactured products in the United States, Klepper and Miller (1995) find a high number of exits with only few firms entering the industry during what is called the 'shakeout phase'.

3.2 Cyclical risk

Regional policymakers are not only concerned about a region's growth prospects, but also try to avoid unnecessary fluctuations in business activities, because they cause inefficiency and welfare losses. Among the sources of instability in regional economic activities (Bolton, 1986) the national business cycle is the most important, but because of increasing globalisation the region can also be affected by unexpected events outside the nation, for example, changes in international energy prices. Because business cycles are unevenly distributed among industries, the region's industrial structure (that is, its mix of different industries), determines the level of instability for the region as a whole. Consequently, the search for the optimal industrial mix has been going on in the literature for quite a while. Suggestions that have been made include the 'ogive' index of diversity, the 'national average' measure, the 'minimum requirements' measure, or the 'percent durable' measure.[2]

We claim that a region dominated by one or more clusters is exposed to an increased risk of cyclical variations in economic activities. This hypothesis will be analysed in more detail, with the methodology of regional portfolio theory. In the next section this methodology and the underlying theoretical approach are discussed.

Regional portfolio analysis—a short discussion of the basic concept [3]

In his seminal work in the early 1970s, Conroy (1972; 1974; 1975) made use of financial portfolio theory [4] to analyse regional industrial diversification. His concept of a regional industrial portfolio has been widely applied in examining the industrial structure of regions, and in making recommendations for efficient diversification policies. In this section we will discuss the basic concept of regional portfolio theory and some of the most significant contributions to its conceptual and empirical development.

Assume that the residents of a region are investors, who draw from the various resources available in the region (natural and human resources, capital) and want to utilise them efficiently in order to maximise their welfare. The resources are transformed into returns (for example, income, employment, or output) by economic activities, which are commonly divided into industrial categories. These industries are the assets in the regional portfolio, equivalent to assets in a financial portfolio. Like financial returns, the returns of regional industries are stochastic. At best, their probability distribution is known. If the residents are not indifferent towards uncertainty (that is, risk neutral), their goal will be to maximise returns, given that a certain level of risk is not exceeded, or minimise risk, given that a certain level of returns

---

[2] For a discussion of the history of the economic diversity and instability literature see Kort (1981). A critical discussion of various diversification measures is also included in Conroy (1974).

[3] This section draws heavily on Fritz (1995).

[4] Portfolio theory was developed by Markowitz and Sharpe in the 1950s and 1960s (for example, see Markowitz, 1952; Sharpe, 1970).

is attained.[5] Because firms cannot be sold or bought by regions like financial assets, but have to be given incentives to change their level of activities or relocate their production, the regional application of financial portfolio theory has to include constraints reflecting this inflexibility. This problem will be discussed in more detail later on.

Different portfolios contain different combinations of industries and can be described by their mean and variance. In this framework, regional policymakers are put in the place of the investor. Based on the region's welfare function they want to maximise regional income or employment and minimise the risk of destabilising fluctuations in income or employment levels, caused by regional, national, or international business cycles. Their goal is to diversify the regional portfolio efficiently, that is, choose the optimal industrial mix, which coincides with the welfare-maximising solution if the welfare function is quadratic and returns are normally distributed (for example, see Levy and Markowitz, 1979; Sharpe, 1970).

The policymakers' decision problem can be described by the following optimisation problem:

$$\underset{w_i}{\text{maximise}} \sum_i w_i r_i - \lambda \sigma_P , \qquad (1)$$

where r_i is the expected rate of return of industry i, w_i is the relative weight of industry i in the portfolio; λ is a parameter characterising degree of risk aversion with $\lambda = 0$ for risk-neutral preferences, and σ_P is the portfolio variance, which is a weighted sum of individual industries' variances and covariances:

$$\sigma_P = \sum_{i,j} w_i w_j \sigma_{ij} = \sum_i w_i^2 \sigma_i^2 + \sum_{i,j \neq i} w_i w_j \sigma_{ij} . \qquad (2)$$

From equation (2) it is obvious that a regional diversification strategy aimed at attracting stable industries (that is, industries with low variances) will lead to the risk-minimising portfolio only if all industries are stochastically independent. Such an assumption seems highly unrealistic, given that a regional economy is a socioeconomic system whose various activities and actors are linked in many different ways. Even if a regional economy without any intraregional linkages existed, stochastic dependence would result from these industries' dependence on the national economy. Consequently the performance of regional industries will vary according to national business cycles and will be procyclical for some, countercyclical for others.

The different portfolios can be projected into a return–variance space. The efficiency frontier is defined as the locus of all portfolios whose expected return cannot be increased without increasing its variance or whose variance cannot be reduced without lowering its expected return. The points on the efficiency frontier are the solutions to the optimisation problem of equation (1). The optimal portfolio is the one that maximises regional welfare or comes

[5] This approach goes back to Markowitz (1952) and follows that of a rational investor, who maximises returns and minimises risk.

reasonably close to the welfare maximum if the welfare function is not quadratic or if the normality assumption does not hold. It is located where the welfare indifference curve is tangential to the efficiency frontier.

4.1 Conceptual and empirical issues in the application of regional portfolio theory
In the application of the portfolio framework to the analysis of regional industrial diversification, several conceptual and empirical problems surface. Recent workers have addressed many of these difficulties and sought to improve the theoretical and methodological basis of regional portfolio theory (Bolton, 1986; Hunt and Sheesley, 1994; McKillop, 1990; Schoening and Sweeney, 1992; Sherwood-Call, 1990; Siegel et al, 1995).

Many of the issues arise out of the following questions:
(1) What are the returns to regional assets?
(2) What is the correct measure of the risk associated with expected returns?
(3) Is portfolio variance a good indicator for observed regional economic instability?
(4) How can the limitations to arbitrary manipulation of industrial activity levels be implemented conceptually and/or empirically?

Another issue is related to the fact that the shares of regional industries cannot be changed independently, because of interindustry linkages and other general equilibrium effects, representing the region's economic structure. These structural links restrict the solutions to regional portfolio models, because altering the shares of individual industries will indirectly affect other industries' shares in the portfolio. Different authors (for example, Cho and Schuermann, 1980; Hunt and Sheesley, 1994) have tried to cope with this problem by adding interindustry linkages (expressed in an input–output model, a system of sectoral multipliers, or even a general equilibrium framework) to the portfolio analysis.

Based on the financial portfolio literature, there is no clear justification for or against the existence of a risk–return trade-off and thus a concave efficiency frontier for regional portfolios. Most empirical examples of efficiency frontiers in the literature, however, show the 'right', concave, shape (for example, Brewer, 1984; Hunt and Sheesley, 1994; Lande, 1994; St. Louis, 1980), suggesting that cyclical industries exhibit higher growth rates than more stable industries.

In the next section, the regional portfolio framework is applied to analyse the risk implications of a cluster-oriented regional policy.

5 An analysis of cluster-induced cyclical risk in the regional portfolio framework
Assume a regional economy whose firms belong to one of n different industries. Some of these industries, say k to l, are characterised by a high level of interindustrial linkages, based either on sales and purchases of intermediate goods or on other forms of cooperation and interrelatedness. This subset of industries can therefore be defined as a cluster.

If regional decisionmakers want to pursue a cluster-oriented policy, they will attempt to promote the existing network of industries. Three policy options are available for this purpose: strengthening existing linkages, extending the number of participating firms, or adding other industries to the cluster. The decisionmakers may implement one of the three, two of them, or all three together. No matter which strategy they follow, their goal is to promote the cluster by increasing its share in total economic activity in the region. This implies changing the composition of the regional industrial portfolio in favour of those sectors participating in the cluster.

In terms of regional portfolio theory, interindustrial linkages correspond to interindustrial covariances. Thus a strong link between two industries will likely result in a positive value of covariance of the returns of these industries in the portfolio. This implies that the levels of their economic activities will be positively correlated and their respective business cycles will be similar. The extent of the comovement of these two cycles will, however, depend on the type of link between these industries, with a supplier–buyer relationship implying a higher level of positive correlation.

Consequently, whereas a risk-minimising regional planner tends to diversify his or her portfolio by increasing the shares of those industries whose variances are low and/or whose covariances are negative (Markowitz, 1952; see also Bolton, 1986), a cluster-oriented policymaker will increase the share of those industries in the regional portfolio that are closely interrelated, that is, that have positive covariances. It will therefore likely result in an increase in the level of the portfolio variance. If one accepts the portfolio variance as an indicator for cyclical variations, the region will thus be exposed to a higher level of cyclical risk. The more a regional economy is dominated by a cluster and the higher is the interrelation within this cluster the more risk is associated with the regional portfolio.

On the other hand, clusters are thought to be more dynamic and successful than isolated industries, which is the very reason for the pursuit of cluster-oriented policies. The average growth rates of the industries within the cluster should therefore be above the regional average. As a result, assuming that growth is a reasonable indicator for returns on regional assets,

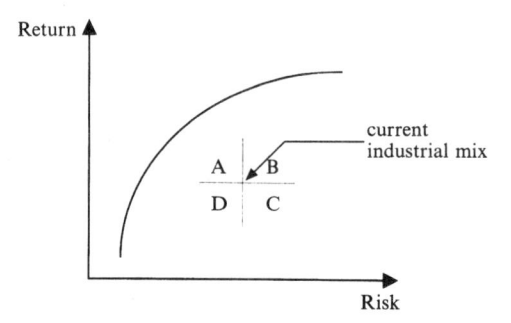

Figure 1. Risk–return trade-off.

the regional decisionmaker may be confronted with a risk–return trade-off as illustrated in figure 1. Given the region's original industrial mix as shown in the figure, following a cluster strategy will shift the location of the region in the return–risk space towards the top right (area B). Assuming that return and risk are the only variables in the regional welfare function, the net welfare effects of a cluster policy remain undetermined, depending on the (aggregated) regional risk preferences.

These results, however, depend on the assumption that networking and cooperation within a cluster do not affect the level of the variances for each cluster industry. It may be more realistic to assume that industries are to some extent able to reduce their individual exposure to cyclical variations in the level of their business activity. In that case a cluster-oriented strategy may very well lead to a reduced regional portfolio variance. Furthermore we assume that cluster-oriented policy addresses only a small number of different clusters. Of course it would be possible to diversify the regional portfolio and reduce the risk associated with cluster-oriented policy by means of promoting different clusters in a region. Without an empirical estimation of the portfolio variances of actual regions, the policy conclusion remains unclear.

Given that not all firms within one industry may be part of a cluster, the measured covariances between the industries to which the firms participating in the cluster belong to may give an inaccurate estimation of the links between them. One way to avoid this measurement error is to treat a cluster as one sector, that is, as one asset in the regional portfolio. Additionally, treating the cluster as one asset in the portfolio enables the regional planner to diversify the regional industrial structure optimally without giving up the cluster orientation.

It would be interesting to discuss briefly the implications of another regional strategy which may take advantage of the dynamism of clusters and networks but at the same time avoids some of the potential negative effects just mentioned. This policy is characterised by efforts to increase the level of internationalisation of the regional economy. In this strategy, industries are encouraged to participate in different interregional, national, or international clusters in order to raise their competitiveness. If successful, these industries will exhibit high growth rates (equivalent to returns), without necessarily increasing the shares of industries with positive covariances in the regional portfolio. As a result, net welfare may undoubtedly increase. This case is represented in the figure by a shift from the original position in the risk–return space to a point farther to the top left (area A in the graph).

The analysis provided here does not deal with structural risk, even though applying regional portfolio analysis to that type of risk seems reasonable as well. Nevertheless, the analysis above does not take into account the potential interdependency of structural and cyclical risk. Regional product cycles, for instance, may influence cyclical variations; decline in the economic activities of the cluster will also affect the regional business cycles. Using the regional portfolio methodology to analyse cyclical risk, one should control for structural risk. Further research on this matter is necessary.

6 Conclusions

In the analysis above, which is still at a very preliminary stage, we have not attempted to discredit cluster-oriented policies. Instead, we have tried to emphasise the positive aspects but also to call attention to the possible destabilising effects brought about by the concentration of resources on a group of strongly linked industries. All of these different aspects have to be examined before any decision about the appropriate mix of regional policies to be applied is taken.

Industries which engage in active cooperation and networking instead of fighting on their own may undoubtedly increase their competitiveness and thus increase the welfare of the regions in which they are located. Nevertheless, a regional policymaker should be aware of the potential long-run structural consequences as well as the short-run instability implications a cluster strategy may have. Only then can he or she attempt to react to these negative effects, perhaps by adjusting the specific cluster-oriented measure in order to make it more immune to the type of risk mentioned.

References

Bolton R, 1986, "Portfolio analysis of regional diversification: some empirical results", paper presented at the meetings of the North American Regional Science Association, Columbus, OH, November; copy available from the author, Department of Economics, Williams College, Williamstown, MA

Brewer H L, 1984, "Regional economic stabilization: an efficient diversification approach" *Review of Regional Studies* **14** 8 – 21

Cho D W, Schuermann A C, 1980, "A decision model for regional industrial recruitment and development" *Regional Science and Urban Economics* **10** 259 – 273

Conroy M E, 1972 *Optimal Regional Industrial Diversification: A Portfolio-analytic Approach* unpublished PhD dissertation, Department of Economics, University of Illinois at Urbana-Champaign, IL

Conroy M E, 1974, "Alternative strategies for regional industrial diversification" *Journal of Regional Science* **14** 31 – 46

Conroy M E, 1975, "The concept and measurement of regional industrial diversification" *Southern Economic Journal* **41** 492 – 505

Fritz O, 1995 *Three Essays on Industrial Pollution Generation and Structural Change in a Regional Economy* PhD dissertation, Department of Economics, University of Illinois at Urbana-Champaign, IL

Hunt G L, Sheesley T J, 1994, "Specification and econometric improvements in regional portfolio diversification analysis *Journal of Regional Science* **34** 217 – 235

Hutschenreiter G, Peneder M, 1994, "Ziele und Methoden der Clusteranalyse wirtschaftlicher und innovativer Aktivitäten" *WIFO-Monatsberichte* **67** 617 – 623

Klepper S, Miller J H, 1995, "Entry, exits, and shakeouts in the United States in new manufactured products" *International Journal of Industrial Organization* **13** 567 – 591

Kort J R, 1981, "Regional economic instability and industrial diversification in the US" *Land Economics* **57** 596 – 608

Lande P S, 1994, "Regional industrial structure and economic growth and instability" *Journal of Regional Science* **34** 343 – 360

Levy H, Markowitz H M, 1979, "Approximating expected utility by a function of mean and variance" *American Economic Review* **69** 308 – 317

McKillop D G, 1990, "Industry risk factors and their implications for diversification within the Northern Ireland regional economy" *Applied Economics* **22** 301 – 311

Markowitz H M, 1952, "Portfolio selection" *Journal of Finance* **7** 77 – 91

Markusen A R, 1996, "Sticky places in slippery space: a typology of industrial districts" *Economic Geography* **72** 293 – 313

Marshall A, 1920 *Principles of Economics* (Macmillan, London)

Meijboom B R, Rongen J M J, 1995, "Clustering, logistics, and spatial economics", mimeo, Faculty of Economics and Business Administration, Tilburg University, Tilburg

Penttinen R, 1994, "Summary of the critique on Porter's diamond model: Porter's diamond model modified to suit the Finnish paper and board machine industry", DP-462, The Research Institute of the Finnish Economy, Helsinki

Porter M E, 1990, "The competitive advantage of nations" *Harvard Business Review* **68** (2) 77 – 93

Schoening N C, Sweeney L E, 1992, "Proactive industrial development strategies and portfolio analysis" *The Review of Regional Studies* **22** 227 – 238

Sharpe W, 1970 *Portfolio Theory and Capital Markets* (McGraw-Hill, New York)

Sherwood-Call C, 1990, "Assessing regional economic stability: a portfolio approach" *Federal Reserve Bank of San Francisco Economic Review* Winter, 17 – 26

Siegel P B, Johnson T G, Alwang J, 1995, "Regional economic diversity and diversification" *Growth and Change* **26** 261 – 284

St. Louis L V, 1980, "A measure of regional diversification and efficiency" *Annals of Regional Science* **14** 21 – 30

Tichy G, 1992, "Technologiepolitik, Industriepolitik und Wettbewerbsfähigkeit" *Wirtschaftspolitische Blätter* number 4, 408 – 415

Tichy G, 1995, "Die Wirtschaftspolitische Bedeutung ökonomisch-technischer Cluster Konzepte", in *Regionale Innovation: Durch Technologiepolitik zu neuen Strukturen* Ed. M Steiner (Leykam, Graz) pp 89 – 103

Tichy G, 1997, "Are today's clusters the problem areas of tomorrow?", in *Competence Clusters* Ed. M Steiner (Leykam, Graz) pp 94 – 100

Networks, Innovation, and Industrial Districts: The Case of Scotland

M Danson, G Whittam
University of Paisley

1 Introduction

In this chapter we explore the extent of networking within the Scottish economy and its significance for innovation in products, processes, and policy formation and implementation. The principle of networking between the main economic players in the Scottish economy has been accepted over a long period (Morison, 1987), and has in many ways been the progenitor of the partnership approach to delivering structural funds across the European Union (EU) (Danson et al, 1997). Apparently accepting the philosophy of networking and partnership as a legitimate and appropriate strategy to improve the performance of the economy, the development agency for Scotland, Scottish Enterprise, has sought to establish networking through its policy statements, objectives, and operations. However, by way of contrast to the accomplishments in interagency linkages and relationships, Scottish Enterprise has had only limited success in achieving these aims as they apply to sectors and firms. This is somewhat surprising given the strength of the corporate consensus approach to regional policy in Scotland, with many other areas adopting the innovative Scottish partnership model as the way to organise local and regional development strategies (Danson et al, 1997). Furthermore, there are historical precedents which demonstrate the advantages which can arise from production within industrial districts in the Scottish context (Slaven, 1975). In the current period, the successful networking model developed in Scotland, and illustrated in the Strathclyde European Partnership as outlined in section 3, has at its core a framework of accountability. This is promoted through the established clusters and networking institutions being present and active in the partnership, and by their collective moulding of an inclusive approach. But an examination of the approach adopted by Scottish Enterprise to develop networking relationships reveals that certain key aspects which are essential for the establishment of value-added partnerships (Johnston and Lawrence, 1988) are missing.

We specifically examine the arrangements within the context of the Scottish economy because Scottish Enterprise has recently identified networking, particularly amongst small and medium-sized enterprises (SMEs), and the promotion of clustering amongst firms within key sectors as ways to achieve improved company performance, growth, and hence job creation. This twin approach on clustering and networking reflects the dual strategy to promoting economic prosperity adopted by Scottish Enterprise. On the one hand, there is support for new firm formation via the birth-rate strategy and, on the

other, the attempt to attract inward investors. The rationale for promoting networks between SMEs is made clear in the following quotation:
"Networks are important: many of the solutions will be found in the actions of individual entrepreneurs, backed by their networks of family and friends. An important focus of action for the strategy is to improve the effectiveness of these networks and to make potential entrepreneurs more aware of what they can do themselves to achieve success. Part of this involves improving the support given by the formal support networks in the private and public sectors" (SE, 1993a, page 4).

The resulting Business Birth Rate Strategy, for instance, acknowledges that there are problems with existing networking agencies within the Scottish economy and seeks ways to encourage the effectiveness of both formal and informal networking arrangements (SE, 1993a, page 8). Beyond SMEs, both in traditional and in new industries, Scottish Enterprise has promoted the creation of employer and agency associations to encourage networking; in a way recognising market failure in establishing such relationships. In terms of clustering, Scottish Enterprise has identified thirteen broad clusters as priorities for action. It is envisaged that, by concentrating on the development of these clusters, Scotland will develop economies of agglomeration in applied technologies, specific high-tech skills, suppliers of key components, institutions, etc, which will attract the generalised factors such as capital, raw materials, and scientific knowledge which are highly mobile. Although many regions and countries are noted for specialisation in particular goods and services, what can give these regions and countries a competitive edge is the method of organising production. The promotion of cooperation between firms has long become accepted as a policy objective, whereby firms can develop linkages, either in networking arrangements, industrial districts, or clusters, to achieve a competitive advantage. Where clusters are not well formed there is not the possibility of firms being able to exploit these advantages of organisation.

Within the context of these attempts to improve new firm formation rates and to promote clustering amongst the key sectors, an understanding of the essential components of what makes networking successful is needed. In this chapter we are focusing on the issues of the institutional frameworks which are seeking to organise the networking and clustering between firms and the relevant agencies. As previously noted, there is a successful partnership model of interagency linkages within the Scottish economy but this successful model has not been adopted by Scottish Enterprise to facilitate the development of networking and clustering between firms and agencies. The chapter is therefore organised on the following lines. In section 2 we provide a brief history of the Scottish economy, focusing on the significance of the past for present-day policy proposals. In section 3 we identify the significance of networking both between agencies and between enterprises. In the fourth section we analyse how successful Scottish Enterprise and the other main actors in the Scottish economy have been in attempting to establish networking arrangements. Examples of networking in new industries in Scotland are covered in the fifth section. In conclusion, policy recommendations are suggested in an attempt to ensure the future success of networks within the Scottish economy.

2 Scotland: on the periphery of Europe

Although the period since the United Kingdom joined the EU has refocused the trade, industry, and so the regions of the British Isles, in many ways this was merely exacerbating a much longer relative decline in Scotland, Wales, and the north of England. Indeed, over "most of this century Scotland has been declining relative to the rest of the United Kingdom and, by extension, the rest of Europe" (Danson, 1991, page 89). As one of the first industrialised regions of Europe and so of the world, Scotland has experienced both an early period of growth and a long history of stagnation. The growth and development of the economy were built on strong networks, however. According to Slaven (1975, page 182, describing the early part of this century):

"A community of interests was growing among steelmakers and shipbuilders, and this was a link more strongly developed at a later date. ... The demands of the shipyards boosted the growth of steel and dictated the changing production patterns of the pig-iron and malleable-iron producers. Marine engineering and shipbuilding lay at the centre of a complex concentration of heavy engineering and finishing trades."

Behind the development of the Scottish economy, and of Clydeside in particular, were factors associated with the British Empire. Britain's role as an imperial power, based on naval supremacy, prompted the establishment of the coal and steel industries of central Scotland. The regional economies, and through linkages and migration the economy of the rest of the nation, became inextricably dependent on the trading and military position of the United Kingdom as a whole. A coherent and integrated Clydeside experienced very significant growth in industry and population in the years up to 1914. Records on shipbuilding tonnage (Slaven, 1975) confirm that this area was at the heart of the Empire in terms of industrial output and importance. The First World War was perhaps the watershed, although the seeds of external destruction may have been inherent before then, the decline of British power and controlled imperial markets exposed a rapid and deep structural imbalance in the Scottish economies. And, because of the close networks and linkages between sectors and corporations, these forces of decline spread rapidly through the regional economy. Massive unemployment, poverty, deprivation, and emigration marked the period up to the next World War, with Glasgow suffering the worst urban slums in the history of the planet (Damer, 1990). Since 1930 this industrial legacy has ensured Scotland has been subject continuously to a broad set of economic policies targeted at relieving the worst effects of the rundown and closure of the staple industries—steel, coal, shipbuilding, heavy engineering, and textiles.

Having struggled to maintain former glories in output and trade through most of the century, Scotland—and the rest of the periphery of Britain—has been in almost continuous benefit of various forms of regional development aid. Regional economic policies, based on restructuring the old industrial areas, and the decongestion of overcrowded housing and manufacturing areas, have been increasingly concerned with promoting foreign direct investment and withdrawing the state from direct involvement in production. This has led to the degree of external control and ownership of the industries of Scotland progressively increasing, at times through merger and takeover, through nationalisation and privatisation, at other times through differential rates of decline and growth of native and foreign companies. Research suggests that such changes, in complex ways, put a relative brake on the rates of new firm formation and indigenous development (Ashcroft et al, 1987). Concomitantly, output, trade, and investment have become more narrowly dependent on a few key sectors.

With the regional policy focus on the attraction of inward investment and with low rates of endogenous growth, a branch-plant economy has been created to replace the former heavy industrial clusters. With few local supplier or purchaser linkages, and an absence of many of the higher business functions of R&D, finance, marketing, corporate strategy, etc, branch plants offer jobs to the local economy, and little else. In particular, there are minimal relations established or developed between inward investors and the local SME sector. Compared with the strong inherent linkages of the traditional industrial complex, the branch-plant economy has failed to network to anything like the same degree.

If we consider the leading sectors, electronics (and computers especially) and whisky have accounted for over half of nonoil manufacturing exports from Scotland (SCDI, 1997), and about 50% of all manufacturing investment in recent years. Both these sectors are dominated by overseas companies, with over 90% of output from non-Scottish firms. Over a quarter of manufacturing employment in Scotland is in overseas-owned plants, and much of the rest is controlled by UK corporations with their headquarters in the southeast of England. With a truncated internal range of higher order functions, and few business-service linkages into the local economy (Fothergill and Guy, 1990), this degree of domination is often blamed, in a simple way, for the massive restructuring of the Scottish economy since 1979.

Since 1980, 40% of all Scottish manufacturing jobs have been lost. In key traditional sectors such as coal, steel, and engineering the decline in employment was heavier, with the replacement of some of these jobs in electronic and electrical engineering. In many sectors (agriculture, fishing, energy and water, manufacturing, and construction) employment is historically at its lowest levels in Scotland, with jobs for both men and women disappearing. Only services are showing any growth over time, and then only in part-time work, and typically in sectors related to simple consumption expenditure alone.

In such past-dominated narrow economies, new firm formation and SMEs have become the panaceas in the mission statements of many agencies, and in particular of the EU regional economic development and regeneration programmes. The business birth rate has undoubtedly increased since 1980 (SE, 1996). However, questions have been raised over the ability of companies dependent on the local market to reverse long-term regional decline on their own. Entering this debate, Malecki and Nijkamp (1988), for example, have suggested that uneven development is endemic, with no possibilities to overcome metropolitan core bias through compensation of the periphery, although Vaessen and Keeble (1995) appear to see enough examples of the counterfactual successful SME in the periphery to argue that regional divergence need not be inevitable. In practice, there has been but limited success in replacing traditional major manufacturing employers. Further, with more prosperous areas of the United Kingdom tending to maintain higher rates of business start-up (Ashcroft and Love, 1996) and with real unemployment rates in the problem regions persistently significantly higher than the UK average (Beatty et al, 1997), this record suggests that greater levels of incomes and wealth are not incompatible with the creation of conditions for enterprise development. Analysis of the nature and form of networking and of the wider business environment is also of critical relevance in determining such comparative performances in regeneration.

Faced with the failure of the market to create formal linkages between enterprises in the Scottish economy, Scottish Enterprise has been promoting the establishment of networking and clustering to fill this gap. In essence, a key question for the success of the new firm strategy and of the inward investment strategy, separately and collectively, and of these direct interventions in the interrelationships between companies is whether they are conducive to the creation of networking and industrial districts. Superficially based on the best-practice models identified elsewhere in Europe and North America in regional economic regeneration strategies, they rely on networking and partnership. It is to a consideration of these themes that we now turn.

The gains that are to be achieved through operating as a network or a more spatially specific industrial productive system (that is, an industrial district) have been identified as arising from competitive external economies, exogenous external economies, and cooperative external economies (Oughton and Whittam, 1997). Competitive external economies arise from internal economies which, because of competition, are passed on in lower costs and prices. Exogenous external economies arise out of agglomeration economies and what Marshall termed 'industrial atmosphere'. Cooperative external economies are associated with the benefits of belonging to the network or industrial district or industrial cluster. It is generally accepted that a crucial ingredient for the success of a policy aimed at achieving the above economies generally, and cooperative external economies specifically, is trust and cooperation. The way that trust and cooperation can be developed is by a sense of belonging to the particular network, industrial district, or cluster.

"A carefully nurtured collective identity can potentially provide the social fabric which sustains cooperation in an industrial district as in a corporation" (Best, 1990, page 237). The identification of the industrial district encompassing firms within a wider community as a whole, "as in a corporation", is a point identified by Becattini in his definition of an industrial district:

"I define the industrial district as a socio-territorial entity which is characterised by the active presence of both a community of people and a population of firms in one historically bounded area ... The most important trait of local community is its relatively homogeneous system of values and views ..." (1990, pages 38–39).

Until such a time as trust and cooperation can be assured, enforcement mechanisms can be utilised. In the context of networking, member firms have to join. Firms breaking the rules of membership can be excluded. Belonging to a network suggests a commitment to working in a collective manner. To encourage membership, firms need to be aware of the benefits to be achieved from belonging: "organisations that are not able to make membership compulsory must also provide some noncollective goods in order to give potential members an incentive to join" (Olson, 1971, page 17).

In this section we have discussed the specific situation of the Scottish economy, existing as it does on the periphery of Europe. We have noted the industrial districts of its past and commented upon an essential ingredient of successful networking. Through promoting networking it is hoped to establish cooperative relationships. This can be achieved through the promotion of trust and the establishment of interfirm relationships and, drawing on the experience of the Third Italy, for instance, the business development agencies in Scotland have introduced policy delivery mechanisms which can assist in the development of trust and cooperation. This idea of examining firms within a framework of institutions and organisations has suggested to Scottish Enterprise, with its objective of establishing networks, that it needs to develop a collective approach to the establishment of interfirm relationships. We now examine the recent experiences of what in many ways can be identified as an 'ideal type' of networking which is to be found in the Scottish economy: the Strathclyde European Partnership model. In the following section we will look at the efforts of Scottish Enterprise to establish such networking between firms within the Scottish economy.

3 Networking within Scotland: the Scottish partnership model
In policy-delivery terms, Scotland differs from the rest of the United Kingdom and indeed other European regions in several ways which may have impacted on the success of the partnership model to date (McCrone, 1992; Paterson, 1994). Persistently different from the dominant UK model because of the presence of the Scottish Office, the development agencies, and other organs of the state, there is a number of infrastructural and implementational ways in which policy is created and delivered in Scotland which marks it out from the practices in England, especially since 1980. In the wider EU context,

according to Danson et al (1997) there are three differences in the Scottish approach to delivering European integrated programmes. First, the responsibility of the implementing authority has been devolved to the Scottish Office rather than being overseen by a department of central government in the capital. This relative autonomy has facilitated better relationships with local and regional organisations and given the regional partnerships more scope to develop their own momentum and operating procedures. Second, powerful regional authorities have been evident, especially in the form of Strathclyde which was created in 1975, but abolished in 1996. In addition to this, Strathclyde Regional Council and the City of Glasgow Council were also proactive in the formation of alliances with other European regions to promote common interests and to lobby for appropriate policy reforms as well as being represented on the main bodies in structural fund negotiations. In other words, Scotland has been actively involved in accessing the structural funds and negotiating their position and has built up a great level of experience and knowledge in this arena over the past twenty-two years of European regional development funding. The historical tradition of partnerships for urban and regional development within Scotland and in particular within Strathclyde Region, which includes Glasgow, has been evident since the 1970s when the Glasgow East Area Renewal scheme was created to be followed by Task Forces, Area Initiatives, etc (see Moore and Booth, 1986; Morison, 1987), and then by the National Programme of Community Interest. These initiatives were the first of their kind in Scotland, where decisions on projects for Glasgow were taken locally by a partnership comprising the key agencies for regional development, including the Regional Council, the Scottish Office, and the European Community.

Indeed, the development agencies in Scotland are themselves essentially networks of local enterprise companies. These lower level bodies typically establish ad hoc and more permanent partnerships internally to their areas, coming together with other local agencies to create economic development and training vehicles for specific parts of their territory or sectors. Two examples of such initiatives are the Partnerships for Priority Areas (for severely socially and economically disadvantaged communities) and the Ayrshire Economic Forum (for addressing economic performance of a county).

The concept of a more decentralised decisionmaking partnership for regional development was extended in 1988 when Strathclyde secured an Integrated Development Operation (IDO). This was the first European-funded integrated Objective 2 economic development programme approved for a five-year period in the United Kingdom. A novel feature of the IDO was the creation of a separate secretariat or Executive to manage the programme within the region. The IDO enabled decentralised decisionmaking on policy issues as well as on project applications by a committee. The committee consisted of agencies with a role in economic development of the region and was chaired by the Scottish Office.

An indicator of the IDO's success is the further three independent executives introduced in Scotland to manage the 1994–99 Single Programming Documents in Highlands and Islands, Eastern Scotland, and Dumfries and Galloway. Strathclyde continued to receive structural Objective 2 funding in the form of the 1994–99 Operational Programme which continued to be managed by the Executive. Following the reorganisation of local government in 1996 the Programme Executive was made a company limited by guarantee and is now known as The West of Scotland European Partnership. Since the creation of the IDO, membership of the committees and their advisory groups has been widened to reflect an expanding partnership and to include independent technical experts.

The European Commission is considered (Bennett and Krebs, 1994) to be one of the main promoters in the development of networking behaviour, with the aim of achieving the goal of European integration and cohesion. As a means of encouraging this behaviour many of their initiatives and programmes insist, as a prerequisite, that the participating organisations must be involved in a partnership or network (Cooke and Morgan, 1993). This is evident, as highlighted earlier, in the arrangements for regional policy funding and for accessing structural funds through partnership. The aim of those arrangements is to harness the energy, skills, and resources of the key actors in the regional development environment and to develop and implement solutions to increase cohesion across the European Union.

Figure 1 (over) is an outline of the structure and functions of the West of Scotland Partnership and indicates the internal and linking networks and relationships established in this approach. Almost 200 organisations have accessed structural funds in the current and previous programmes. These organisations are referred to as partners and are involved in a locally based decisionmaking process. The partners contribute 50% of the management costs of the programme and the European Commission contributes the other 50%. The partners categorised by sectoral or interest groups are:

European Commission DGXVI and DGV,
The Scottish Office Industry and Education Department,
Local authorities,
Scottish Enterprise Network,
Universities,
Strathclyde Further Education Partnership,
Other colleges,
Local Enterprise Trusts,
Local Initiatives and Affiliated Bodies,
Area Tourist Boards,
Charitable and voluntary organisations—employment and training,
  —business development and infrastructure,
  —culture, environment, heritage, and tourism,
National agencies or organisations,
Regional agencies or organisations.

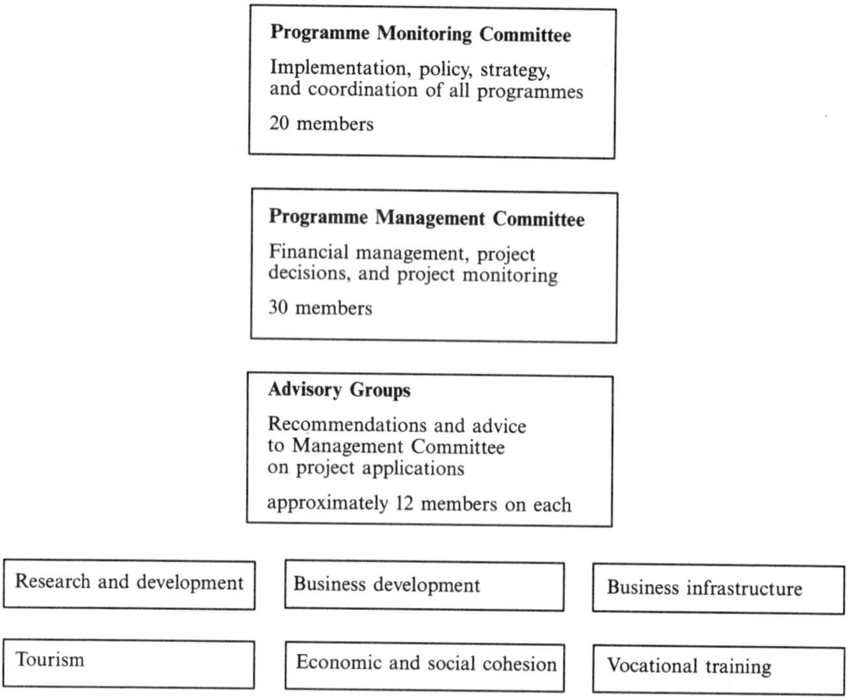

```
┌─────────────────────────────────────┐
│ Programme Monitoring Committee      │
│ Implementation, policy, strategy,   │
│ and coordination of all programmes  │
│ 20 members                          │
└─────────────────────────────────────┘

┌─────────────────────────────────────┐
│ Programme Management Committee      │
│ Financial management, project       │
│ decisions, and project monitoring   │
│ 30 members                          │
└─────────────────────────────────────┘

┌─────────────────────────────────────┐
│ Advisory Groups                     │
│ Recommendations and advice          │
│ to Management Committee             │
│ on project applications             │
│ approximately 12 members on each    │
└─────────────────────────────────────┘
```

| Research and development | Business development | Business infrastructure |
| Tourism | Economic and social cohesion | Vocational training |

Figure 1. Committee structure 1994–96: West of Scotland Partnership.

The long history of collaborative working (Moore and Booth, 1986; Morison, 1987) and the formalised structure of the European Partnership which necessitated the formation of relationships between agencies, politicians, and organisations within Scotland, has the potential to realise the benefits of networking. Despite a recognition that autonomous regional development agencies offer a number of advantages over alternative forms of local and regional economic interventions (Halkier and Danson, 1997), including being able to intervene and interact with the economy at the most appropriate jurisdictional level (Armstrong, 1997), the Scottish agencies have been involved in partnerships and networks from their inception. Thus, although the benefits of avoiding splitting a number of responsibilities between different departments or quasi-government agencies—such as the provision of sites, attraction of industry, environmental improvement, sector strategy, and urban development—has been long recognised and the perceived advantages of locating these functions within a single organisation to realise synergies (Moore and Booth, 1986, page 119) has been fully appreciated, these benefits and advantages are increasingly claimed for partnerships between these very bodies (Danson et al, 1997). In other words, experience of networking between policy bodies has promised and delivered significant gains to the Scottish economy.

4 The promotion of networking and clusters by Scottish Enterprise

As suggested above, Scotland could be said to be an export-oriented economy, now relatively protected from UK business cycles, but open to new sensitivities—locked into the supply needs of multinational oligopolies in the electronics and oil sectors, in particular. It is clear that the future relies on European markets, and also, consequently, in attracting foreign direct investment from North America and the Pacific Rim as locations for entry into the EU. Exports surveys for Scotland show the degree to which the historical patterns of trade have been transformed. According to a well-established survey (SCDI, 1997), two thirds of Scottish exports go to Western Europe. Electronics account for 49% of all sales overseas, most going to Europe and to a lesser extent the Middle East and Africa. Globalisation of production in essence means multinational enterprises arranging a configuration of plants across the world which meets their needs to supply in and into a number of trading blocs. The deepening reliance of Scotland on the attraction of such titans means competing for highly mobile investment, with this very competition between regions and states threatening to heighten the propensity of such capital to be mobile. Without a counteracting, long-term sustainable development of indigenous companies, it is recognised that such peripheral economies will progressively lose further control over their own destinies: the development of their underdevelopment.

In delivering the twin supply-side strategies of inward investment and new firm formation to address regional development and regeneration, regional development agencies and local authorities collectively arrange for property investments, improvements to the image of the area, training, technology transfer, information technology, and telecommunications expenditures to favour investment, often through joint local, UK, and EU interventions. Confirming such collective actions, recent research (Danson et al, 1997) has demonstrated the strength of the corporate consensus approach to regional policy in Scotland, with many adopting the Scottish partnership model as the way to organise local and regional development strategies. At the political and bureaucratic levels, as described above, networking and partnership are accepted as valuable features of economic development strategies.

Given this culture, to what extent has this experience been applied to the implementation of the business birth rate and sectoral strategies? Addressing birth rate requires a consideration of the original and subsequent related research into the low rates of new business start-ups in Scotland. As part of the overall study on this subject, Scottish Enterprise sponsored research on specific sectors which were considered to provide significant lessons or to have a key role in the economy. Analysis of clothing and textiles showed that the popular view of company failures and declining employment was disguising the actual and potential contributions of new firms to the industry in Scotland. However, their role was restricted by a lack of innovation and dynamism. Evidence from Italy and Germany, with their strong networking traditions, suggested there was the potential for a more significant contribution to the economy, with development through a growth in the number of new, small firms rather than in the size of existing companies.

Scottish Enterprise has identified textiles as an industry where clustering should be encouraged so that the industry can compete. At the local level Enterprise Ayrshire has been instrumental in trying to implement this policy via the Ayrshire Textile Group (ATG). The textile industry in Ayrshire now represents 25% of Ayrshire manufacturing industry, employing 9000 people and exporting £60 million per annum. In 1991 a partnership arrangement between Enterprise Ayrshire and the textile sector was established. This led to the creation of the ATG in 1992 with the establishment of a programme of assistance leading to the ATG acquiring its own premises in 1994. The commitment to working in partnership with the industry reflects the ethos of Enterprise Ayrshire's approach to local economic development: "it seeks to develop and build relationships with companies rather than superimposing a structure onto a sector" (ATG, 1994, page 3). The ATG appears initially to have been successful in the "delivery of information advice and programmes in response to proven needs ... with more than 60% of local industry participating in some meaningful way" (page 4). This has led to product development, moving products more upmarket, increased diversification, and marketing Ayrshire textiles as a group. However, there has been a failure to take the project forward to develop a 'stand-alone resource centre'. To become 'free standing', it is envisaged that the ATG must recover all its operating costs. In this context, Johnstone and McLachlan (1996) found textile companies reluctant to participate fully in the network which became an obstacle to realising economies of scale and scope, with "a low level of trust and the strongly adversarial nature of the [local] sector" (page 755) a significant factor. Unwillingness to pool resources, concern over allocation of orders within the network, apprehension over cooperation, and a failure to communicate were all cited as contributory reasons for the lack of a full commitment to the project by the member firms. The need for training in networking protocol and processes, the key role of the facilitator, and the benefits of contractual agreements between members were identified as necessary elements if the local industry was to create a successful cluster. It would appear that although ATG has attempted to develop a partnership the companies do not see the group as theirs. There is a tacit recognition of this problem in one of the reasons sought for the establishment of the freestanding facility: "The Ayrshire Textile Industry must view the ATG and its Resource Centre as their own" (ATG, 1994, page 13). This is one of the essential features of the 'ideal model' of the Strathclyde European Partnership.

A further attempt at delivering the clustering objective envisaged by Scottish Enterprise by a local enterprise company is again to be found in Ayrshire: the Engineering Ayrshire (EA) initiative. Engineering is the largest manufacturing sector in Ayrshire with over 180 companies employing over 14 000 people. It has a turnover of £1.9 billion and generates £1.3 billion in export sales (EA, 1997, page 6). The original rationale for the establishment of the group arose because of the perceived lack of sales outside the west of Scotland. To this end EA has organised trade missions, launched a public procurement initiative, and provided individual members with export sales and support to improve product or process. The EA group is now seeking to widen its activities in order to capitalise on opportunities presented by changing market conditions. Of significance and giving grounds for optimism, vis-à-vis the possibility of the group becoming self-financing and realising key objectives such as developing the skills base in a collective manner, is the structure of the organisation. Unlike the ATG, the aims of the EA state it should "be an organisation run by its members for the benefit of its members and the Ayrshire engineering and electronics sector" (EA, 1997, page 9). Within its strategic objectives there is a responsibility placed on the membership to work in partnership with the relevant organisations, such as Enterprise Ayrshire, local councils, and trade associations and to attend relevant seminars and programmes. The group is managed by a steering committee, comprising senior representatives from six companies and two Enterprise Ayrshire executives, which meets monthly. A full members' meeting is held at least once a year. By placing obligations and responsibilities onto the membership and by having an accountable framework for decisionmaking of the group as a whole, it is hoped that the problems associated with the ATG in attempting to develop its strategy will be overcome. The essentials of an ideal network, outlined in the introduction, can be seen to be reflected in the EA clustering initiative.

5 Networking in new industries: examples from Scotland
Looking at newer industries and at high-tech and academic spin-offs especially, the economic consultants PACEC (SE, 1993b, pages 67–71) reported minor differences in the factors explaining low levels of new firm formation in these specific sectors and in the overall economy. The research base within industry was perceived as inadequate, with few organisations acting efficiently as incubators for spin-offs and new high-tech companies. The infrastructure and superstructure were deemed deficient in these respects, with poorly developed and weak specialist resources, social networks, and information flows within the high-tech community. The lack of demand for high-tech companies was highlighted (with the exception of the North Sea oil industry), constraining development of a home base, and putting firms at a cost disadvantage in export markets. In comparison with other areas, notably Massachusetts, policy support critically was seen as less extensive. Problems in establishing a high-tech firm are compounded by the actual process of formation and growth, with the lack of support and ambition at home stunting development opportunities. All this seemed minor compared with the situation of academic spin-offs, the authors argued, where major restrictions were identified in Scottish higher and further education institutions and in their business, funding, and cultural environments in promoting new firm formation. Entrepreneurship was perceived as alien to academic traditions,

risky and stigmatising, so that significant barriers to entry into the market place had been erected. Westhead and Cowling (1995) also looked at high-tech business start-ups and questioned the certainty of such conclusions. They believe that the debates on the major factors are still unresolved. In a US study, Muniak (1994) has argued recently that there are essential conflicts between the promotion of regeneration policies, based on high tech, at the urban and at the national level. In the same vein, and as seen here, much of the policy and analysis in the United Kingdom seems to suggest that the needs of local and of national economies may not be coincidental with regard to the encouragement of high-tech and academic entrepreneurs. Policies to improve new firm formation in the country as a whole may lead to differential effects across regions, perhaps providing strong virtual and vicious cycles of uneven development.

These various studies on high-tech and academic spin-offs and start-ups suggest that networking, partnership, trust, and cooperation were poorly developed or actively discouraged in Scotland. The role of branch plants in diminishing the ability of entrepreneurs to set up their own businesses was noted frequently, as more unexpectedly was the scepticism and obstruction of the business development and academic establishments.

In a related sector, Clarke (1996) reported on the Scottish software industry. This displays many of the characteristics of a high-tech sector requiring peculiar forms of support, within a general programme of assistance. Although sector-specific programmes are seen positively, he questioned the concentration of state business development agencies on a minority of export-oriented product-based companies. In like manner, Turok (1993a) examined the development of the printed circuits industry in Scotland. Unlike much of the electronics sector, it is dominated by indigenous enterprises which are growing strongly. He argued the main reasons for this buoyancy are the local concentration of firms within Scotland, the avoidance of direct competition between companies, the particular demand conditions facing the sector (based as they are on batch production of customised products for technically close customers), and the need to satisfy just-in-time manufacturing processes. However, although Turok recommended closer connections between enterprises and so the growth of a Scottish industrial cluster in electronics, as the sector tends to supply original equipment manufacturers (OEMs) in a dedicated chain with few external economies obvious from enhanced linkages between companies, it is not clear how agglomeration economies would be promoted through improved networking and, by extension, why the industry would support such moves.

To encourage more local sourcing Scottish Enterprise has developed a Strategic Alliance with the major electronics firms who are organised through the Scottish Electronics Forum. The objective is to provide "vital support and infrastructure" (SE, 1998) so that the industry can maintain its competitive edge. This is to be achieved by ensuring the electronics industry exploits developing markets by developing more knowledge-intensive activities enabling the development of 'a value-added logistics hub'. The Alliance identifies four objectives to achieving this aim: commercialising academic and industrial research, increasing the level of Scottish content in electronics products, developing the skills base of the Scottish workforce, and attracting R&D investment into Scottish plants (SE, 1998). This Alliance has now been adapted to fall in line with the clustering strategy adopted by Scottish Enterprise, as outlined earlier. This has resulted in a group of singularly functional, operational departments combining to form the Information Industries Group. The information industries cluster team has a dominant role in terms of working closely with the Electronics, Optoelectronics, Software, Multimedia, and Information Society partners. The problems facing this particular cluster in terms of the electronics industry concern job creation, local sourcing, and technology transfer. "The Scottish electronics sector's total output has nearly doubled in the past decade. But total direct employment today, despite some growth in the late 1980s is roughly where it was in the mid 1980s ... . Nearly 80% of all components and sub-assemblies which go into Scottish-produced electronics output is sourced outwith Scotland" (Young, 1997, page 14).

Although there is a low level of indigenous employment and output, in a study of linkages within the electronics industry in Scotland, Turok (1993b) suggested that there are potential opportunities for much higher levels of local sourcing by the foreign-owned OEMs. However, where these do exist, he described them as corresponding to a 'dependency model' rather than to a 'self-sustaining economic development scenario'. Linkages are seen as weak or based on dependency; therefore the industry is not embedded into the regional economy. With few high-tech linkages between indigenous supplier and OEM, there are no effective clusters, as connections are founded on market strengths, not on trust and cooperation. Few cases of transfers of staff or technologies have been recorded. Classically, the retention of higher order functions (R&D, marketing, strategic management, financing, etc) elsewhere limits the ability of the local economy to establish business services externally to the multinational enterprises themselves, and to furnish spin-offs, transfers, and the other components of a successful milieu. Not least the failure to acquire a broad range of skills and contacts through networking in the Scottish economy is significant in depressing the strategies for self-sustaining development based on the formation of new firms. Thus Clarke and Turok confirm the structural obstacles to the promotion of new enterprises and networking which strategies based on encouraging individual entrepreneurs to start up businesses are unable to address.

In a similar vein, research on the North Sea oil industry has demonstrated that Scottish companies tend to be restricted to the peripheral, low-value-added areas of the sector, such as catering, supplies, labour-only contracts, and maintenance (Lipka and Howie, 1993). Unable to integrate forward because of a lack of market strength and too low down the value chain to establish an export-oriented sector of any note, there are few opportunities to be realised in cooperation in an industrial cluster for these companies.

One model to address such problems is corporate venturing. This covers assistance to entrepreneurs or SMEs to exploit new ideas, corporate restructuring through staff creating new independent businesses, and the development of new profit centres and subsidiaries to exploit new products, processes, and markets. It was suggested (SE, 1993a) that Scotland is poorly provided with such approaches to regeneration and restructuring, which is unfortunate because, by reducing risk and establishing new ventures with better prospects, they offer an effective, additional element in promoting new firm formation, especially in the high-tech area.

A perennial problem facing new entrepreneurs in Scotland, and in the United Kingdom more generally, has been financial support. Considering the extensive research on financial aspects of supporting new firms and networks, the Scottish Enterprise (1993a) findings proposed an attempt to close the widening equity and financial gap with a move to the provision of risk finance by the public sector and by the private investor, the so-called 'business angel'. Some of the chapters on finance in *Scotland's Business Birth Rate* (entitled "Finance—the big issue?", "Informal investment—a neglected source?", and "Finance—the ongoing debate") suggested that little has been resolved. The equity gap is contrasted with the availability of money: the question of access is a major and persistent feature in the Scottish business environment, and to a greater extent than elsewhere. Given the significant power and size of the Scottish financial sector, the invention and introduction to the world of investment and unit trusts, and the strong history of (savings) banking, there is an argument that "improvements in the behaviour and performance of both lenders/investors and actual/potential business founders" can and must be encouraged (SE, 1993b).

Although many of the factors influencing enterprise creation and regional economic development raised in the Scottish Enterprise enquiry are common to most communities, with the arguments well rehearsed elsewhere, the identification of entrepreneurial potential and cultural issues as significant does add an extra dimension to our understanding of the processes involved in new firm formation. The research directly challenges the business development community in Scotland to address its own behaviour and attitudes to potential entrepreneurs. It has been argued that persistent unemployment, poverty, and deprivation, against a relative economic decline of eighty years duration (Danson, 1991), have thralled Scots to the philosophy of collective intervention to generate jobs and, as argued here, the strategies of Scottish Enterprise, with their denial of the relevance of trust and cooperation, have exaggerated rather than addressed this view.

Although further analysis by Scottish Enterprise (SE, 1993b) showed a lower pool of self-confessed, potential entrepreneurs, there was still a substantial proportion of the adult population who believed they had the ability and desire to run their own business. This analysis suggested that there was no anti-enterprise or dependency culture; rather, with much unexploited potential, there seemed to be a barrier to converting desire into reality.

Such research is indicative of a wider growing interest in the concept and importance of social networks in the entrepreneurial creation process. Recognising this, Scottish Enterprise addressed the role of networking directly. This revealed Scotland as having a more extreme set of rules and social mores in determining support and advice. Those who would have most difficulty establishing a business elsewhere (the young, women, and the working class) appear to face higher hurdles than in the southeast of England and beyond. Lack of security and of alternative employment in the event of failure, and the effects on family life of creating a new firm are the major concerns of potential entrepreneurs according to the Scottish Enterprise report (1993b, page 24). The report believed these problems could be overcome in some instances by wider discussions with informed contacts and with existing entrepreneurs. This seems all the more important given the less than positive attitude of business development organisations. Unfortunately, the low level of entrepreneurship identified above becomes self perpetuating in these circumstances, compounding the earlier barriers to enterprise.

Knowing an entrepreneur is an important 'rocket' in the process of converting potential into actual business creators; those who are most dissuaded from making the transition are least likely to have a set of relevant contacts. In cross reference to the details on financial support, the greater severity of funding problems in Scotland is associated with the lower penetration of the ideology of the property-owning democracy. With the UK funding bodies' preference for mortgages on the borrower's home to be used as collateral, there is an additional barrier to establishing the company where owner-occupation of housing has traditionally tended to be low.

McNicoll (1996) and Whittam and Kirk (1996) have discussed this set of elements in more detail. McNicoll on behalf of Scottish Enterprise concluded that positive encouragement to progress the idea of becoming an entrepreneur would often be sufficient, within the environment of all the other hurdles and obstacles, to prevent most potential new ventures from failing to reach the starting line. In a contemporary study, Struthers et al (1996) show the influence of organisations and individuals within current and former networks to be significant in the processes of new firm formation and business development. This discussion on the role of information and networks is extended by Devins (1996), when he considers measures taken to overcome market failure in business advice and knowledge. The dual need to monitor and evaluate advice programmes and services, and to know the companies supported, are seen as essential to the long-term health of the new firm sector.

Harrison and Leitch (1996), in considering Northern Ireland experience, argue for the establishment of entrepreneurial teams to progress business ideas and opportunities, positively assembling collection of talent and expertise, rather than an approach which passively supports the market. The consensus, here as earlier, is for intervention to overcome market failure in labour, capital, land, and knowledge sectors, but to address the establishment of trust and cooperation directly.

6 Conclusion

In this chapter we have identified that the economic development of the Scottish economy suggests there is scope for the establishment of networking arrangements once again, highlighting the parallels between the industrial clusters of the last century and the success of the Strathclyde European-Partnership model today and its adoption as 'an ideal type'. However, attempts to create sector clusters explicitly to mirror these partnerships have encountered problems. In the case of the Scottish Electronics Forum, there is little accountability to the indigenous industry, domination by foreign multinational companies, and hence a power imbalance; this raises questions over whether Scottish Enterprise or any other players from within Scotland have control over this key sector. The Ayrshire Textile Group could be considered to be failing because the companies do not see it as their own, with a dependency tradition being reinforced. In the same county there is optimism for the Ayrshire Engineering group because of the commitment to involve members from the beginning rather than to accept passive membership, while obligations placed on being a member strengthen these relationships. In this latter association, it can also be argued that there is more of a commitment from the sector because of the high rate of inclusion of indigenous enterprises.

With regard to the business birth strategy in Scotland, this has tended to be passive in the promotion of trust and cooperation. Networking is identified as important but is to be encouraged at the level of the individual entrepreneur on an atomistic basis. This contrasts sharply with the extensive and successful model of regeneration established between the regional economic development agencies and organisations themselves, and with the industrial clusters of traditional sectors which created the 19th-century Scottish economy.

From the experiences with these two general approaches to economic development, we would argue that, to ensure the success of potential networks, Scottish Enterprise should move from trust based on contracts to trust based on goodwill within the networking arrangements. In order to achieve this it is necessary to develop a group approach to firm organisation, which draws on the experience from these experiments and from elsewhere. This requires the involvement of all potential firms in the decisionmaking process to ensure that the needs of the firms are being catered for within the networks. This collective approach is required because the services being delivered within the network of firms will consist of elements of public goods such as information. We have noted that punishment strategies exist within networking arrangements but, rather than prohibit trust, these strategies can enhance trust, acting as a safeguard mechanism. With an economy dominated by branch plants, however, it is unclear how multinational enterprises can be encouraged to adopt more cooperative strategies, and so become more embedded into the Scottish economy.

References
Armstrong H, 1997, "Regional-level jurisdictions and economic regeneration initiatives", in *European Research in Regional Science, 7. Regional Governance and Economic Development* Eds M Danson, S Hill, G Lloyd (Pion, London) pp 26-46
Ashcroft B, Love J, 1996, "Employment change and new firm formation in GB counties: 1981-89", in *Small Firm Formation and Regional Economic Development* Ed. M W Danson (Routledge, London) pp 17-35
Ashcroft B, Love J, Schouller J, 1987 *The Economic Effects of Inward Acquisition of Scottish Manufacturing Companies 1965-1980* Industry Department for Scotland, 5 Cadogan Street, Glasgow G2 6AT
ATG, 1994 *Business Plan* Ayrshire Textile Group, 16 Nelson Street, Kilmarnock
Beatty C, Fothergill S, Gore T, Herrington A, 1997, "The real level of unemployment", Centre for Regional Economic and Social Research, Sheffield Hallam University, Sheffield
Becattini G, 1990, "The Marshallian industrial district as a socio-economic notion", in *Industrial Districts and Inter-firm Co-operation in Italy* Eds F Pyke, G Becattini, W Segenberger (International Institute for Labour Studies, Geneva) pp 37-51
Bennett R, Krebs G, 1994, "Local economic development partnerships: an analysis of policy networks in EC LEDA Local Employment Development Strategies" *Regional Studies* **28** 119-141
Best M, 1990 *The New Competition: Institutions of Industrial Restructuring* (Polity Press, Cambridge)
Clarke T, 1996, "New software companies in Scotland: growth constraints and policy implications", in *Small Firm Formation and Regional Economic Development* Ed. M W Danson (Routledge, London) pp 189-201
Cooke P, Morgan K, 1993, "The network paradigm: new departures in corporate and regional development" *Environment and Planning D: Society and Space* **11** 543-564
Damer S, 1990 *Glasgow: Going for a Song* (Lawrence and Wishart, London)
Danson M, 1991, "The Scottish economy: the development of underdevelopment" *Planning Outlook* **34** 89-95
Danson M, Fairley J, Lloyd G, Turok I, 1997, "The governance of European structural funds: the experience of the Scottish Regional Partnerships", paper 10, Scotland Europa, Brussels
Devins D, 1996, "The use of external advice by new and established SMEs: some survey evidence", in *Small Firm Formation and Regional Economic Development* Ed. M W Danson (Routledge, London) pp 161-188
EA, 1997 *Three Year Plan 1997-2000* Engineering Ayrshire, 17/19 Hill Street, Kilmarnock, KA3 1HA
Fothergill S, Guy N, 1990 *Retreat from the Regions: Corporate Change and the Closure of Factories* (Jessica Kingsley, London)
Halkier H, Danson M, 1997, "Regional development agencies in Western Europe: a survey of key characteristics and trends" *European Urban and Regional Studies* **4** 243-256
Harrison R, Leitch C, 1996, "Whatever you hit call the target: an alternative approach to small business policy", in *Small Firm Formation and Regional Economic Development* Ed. M W Danson (Routledge, London) pp 223-247
Johnston R, Lawrence P R, 1988, "Beyond vertical integration—the rise of the value-adding partnership" *Harvard Business Review* July-August, 94-101
Johnstone R, McLachlan A, 1996, "The Ayrshire knitwear sector: from competition to collaboration", paper presented to Institute for Small Business Affairs National Conference, Birmingham; copy obtainable from University of Paisley, Paisley PA1 2BE

Lipka J, Howie D, 1993, "Contracts awarded to North Sea oil and gas contractors: a comparison of the UK and Norwegian sectors", STUC – COMETT Partnership, University of Paisley, Paisley

McCrone D, 1992 *Understanding Scotland: The Sociology of a Stateless Nation* (Routledge, London)

McNicoll A, 1996, "Social networking: a comparative behavioural study between would-be entrepreneurs in Scotland and Boston, Massachusetts", in *Small Firm Formation and Regional Economic Development* Ed. M W Danson (Routledge, London) pp 99 – 107

Malecki E, Nijkamp P, 1988, "Technology and regional development: some thoughts on policy" *Environment and Planning C: Government and Policy* **6** 383 – 399

Moore C, Booth S, 1986, "Unlocking enterprise: the search for synergy", in *The City in Transition: Policies and Agencies for the Economic Regeneration of Clydeside* Eds W Lever, C Moore (Clarendon Press, Oxford) pp 76 – 119

Morison H, 1987 *The Regeneration of Local Economies* (Clarendon Press, Oxford)

Muniak D C, 1994, "Economic development, national high technology policy and America's cities" *Regional Studies* **28** 803 – 810

Olson M, 1971 *The Logic of Collective Action* (Harvard University Press, Cambridge, MA)

Oughton C, Whittam G, 1997, "Competition and cooperation in the small-firm sector" *Scottish Journal of Political Economy* **44**(1) 1 – 31

Paterson L, 1994 *The Autonomy of Modern Scotland* (Edinburgh University Press, Edinburgh)

SCDI, 1997 *Export Survey* Scottish Council Development and Industry, 23 Chester Street, Edinburgh EH3 7ET

SE, Scottish Enterprise, 120 Bothwell Street, Glasgow G2 7JP
  1993a *Improving the Business Birth Rate: A Strategy for Scotland*
  1993b *Scotland's Business Birth Rate* with Scottish Business Insider
  1996 *Business Birth Rate Strategy: Update*
  1998 *A Strategic Alliance: Scotland's Future in Electronics* with Scottish Electronics Forum

Slaven A, 1975 *The Development of the West of Scotland: 1750 – 1960* (Routledge and Kegan Paul, London)

Struthers J, Young A, Wylie J, 1996, "Networks and new enterprise development in Russia: a case study of the Yaroslavl region", in *Small Firm Formation and Regional Economic Development* Ed. M W Danson (Routledge, London) pp 144 – 160

Turok I, 1993a, "Making connections: the growth of a Scottish industrial cluster in electronics", Strathclyde Papers in Planning 24, University of Strathclyde, Glasgow

Turok I, 1993b, "Loose connections: foreign investment and local linkages in 'Silicon Glen'", Strathclyde Papers in Planning 23, University of Strathclyde, Glasgow

Vaessen P, Keeble D, 1995, "Growth-oriented SMEs in unfavourable regional environments" *Regional Studies* **29** 489 – 506

Westhead P, Cowling M, 1995, "Employment change in independent owner-managed high-technology firms in Great Britain" *Small Business Economics* **7** 111 – 140

Whittam G, Kirk C, 1996, "The business birth rate, real services and networking: strategic options", in *Small Firm Formation and Regional Economic Development* Ed. M W Danson (Routledge, London) pp 132 – 143

Young A, 1997, "The best is yet to come" *The Scottish Electronics Review* 14 – 17

Learning with Clusters: A Case Study from Upper Styria

M Steiner, C Hartmann
Department of Economics, University of Graz and Joanneum Research, Graz

1 Introduction
Clusters as units of analysis presume a high density of interaction, cooperation, and creation and diffusion of knowledge. In this chapter the cluster concept will be interpreted from the perspective of 'learning organisations' to give a description of joint knowledge creation and technology spillovers that may take place within clusters. A special focus is put on the different forms and organisational aspects of joint learning. The recent literature (for example, Coombs et al, 1996; Dodgson, 1996; Senge et al, 1995; Senker and Sharp, 1997) suggests that this new approach may offer new findings about clusters and networks of small and medium-sized enterprises (SMEs) that go beyond the well-known ideas of learning by doing and knowledge spillovers.

The emerging materials and metal cluster in Upper Styria (an old industrial area formerly dominated by nationalised steel mills and coal and ore mines and now trying to regain socioeconomic momentum) will form the empirical basis for the interpretation of clusters as learning organisations. The results of a questionnaire will serve to assess the current learning behaviour of the firms within the cluster. Supplementary qualitative interviews were undertaken to reveal more details about joint learning activities between firms and to make the particular mechanisms that lead to knowledge spillovers more explicit.

2 Clusters as learning organisations
As a starting point for a definition of cluster it is still useful to refer to Marshall (1920) and his threefold dimensions of cluster-forming effects: clusters are sets of complementary firms (in production *and* service sectors), public, private, and semipublic research and development institutions, which are interconnected by labour market and/or input–output and/or technological links. They are highly competitive because these links generate a situation which combines the advantages of both the market mechanism and the direct control structures of a single organisation. First, because there are many different firms within a cluster serving many different markets within and outside the cluster, which keep the forces of competition alive and guarantee flexible and efficient handling of activities. Second, because the interconnections of the agents within a cluster allow for close coordination of activities, the development of strong long-term complementarities, and the avoidance of external effects (that is, external to the cluster).

This raises the question of to what extent this competitiveness is automatically created by the existence of these threefold effects or whether conscious efforts are needed to maintain and develop the competitiveness of clusters. These efforts may be pursued from outside the cluster, for example, from policy institutions with their goal orientation and consequent instrument use (for a treatment of policy issues to enhance learning in clusters see the contribution by Cullen in this volume), but they can also be generated from inside the cluster as a coordinated attempt by the members of the cluster to improve their relations and links. Hence, clusters can be regarded as learning organisations, and concepts of learning can be applied to cluster analysis.

2.1 The nature of learning

The concept of learning has changed considerably in recent years. For a long time it was primarily considered as an adaptive response by an organism to a change in the environment. According to an essentially behaviourist – reductionist perspective, learning was a linear process and something that has to start from the level of the individual so that learning in a social context can be understood as the aggregate of individual behaviours.

As Cullen (1998, page 4) argues, conventional models of organisational learning still retain elements of these positions, taking as a starting point an 'information-processing' model or 'black box' conceptualisation of learning, in which information is converted into knowledge and then action. Applied to the concept of organisational learning, it can be understood as a collective and purposive strategy to achieve the goals of the firm; it can furthermore be extended to the notion of clusters as learning organisations with common goals and shared agendas.

Yet learning cannot only be regarded as a process leading to changes in capabilities and competencies; it also has to be considered as a social process of ongoing development embedded in a sociocultural (regional) context. Learning then becomes essentially a communicative process rather than a cognitive performance, requiring new thinking about the nature and forms of the transmission and dissemination of knowledge within a social and organisational context, such as the firm or a cluster (Cullen, 1998, page 5).

2.2 Agglomeration economies and learning by doing as basis of analysis

The idea of learning by doing and positive externalities was raised again by Romer (1986) but goes back to Arrow (1962). Technological know-how is conceived as a factor of production; learning by doing works through each investment of the firm. Specifically an increase in a firm's capital stock leads to a corresponding increase in the stock of its knowledge. This process reflects Arrow's position that knowledge and productivity gains come from investment and production, a formulation that was inspired by the empirical observation of large positive effects of experience on productivity in airframe manufacturing, shipbuilding, and other areas. The second key assumption is that each firm's knowledge is a public good that any other firm can access at zero cost. In other words, once discovered, a piece of knowledge spills over instantly across the whole industry or region. This assumption implies that a change in each firm's technology corresponds to the regional economy's overall learning and is therefore proportional to the change in the aggregate capital stock (Barro and Sala-i-Martin, 1995).

2.3 Organisational learning

Whereas learning by doing and agglomeration economies focus mainly on the issue of productivity gains, organisational learning deals rather with the question of the firm's innovation activities. Learning in this context is regarded as an effort that is pursued actively and strategically. Learning by doing may be considered as something that is carried out passively in parallel with the acquisition of new technologies whereas organisational learning follows clear decisive objectives and may even lead through innovations to the development of new technologies.

> "Learning organizations are organizations where people continually expand their capacity to create the results they truly desire, where new and expensive patterns of thinking are nurtured, where collective aspirations are set free, and where people are continuously improving their personal capabilities" (Senge, 1990, page 15).

Of particular importance seems to be the interplay of the individual, the team, and the organisation as a whole. Organisational learning is the outcome of three overlapping spheres of activity—individual, team, and system learning. All three kinds of learning take place simultaneously. Individual learning takes place each time an individual reads a book, performs an experiment, or gets feedback from workmates or colleagues. Team learning takes place when two or more individuals both learn from the same experience or activity. Team learning may involve new ways to address the team's responsibilities, or it may involve some aspect of the interaction between the members of the team themselves. System learning takes place when the organisation develops systemic processes to acquire, use, and communicate organisational knowledge (Dixon, 1995).

All these definitions have several characteristics in common. First, learning is conceived as something that is deliberately pursued by the organisation and its members. Organisational learning therefore seems to be something that has actively to be achieved. Second, the learning process is considered as continuous. Third, learning is depersonalised. It is not a person or an elite (the owner or the top management) who is learning (even when he or she is learning for the organisation); organisational learning is a change in the knowledge of the whole organisation (Staehle, 1991, page 844).

2.4 How do organisations learn?

Learning is a conscious attempt on the part of the organisation to retain and improve competitiveness, productivity, and innovativeness in uncertain technological and market circumstances. The greater the uncertainties, the greater the need for learning. Organisations learn in order to improve their adaptability and efficiency during times of change (Dodgson, 1993).

To correspond to concepts developed by Piaget (1985) Argyris and Schon (1978) have described three basic types of organisational learning:

Single-loop learning (SLL) SLL occurs when errors are detected and corrected and organisations carry on with their present policies and goals. According to

Dodgson (1993), SLL can be equated to activities that add to the knowledge base or firm-specific competencies or routines without altering the fundamental nature of the organisation's activities. SLL has been referred to as lower level learning by Fiol and Lyles (1985) and adaptive learning or coping by Senge (1990).

Double-loop learning (DLL) DLL occurs when, in addition to detection and correction of errors, the organisation is involved in the questioning and modification of existing norms, procedures, policies, and objectives. DLL involves changing the organisation's knowledge base or firm-specific competencies or routines (Dodgson, 1993). DLL is also called higher level learning by Fiol and Lyles (1985) and generative learning (or learning to expand an organisation's capabilities) by Senge (1990).

Deutero-learning (DL) DL occurs when organisations learn how to carry out SLL and DLL. The first two forms of learning will not occur if the organisation is not aware of the fact that learning should take place. The awareness of ignorance motivates learning (Nevis et al, 1995).

2.5 What do organisations learn?

According to Dodgson (1993), organisational learning occurs when an organisational knowledge base, firm-specific competencies, and routines are developed. Knowledge bases are created by acquiring, storing, interpreting, and manipulating information both from within and outside the organisation. Thus the question arises, which types of information may contribute to such knowledge bases? According to various authors (Belden et al, 1993; Dodgson M, 1991; Pedler et al, 1991; Weisboard, 1987) three different types of knowledge to be acquired through organisational learning may be described:

Technological learning deals with the acquisition and interpretation of know-how associated with new technologies. These technologies may either be acquired by the organisation from outside or developed within. Technological learning occurs typically in R&D efforts by the company (Dodgson, 1991).

Management learning deals with the improvement of the existing organisation of the processes of value creation by the firm. Management learning may be focused either on the further improvement of an existing management system or on the development of new forms of management within the organisation (Pedler et al, 1991).

Marketing learning deals with the acquisition of know-how about the needs and wants of the clients of the firm. Such knowledge is typically acquired through salespersons in direct contact with the clients (Belden et al, 1993).

2.6 Interfirm links and organisational learning

External links, with customers, suppliers, and other sources of information, are critical to the innovation process. Indeed, it is argued that these links are becoming more central in the innovation process (Rothwell, 1992). Interfirm links provide an opportunity to observe novelty through the approaches of the partners, and can stimulate reconsideration of current practices. They can be an antidote to the 'not-invented-here syndrome', and learning vicariously can help prevent the repetition of mistakes.

The primary motive ascribed to learning through interfirm links is to deal with technological and market uncertainty (Chesnais, 1988; 1996; Ciborra, 1991; Mody, 1990). Ciborra argues that alliances are the institutional arrangement that most efficiently allows firms to implement strategies for organisational learning and innovation:

"The alliance brings into the corporation new expertise concerning products, marketing strategies, organizational know-how, and new tacit and explicit knowledge. New management systems, operating procedures and modifications of products are the typical outcomes of this incremental learning" (1991, page 59).

These links can assume two (partly complementary) forms.

2.6.1 *Interfirm links and the concept of the value chain*

According to Porter (1985) or Galbraith (1983) the whole process of the production of a good may be described in terms of a value chain reaching from the provision of raw materials to market distribution and retailing. Each firm may be located in this value or industry chain depending on its prevailing core competencies. The concept of the value chain offers a way to describe interfirm links expressed as vertical or horizontal relations along this chain.

Downstream links Links to other firms that act as clients may be described in general as vertical links and in particular as downstream links or supplier–client relations (Galbraith, 1983). The quality of the link may reach from simple input–output relations to joint R&D efforts described as technological links (Dodgson, 1996).

Upstream links Links to other firms that act as suppliers may be described in general as vertical links and in particular as upstream links or client–supplier relations (Galbraith, 1983).

Links in one stage of the value chain Links to firms that have similar core competencies may be described as horizontal links (Galbraith, 1983). Such links may consist of the joint use of machinery between the partners or several forms of the sharing of production capacities.

2.6.2 *Interfirm links and the concept of the network*

A complementary perspective to the value-chain concept is the network approach (Cooke and Morgan, 1991). Networks may be perceived as "intermediary" between market and hierarchies (Williamson, 1985). Or as Imai (1990, page 158) puts it:

"Networks can be viewed theoretically as interpenetrated forms of market and organization. Empirically they are loosely coupled organizations having a core of both weak and strong ties among constituent members. Cooperative relationships among firms are the key linkage mechanism in network organizations."

These cooperative relationships may be organised along a specific value chain but need not have such a specific structure (Saxenian, 1996). Thus the activities carried out within networks may reach from the joint production of goods and services to activities such as strategic R&D efforts.

Using these conceptual elements of different forms of learning we will in the next section use the case of the emerging cooperation between firms in the materials and metal cluster of Upper Styria to answer the following questions:

(1) What role do learning activities such as learning by doing play at the enterprise level? Are spillover effects present, as presumed in the work of Arrow (1962) and Romer (1986)? How do they manifest themselves?

(2) To what extent do deliberate learning activities exist between the different firms in addition to knowledge spillovers? How may those learning activities be categorised (SLL or DLL)? What are the contents of such learning efforts?

(3) Are there hints for clusters or SME networks acting as learning organisations? What kind of interfirm links could foster which kind of learning activities?

3 Clusters as learning organisations: the case of the materials and metal cluster in Upper Styria

The case study focuses on firms within the materials and metal-processing sector in Upper Styria. Because the region was for decades (if not centuries) dominated by large firms and the SMEs of the region were either integrated with or strongly dependent on these large firms there have only recently been attempts made to begin links and cooperation among SMEs. There was, therefore, no predefined 'cluster'. To discover the slowly emerging forms of cooperation between these firms we had to proceed in three, rather qualitative, steps without being able to use rigorous quantitative methods.

3.1 Steps in the field of research

First step: *identification of cluster-firms*. In order to discover existing SME networks and cooperation in the region, links and joint learning activities between the different firms had to be identified and located. First, 186 firms related to materials and metal in the region were identified through the database of the regional technology-transfer centre. Then ten regional experts (two managers of technology centres, two managers of regional development agencies, two managers of regional technology transfer centres, two management consultants, and two members of the regional chamber of commerce) were interviewed in order to spot firms among those 186 where closer network and value-chain relations exist. In this way we were able to identify thirty-eight firms that are embedded in regional networks and interfirm links ('mapping' of the cluster).

Second step: *qualitative interviews in the firms*. Fifteen out the thirty eight identified firms were visited and interviewed personally. The first aim of these qualitative interviews was the identification of projects related to learning activities in the cluster. The second aim was a deeper investigation into the networks and interfirm links reaching through the cluster. As a result two network structures and thirteen links between the thirty-eight firms were identified.

Third step: *the baseline survey.* In order to get results on a quantitative basis a 'baseline survey' was prepared and carried out (the questionnaire comprised twenty pages and thirty-three questions). The questionnaire was first sent to the thirty-eight cluster firms identified in the first step of our analysis. These sample firms were then questioned through structured telephone interviews.

3.2 A brief characterisation of the cluster
3.2.1 *The background*
The northern part of Styria is characterised by a centuries-old industrial tradition based on rich deposits of iron-ore and coal. The region may be characterised as an old industrial area. Of the 159 071 people employed (according to the 1991 Census) 6% work in the primary sector, 43% in the secondary sector (the Styrian average is 38%), and 51% in the tertiary sector. It was dominated by nationalised industry, guaranteeing stable employment until 1986. Decreasing competitiveness led to far-reaching structural changes such as subdivision into companies and a shift towards concentration on manufacturing. The restructuring of the nationalised industry led at the beginning of the 1990s to several spin-offs but also to a decrease in the number employed and to a decrease in the average size of firms in the region. But large companies in iron and steel industries still dominate the scene.

3.2.2 *Member firms and organisational structures of the cluster*
Between those thirty-eight firms identified through our investigations as members of the cluster several overlapping links and cooperative structures could be located. As a major result of our investigations we have been able to identify particular interfirm links within the cluster. These links may be categorised in part as links according to the value-chain concept and in part as links according to the network concept.

Interfirm links and the value-chain concept Thirteen links between these firms were identified within the value-chain concept.
(a) Eight links between the firms may be categorised as *vertical* (along the value chain). The eight vertical links can be divided into two groups. In five cases firms act as suppliers (upstream firms) to clients in the region. In three cases firms (downstream firms) cooperate with suppliers in the region.
(b) Five links between the firms can be categorised as *horizontal* cooperation (at one stage of the value chain). They can be divided into two groups. There are three cases of joint usage of machinery—a machine is bought and jointly used by different firms. In two cases of production, capacities are systematically shared between the partners.

Interfirm links and the network concept There have been two regional interfirm link systems identified within the network concept.

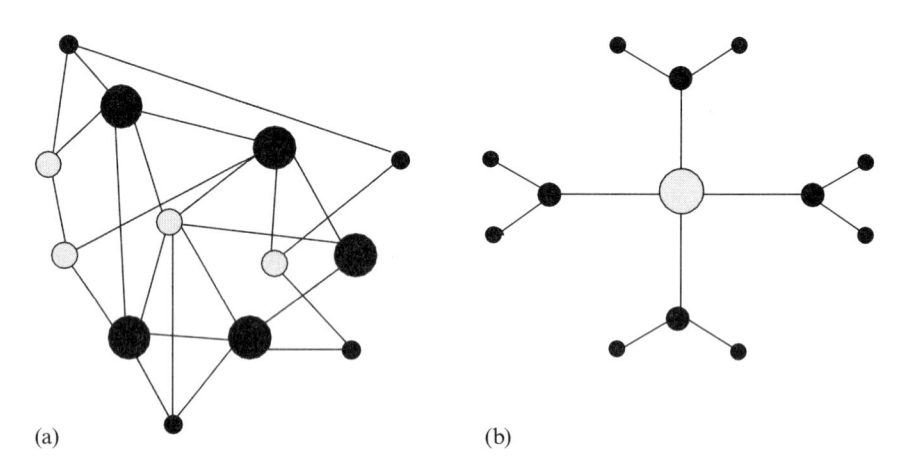

Figure 1. Types of network structure: (a) the web, (b) the snowflake.

The *first network* (seventeen members) offers general building and construction solutions (for example, the new design of the main square of towns). The structure of this network may be described as a 'snowflake' (figure 1). Several suppliers are centred around the main contractor with direct contact with the clients. Thus most of the communication activities are carried out by this main contractor in the centre of the structure. The following firms are members of the network: the main contractor (a general project manager for building and construction projects), two planning and construction bureaus, three suppliers of raw materials, two locksmiths, three plumbers, four electricians, and two construction firms. Because of the specific structure of the network there are hierarchical relations between the main contractor and the suppliers. Contact in the network is maintained through periodic meetings of its members.

The *second network* (ten members) examined offers system solutions that combine software products and measuring techniques for high-precision metal-processing firms. It consists of six firms specialising in custom-made software solutions and four firms specialising in the development and implementation of measurement systems. The structure of this network may be described as a 'web'. A typical product may be an electronic measuring system that continuously controls the production of seamless tubes in a tube mill. The network will not only implement this control system, it will also provide training and consulting services to the staff. In contrast to the first network there is no general main contractor in this network—every member may act in a particular project as supplier of system solutions. If one firm has a new client, the other network members will be invited to participate in the planned project. Most of the networking activities are performed through tele-cooperation (the individual computer networks of the different partners are linked through a spatially distributed groupware environment).

3.2.3 *Actors and support institutions of the cluster*

The cluster firms are embedded in a rich environment of supporting institutions and intermediaries. Examples of research, technology, and development institutions are the University of Leoben (raw materials and raw materials extraction, basic elements and materials, machine construction), the Christian Doppler Laboratories, Joanneum Research (the leading applied research institution in Styria and the second largest of its kind in Austria) and the Technology Transfer Centre in Leoben. Several training institutions ensure the transfer of new knowledge to the firms.

3.3 Learning at the firm level and knowledge spillovers

First, we focus on the almost classical issue of agglomeration economies and learning by doing. Through our qualitative personal interviews and survey we found possible hints to learning by doing and knowledge spillovers.

3.3.1 *Learning by doing in the cluster*

More than two thirds of the twenty-five regional experts and firms interviewed in the first and second steps of our analysis claimed that knowledge is acquired in the cluster to a large extent through learning by doing and experience. When new technologies and/or products are adopted this learning process is propelled by arising problems—the solution of these problems generates tacit knowledge that forms the basis of future core competencies of the firms. Learning by doing seems to be particularly present in the spin-offs in the cluster. Five start-ups interviewed during the 'mapping' exercise all stated the learning by doing and experience is not only important in the field of product development but also in marketing and management issues.

The empirical evidence of the baseline survey underlines this qualitative result—learning by doing seems to play an important role at the enterprise level within the cluster. Sixteen of the thirty-eight interviewed firms claimed to have built up their own expertise mainly through learning and training on the job. This picture gets even clearer if one looks at the way the staff of the firms has built up its know-how. For twenty-one of the thirty-eight interviewed companies learning by doing is an important source of corporate expertise—they believe that training and learning on the job are important factors for building up their core competence.

3.3.2 *Potential sources of knowledge spillovers in the cluster*

Several potential sources of knowledge spillovers in the cluster were mentioned in the qualitative interviews. The firms claim that regional 'watering holes' (informal meeting places such as bars or restaurants) and thus informal contacts with colleagues and also the incorporation of new workers into the firm play important roles in spillover processes. The management consultants stated a different source—according to them open training sessions and seminars are potentially strong sources of spillovers.

Again the empirical evidence of the baseline survey supports the qualitative statements. Twenty-two of the thirty-eight interviewed firms claimed that belonging to the cluster facilitates the circulation and diffusion of knowledge among the firms. For seventeen of the firms informal relations to other firms in the region are an important source of know-how and information. Seventeen of the firms conceived of the incorporation of new workers as a source of new know-how as important. Thirty of the enterprises send their staff to seminar and training sessions.

Thus not only could the existence of knowledge spillovers be verified but three different sources of such spillovers were identified in this cluster:
1. watering holes (informal meeting places),
2. training sessions and seminars,
3. the incorporation of new workers.

3.4 The cluster as a learning organisation: learning through interfirm links

Second, the issue of cooperative learning and interfirm linkages is dealt with. The different kinds of links that have been identified within the cluster will be examined under the focus of joint learning activities. Each particular type of cooperation will be illustrated with a concrete example.

3.4.1 *Vertical links: learning through upstream and downstream links*

One of our major findings is that all examples of vertical cooperation, except one, obviously fostered or favoured interfirm learning activities. In seven of the eight cases we were able to identify during our visits and personal interviews, projects that consisted of actively pursued joint learning efforts. Those efforts may be classified according to Argyris and Schon (1978) as SLL (three cases) and DLL (four cases).

An example of SLL is the case of a long-term collaboration in strategic R&D between one metal-processing firm that supplies seat components to German car manufacturers. In this case new seat components (hinged joints) for minivans are developed jointly. The supplier is allowed to use the capacities of the R&D department of the car manufacturer. In this project a special focus is put on the usage of new materials instead of brass—an R&D problem both partners became aware of together. These activities add to the firm-specific competencies without altering the fundamental nature of the organisations' activities. Because of the R&D focus of the learning activities the content may be classified as technological learning.

Examples of DLL are continuous improvement projects (total quality management) between a producer of plastics and four recycling firms or between a paper mill and a company specialising in the production of cardboard containers. In these cases various business processes were examined among the partners and whenever necessary redesigned in order to improve the quality of the production processes in each firm. Quality circles have been established at all firm levels of the involved partners in order to implement and to foster the continuous improvement of the quality of the production processes and goods to be produced. In these two cases technological learning (concerning the production processes) and management learning (concerning the general organisation of work) was carried out deliberately and the newly acquired know-how was shared between the partners.

Again the baseline survey brought results that may underline the findings of the qualitative research efforts. Twenty one of the firms regarded suppliers inside the region as an important source of knowledge and twenty of the enterprises regard suppliers outside the region as such a source. The relation with clients is even more important. All thirty-eight firms regard clients in the region as a major source of new knowledge and thirty five of them believe the same of clients outside the region. Twenty seven of the firms maintain collaborative links with other companies in the region. These collaborations are divided into the joint realisation of R&D (two firms), supplier–client relations (fourteen firms), and client–supplier relations (eighteen firms). Collaborative links to firms outside the cluster are maintained by twenty five of the firms. These links are maintained by seven firms in order to realise joint R&D projects. Supplier–client relations are maintained by twelve firms and client–supplier relations by thirteen.

3.4.2 *Horizontal links: learning in one stage of the value chain*
Only in the two cases of capacity management between cluster firms could joint learning activities between the partners be identified. Although learning between suppliers and clients is in the main technological learning, the know-how acquired in this horizontal cooperation consists to a very large extent of management skills. When capacities are shared, an elaborate know-how in coordination and communication is necessary in order to carry out the tasks properly. Mistakes and errors need to be corrected and avoided in future. Thus these learning efforts may be classified as SLL. In the three cases of joint usage of machinery, only learning by doing and potential spillovers could be detected.

3.4.3 *Learning in SME networks*
As already shown above, the two SME networks identified in the region differ in their organisational structure, processes, and in the products and services they offer. But nevertheless they have something in common—both networks specialise in offering system solutions to their clients. System solutions are the joint offering of material goods together with complementary services to the client with a strong focus on the particular problems to be solved (Simon, 1993).

In the first network several learning activities have been identified between the firms. With regard to the content of the efforts they may be divided into technological learning, management learning, and marketing learning.

Technological learning is, on the one hand, focused on the continuous improvement of the services the firms offer. To work together as well as possible each partner needs very good knowledge about the technologies the others use. For example, the suppliers of raw material need to have very good knowledge about the processes used by the locksmiths and plumbers. In these cases existing routines or competencies are enhanced while leaving the network's policies and goals untouched—SLL is present. On the other hand, joint activities are carried out with the objective of creating process and product innovations within the network. These learning activities may be classified as DLL because such innovations are able to change the existing norms, procedures, policies, and objectives of the network.

Joint learning activities between all the network members are focused on the further development of project-relevant management skills. Partners do not only carry out actions to improve the management of joint projects, they also try to improve continuously the performance of the network as a whole. This is done through a total quality management approach in which all existing procedures and routines are carefully examined in order to find potential for further improvements in cooperation. Therefore these efforts may be classified as DLL.

Marketing learning is done by developing future projects interactively with the clients. This is in particular done under the guidance of the main contractor in the network. Direct contact to the market enables the network to react quickly to changes in the needs and wants of potential customers. As a result the routines or firm-specific competencies of the network members may change—DLL is present. The members put much effort into this specific type of learning.

In the second network the learning activities identified between these firms may be divided into technological and management learning:

Technological learning is here carried out in each individual project. Because the network offers no standardised products, every solution offered to a client has to be developed individually. In other words, the network carries out joint R&D projects continuously with a special focus on the particular needs of the client. These projects may be therefore classified as DLL.

Management learning is particularly based on the tele-cooperation the network performs through its spatially distributed groupware system. Although the members feel no need at present to improve the hardware infrastructure further, they focus strongly on the optimisation of the business processes in the network. These learning activities touch on issues such as the flow of documents between network members, application sharing, the joint development of a common electronic knowledge base, and the continuous improvement of network performance in software terms. These learning efforts may again be classified as DLL.

Probably because such elaborate and complex networks are still rather rare in the cluster there is not much evidence from the baseline survey. Only four of the thirty-eight firms claimed to be members of an SME network. Another four enterprises declared they are carrying out joint marketing activities and four were cooperating in public tenders.

4 Learning and unlearning: tentative conclusions
4.1 A summary of the results
At the firm level, learning by doing could be identified in most cluster corporations in the qualitative investigations as well as in the baseline survey. In addition to these findings we could also detect spillover effects that are

present at the cluster level. The most important sources of spillovers we identified were watering holes, training and seminars, and the incorporation of new members into the firms. Thus the above results may support the evidence of the presence of agglomeration economies in the cluster.

At the cluster level we were able to identify different kinds of interfirm links: five upstream links, three examples of downstream cooperation, and five examples of horizontal cooperation could be found. In addition we were able to locate two SME networks within the cluster: one building and construction network seventeen firms and one network offering goods and services for the metal-processing industry of ten firms.

Through these interfirm links (upstream, downstream, or horizontal) the firms were able to carry out SLL and to a certain extent also DLL. With regard to content, technological learning through joint R&D dominated these interfirm activities.

In the networks the member firms were all able to perform DLL activities with such contents as technological learning through intense joint R&D efforts, management learning through the continuous improvement of the routines and procedures carried out together, and marketing learning through the development of new products together with the clients.

4.2 The cluster as a learning organisation

Clusters may be characterised as learning organisations when learning at the interfirm (cluster) level is present. The findings of this case study presume that clusters are indeed acting as learning organisations. In almost all cases of interfirm activities within this specific cluster, learning activities such as SLL and DLL could be detected. As Dodgson (1996) has already pointed out, interfirm links may provide opportunities for 'higher level' or double-loop learning. But the findings of this case study suggest that the specific kinds of links between the firms may also influence whether DLL or only SLL activities will be carried out. Although the horizontal cooperation examined showed only SLL efforts, the upstream and downstream cooperation seemed to foster DLL as well. Thus we may conclude that client–supplier relations and supplier–client links have in this particular cluster a higher potential than horizontal cooperation to enable the partners to carry out DLL.

In this cluster networks seem to have an even bigger ability to bring about DLL activities. In this case study all learning efforts carried out in the examined networks could be classified as DLL. This greater ability of the network to foster DLL activities may lead to the suggestion that clusters may act more as learning organisations the more they have a real network structure.

A final interpretative remark: The cluster of this case study consists of slowly emerging links and cooperation between SMEs in a region long dominated by large and often nationalised companies. The decline of this old industrial area in the 1970s and 1980s was to a large degree caused by an inability to learn new forms of behaviour and a preservation of old ones. The reported attempts to develop a culture of learning among SMEs may be also interpreted as a regional process of 'unlearning' old habits (something that is harder than learning new ones). Therefore, the SME cluster creates for the region as a whole positive spillover and demonstration effects as to the importance of joint learning.

Acknowledgements. This case study has been carried out partly within the TSER project "Developing Learning Organizations in SME Clusters" but also received additional funding from the Federal Ministry of Science and Transport and the Provincial Government of Styria. For help at different stages of the project we would like to thank Markus Gruber, Clemens Habsburg-Lothringen, Ursula Mörtlbauer, and Thomas Jud.

References
Argyris C, Schon D, 1978 *Organizational Learning* (Addison-Wesley, Reading, MA)
Arrow K, 1962, "The economic implications of learning by doing" *Review of Economic Studies* **29** 155–173
Barro R J, Sala-i-Martin X, 1995 *Economic Growth* (McGraw-Hill, New York)
Belden J, Hyatt M, Ackley D, 1993 *Towards the Learning Organization: A Guide* (Belden, Hyatt, and Ackley, St Paul, MN)
Chesnais F, 1988, "Technical cooperation agreement between independent firms, novel issues for economic analysis and the formulation of national technological policies" *STI Review* **4** 51–120
Chesnais F, 1996, "Managerial objectives and technological collaboration: the role of national variations in cultures and structures", in *Technological Collaboration: The Dynamics of Cooperation in Industrial Innovation* Eds R Coombs, A Richards, P P Saviotti, V Walsh (Edward Elgar, Cheltenham, Glos) pp 34–52
Ciborra C, 1991, "Alliances as learning experiences: cooperation, competition and change in high-tech industries", in *Strategic Partnership and the World Economy* Ed. L Mytelka (Frances Pinter, London) pp 55–70
Coombs R, Richards A, Saviotti P P, Walsh V (Eds), 1996 *Technological Collaboration: The Dynamics of Cooperation in Industrial Innovation* (Edward Elgar, Cheltenham, Glos)
Cullen J, 1998, "Developing learning organizations in SME clusters: modelling and recommendations", mimeo, Tavistock Institute, London
Dixon N, 1995 *The Organizational Learning Cycle: How We Can Learn Collectively* (McGraw-Hill, Ottawa)
Dodgson M, 1991, "Technological learning, technological strategy and competitive pressures" *British Journal of Management* **2** 133–149
Dodgson M, 1993, "Organizational learning: the review of some literatures" *Organizational Studies* **14** 195–209
Dodgson M, 1996, "Learning, trust and inter-firm technological linkages: some theoretical associations", in *Technological Collaboration: The Dynamics of Cooperation in Industrial Innovation* Eds R Coombs, A Richards, P P Saviotti, V Walsh (Edward Elgar, Cheltenham, Glos) pp 76–97
Fiol C, Lyles M, 1985, "Organizational learning" *Academy of Management Review* **10** 803–813
Galbraith J R, 1983, "Strategy and organizational planning", in *The Strategy Process* Eds H Mintzberg, J B Quinn (Prentice-Hall, Englewood Cliffs, NJ) pp 141–150
Imai K I, 1990, "Japanese business groups and the structural impedient initiative", in *Japan's Economic Structure: Should it Change?* Ed. K Yamura, Society for Japanese Studies, University of Washington, Seattle, WA, pp 152–165
Marshall A, 1920 *Principles of Economics* (Macmillan, London)
Mody A, 1990 *Learning Through Alliances* (World Bank, Washington, DC)

Nevis E, DiBella A, Gould J, 1995, "Understanding organizations as learning systems" *Sloan Management Review* **37** 235–250

Pedler M, Burgoyne J, Boydell T, 1991 *The Learning Company* (McGraw-Hill, New York)

Piaget J, 1985 *Meine Theorie der geistigen Entwicklung* (Fischer Taschenbuch Verlag, Frankfurt am Main)

Porter M E, 1985 *Competitive Advantage: Creating and Sustaining Superior Performance* (Free Press, New York)

Romer P, 1986, "Increasing return and long-run growth" *Journal of Political Economy* **94** 1002–1037

Rothwell D, 1992, "Successful industrial innovation: critical factors for the 1990s" *R&D Management* **22** 221–239

Saxenian A L, 1996 *Regional Advantage: Culture and Competition in Silicon Valley and Route 128* (Harvard University Press, Cambridge, MA)

Senge P, 1990, "The leader's new work: building learning organizations" *Sloan Management Review* **32** 7–23

Senge P, Kleiner A, Roberts C, Ross R, Smith B (Eds), 1995 *The Fifth Discipline Fieldbook—Strategies and Tools for Building a Learning Organization* (Nicholas Brearley, London)

Senker J, Sharp M, 1997, "Organizational learning in cooperative alliances: some case studies in biotechnology" *Technology Analysis and Strategic Management* **9** (1) 35–51

Simon H, 1993 *Industrielle Dienstleistungen* (Schäffer-Poeschel, Stuttgart)

Staehle W, 1991 *Management* (Vahlen, München)

Weisboard M, 1987 *Productive Work-places* (Jossey-Bass, San Francisco, CA)

Williamson O E, 1985 *The Economic Institutions of Capitalism* (Free Press, New York)

Clusters: Less Dispensable and More Risky than Ever

G Tichy
Austrian Academy of Sciences, Vienna

1 Introduction

Since the earliest history of man clusters of economic activity have proved indispensable for economic success. Chinese silk or Persian carpets, Teutonic swords or trade services of the Hanse, English machines or German dies, Swiss watches or Württemberg's machine tools are examples of regional specialisation in the form of clusters—regional networks of interlocking activities, based on sophisticated systems of division of labour among skilled specialists. Today national innovation systems, science-based industry, concentration on core activities, and outsourcing are increasing the importance of clusters even more. The nightmare of old industrial areas with deserted plants and unemployed workers, however, illustrates the end of at least some of the clusters that began in such a promising way. To some extent the same factors that cause the success of young clusters may cause the demise of old clusters. I will try to investigate these factors to find a medicine to prevent clusters becoming old or a medicine for their rejuvenation.

2 Why clusters are less dispensable today and more risky

Marshall (1920) elaborated three causes for the formation and the superiority of clusters: labour-market effects, input–output interdependence, and technology effects (that is, spillovers).

Labour-market advantages result from the fact that the firms of the cluster demand similar qualifications from their employees. Firms profit from the existence of a labour-market cluster in two ways. First, they need not train employees themselves, but can buy them from a market for skilled labour; such a market can only exist, however, when supply and demand for specialised qualifications are high. This allows the cultivation of specialised qualifications and training for specialised skills in specialised institutions. Formal and informal contacts between specialists will improve the skills further if a critical mass of firms and employees can be reached. Second, firms will profit from technology spillovers, resulting from the transfer of skilled heads from other firms. Workers profit from learning effects as well—higher qualifications result in higher wages—and they profit additionally from thick-market externalities, when the market for their skills is large. To sum up: the labour-market advantage of a cluster is the two-sided network that defines, creates, and supplies skills and produces thick markets and externalities.

That today's highly specialised high-skill industries can buy specialised skills in an efficient labour market is more important than ever before. Even if it pays to educate and train staff within a firm, additional demand for qualifications can be served only after a time-consuming training period, which heavily constrains the firm's flexibility. The concentration of skills in the cluster areas, therefore, is a huge competitive advantage. The theory of the learning curve and of learning by doing (Lucas, 1988) applies to clusters even more strongly than to firms.

The second advantage of clusters, *input – output interdependence*, enables firms to outsource work (processes) which others can execute more efficiently. The higher efficiency of those suppliers rests on two pillars: economies of scale and economies of specialisation. Both result from the concentration of demand for intermediate products within the cluster. Therefore a cluster profits from a second network, a network of exchange of intermediate goods and services, relying on an information network as the basis for this complicated but highly efficient division of labour between firms.

Input – output interdependence was highly important in the preindustrial clusters mentioned in the introduction, but got widely lost in the era of concentration of industry and of conglomerate mergers. Today, however, specialisation and outsourcing are again keywords in modern macroeconomics as well as in business economics. New growth theory emphasises the (external) economies of scale resulting from the existence of highly specialised suppliers (Romer, 1987), and the most modern business economics (over)emphasise concentration on core activities, outsourcing, and just-in-time production (Hammer and Champy, 1990; Roever, 1991). All this implies that a firm's competitiveness relies heavily on the existence of a cluster surrounding it.

Technology spillovers, the third component of clusters, result from the existence of R&D specialists in the region, which can exist because of the concentrated demand for their services, the mobility of these skilled specialists between firms, and from the presence and exchange of tacit knowledge. So a third network, a technology network, supplements the labour-market network and the input – output network. Any one of these three networks supports and improves the other two.

The technology network has become the most important constituent of clusters today. New growth theory emphasises the importance of research for economic growth (Grossman and Helpman, 1991; Romer, 1990), and the modern theory of innovation has scrapped the old linear model (Myers and Rosenbloom, 1996). Innovation needs an integrated and interactive approach that blends scientific, technological, socioeconomic, and even cultural aspects with organisational capabilities in rapidly changing environments. Firm-specific knowledge embedded in its workforce has to be blended with generally accessible knowledge. But this knowledge has to be detected—information may be an almost free good, but not knowledge. This integrated and interactive approach to innovation is facilitated by geographical proximity and frequent face-to-face contacts between people with similar interests, that is, by the existence of a cluster. Clusters can therefore be described as learning organisations (Coombs et al, 1996) and this is completely consistent with the fact that new technologies also arise and diffuse in clusters (National Association of Manufacturers, 1994). Important new technologies cannot be applied step by step but only in a big bang; this gives greenfield investments a head start over existing plants and restricts the adaptability of clusters.

One should mention a fourth factor for the importance of clusters today, namely the *new organisational principles of firms*. The increasing complexity of products and processes, the compulsion to innovate continuously, and the pressure of financial markets to maximise shareholder value sky-rocketed the complexity of management and forced firms to concentrate on their core activities. To be viable, however, such a strategy requires the firm to be embedded in a cluster. Today's clusters are a substitute for yesterday's hierarchical firms, which have proved to be unmanageable. If a cluster works well, firms can profit considerably from the R&D expenditures of other firms (Goto and Suzuki, 1989; Griliches, 1992). Own research and a good network enable firms to utilise foreign R&D as well (Coe and Helpman, 1993; Hutschenreiter, 1995). Torstenson (1997) estimates that domestic research (in Canada) yields markedly higher returns than foreign research, but that the Canadian growth rate nevertheless was boosted by half a percentage point through the transfer of foreign innovations.

This list of arguments demonstrates that all of Marshall's arguments are still applicable today. Furthermore, their importance has increased. The intensive specialisation of industries in the innovation-prone high-tech world of the fading millennium and the organisation of firms make clusters indispensable. Another factor, however, has to be added to Marshall's three advantages: the globalisation of research and the country's R&D policies foster the formation of specialised monopolistic clusters. In previous times clusters formed as a consequence first of accumulation, and later as a consequence of the availability of skills of the population. Today's innovation-dependent highly specialised firms need more: universities, research institutions, specialised suppliers not only of goods but of a wide range of services (legal, software, market research, etc). Clusters therefore tend to specialise in increasingly smaller fields. This inherent tendency is fostered by governments' attempts to build competitive national innovation systems (Nelson, 1993), based on the specialisation of their countries in fields in which they can dominate the market. Government attempts to promote the formation of clusters include national research priorities, research programmes, and science parks.

Clusters are therefore more important for high-tech economies than ever before. They allow workers, firms, research institutions, and countries to profit from specialisation, and the large integrated markets allow more division of labour than ever before. Specialisation increases efficiency but it increases risk as well: a specialist is highly vulnerable to shifts in demand or to innovations rendering his or her skills valueless. If this occurs, the cluster ends. To prevent this and to ensure lasting instead of only temporary profits for clusters, ways must be found to keep a cluster young and flexible.

3 Why clusters tend to age and even go to ruin

That some types of clusters have a limited life span needs no long explanation. Raw-material-based clusters necessarily face problems when the deposits are depleted or the materials are no longer in demand. The coal and steel regions of Europe are some of the best examples of petrified clusters, the transition of formerly dynamic and rich areas to problem areas. Product-based clusters may face problems in a similar way, as the goods produced in the cluster age and are no longer in demand. Clusters based on local skills age, when machines make the skills superfluous or methods are developed to produce the goods with unskilled workers. The chip made precision work and precision tools useless and ruined the machine-tool cluster of Baden-Württemberg, the watch cluster of Switzerland, and several (nondigital) telephone clusters.

The examples show that clusters *can* grow old and petrify. But *must* they go to ruin? Some examples demonstrate that even raw material clusters need not age; they can transform to skill clusters or process clusters. London as a financial cluster is a good example of a successful transformation. It no longer depends on the raw material which gave rise to this cluster, that is the export of own capital; London now supplies skill-based services to foreign capital, to borrowers as well as to lenders. Examples of product(ion) clusters which could prevent their ruin are machine tools in Baden-Württemberg, motor cars around the Great Lakes, or watches in Switzerland. All of them have been 'reclustered' in one way or another.

The evidence suggests that clusters tend to age, to become inflexible, and in the worst cases, unable to adapt to a new environment. Why does this happen? Why can clusters not adapt to new conditions, why can they not develop new products, why can they not adapt their skills, when the old specialisation cannot meet demand any more? The theory of the regional product cycle (Tichy, 1987; 1991) can give an answer. The motto could be: "Learning new skills is easy, unlearning old habits is tough" (Ichniowski and Shaw, 1995).

The theory of the product life cycle (Heuß, 1979) says that goods are created in agglomerations, because of their informational advantages and the concentrated specialised and sophisticated demand. In its early life such a good is produced in the agglomeration in which it was developed. As long as the product is still in its development phase, the production process is (therefore) not yet standardised and can afford the specialised skills of the region. The competitive advantages cause a specialisation of the region; networks of skills, suppliers, and information arise; and the region profits from network externalities. It pays to specialise, and the specialisation advantages drive out other skills; specialisation increases even more. The more the product matures and reaches its final form, the more the production process becomes standardised, and the less necessary are the specialised workers, their skills, and the sophisticated information networks. Low production costs gain importance to allow mass demand. Therefore production is shifted to the cheaper but less skilled periphery and scale economies are utilised.

When the life cycle of the product turns towards its end (that is, when demand stagnates or shrinks), the product either moves to a still cheaper (and even less skilled) region or a process of concentration starts: concentration in the number of competing firms as well as concentration of activities. Klepper and Simons (1996) and Suárez and Utterback (1995) have shown how fast the number of firms shrinks in mature industries[1]. The latecomers, however, die away even faster than the former pioneers, thus revealing advantages of Schumpeterian entrepreneurs. As the number of firms is reduced, sophisticated networks are no longer necessary, as no new information has to be transferred; nor are clusters any longer competitive, compared with vertically integrated firms, as the number of nodes has been drastically reduced. The smaller the networks, however, the less—and the less new and stimulating—information they can provide, the lower therefore the chance of the cluster inventing new products, new processes, or a new organisation. The cluster has aged; the region in which the cluster is located has become a problem area, a region with little endogenous potential to find new dynamics. The older a cluster the more inward-looking it becomes, the more it secludes itself against outside influences, the more it develops 'laager-mentality'.

A most important conclusion is that a cluster is more likely to become a problem area, the more successful it is. The more successful a cluster the more it specialises and concentrates on the relevant product or process, the more it attracts matching skills and drives out others, the more the networks become specialised and lose their ability to transfer other information which would be necessary to develop new products. One can observe that new products, which substitute for old ones, are usually developed and produced by *new* firms. Examples are electronic calculators, computers, and printers, which were not invented by and are not produced by the old office-machine industry; scrap steel neither invented nor produced by the old integrated steel works; heating oil not distributed by the old coal distributers, etc. In all these cases the network of the old cluster transferred only current and outdated information, not the very information central for adapting to the new situation.

Clusters are therefore in danger of petrification, especially if they are successful. This danger becomes the more real, the more the cluster's firms concentrate and the more the networks shrink in the course of this process. A cluster is certainly more dynamic and less likely to petrify than a vertically integrated firm, but old clusters tend towards vertical integration, in an unattainable desire to survive by utilising economies of scale and to save on the cost of inputs. Tendencies for petrification of clusters may arise internally as a result of the dynamics of the regional product cycle (mass production, concentration) or externally as a result of shrinking markets for their products. The latter may be caused by the fading away of former locational

---

[1] In the period since 1910 the number of US car producers shrank from 275 to 3, of tyre producers from 275 (in 1924) to about 30, of television-set producers from 100 (in 1955) to 6, etc (Klepper and Simons, 1996).

advantages, by new technologies, or by shifting demand. In all cases a sclerosis of institutions and persons sets in, of entrepreneurs as well as of workers, of bureaucrats as well as of labour unions. They all tend to do whatever they can to sustain the old structures in the short run, thereby destroying the cluster's basis in the long run. They do not want new people to join their sclerotic club, and thereby they prevent new information and new network nodes.

4 Strategies to avoid the ageing of clusters

Clusters are an important element of our economies' competitiveness and their importance will increase even more in coming years. New management strategies as well as governments' attempts to build national innovation systems both imply concentration on a more and more limited number of activities. In the short and medium term this will increase competitiveness, although at the cost of potential greater risk, resulting from the reduced portfolio. In the long run, however, serious problems may arise, if countries do not succeed in preserving the flexibility of the clusters and a high density of diversified information within the cluster regions. Policymakers for industry and technology must be aware of this problem. However, this does not demand *more* intervention, but *better* intervention, with a sense for the long-term consequences. Some elements of a cluster policy will be sketched below.

4.1 The creation of clusters

A policy which is restricted to rejuvenating old sterile clusters normally comes too late. Policy should try to prevent clusters from becoming sterile, it should help new clusters to form, and it should watch that newly formed clusters are constructed in a way that promises growth potential, flexibility, and a long life. Many clusters arise out of indigenous market forces, but others are the products of policy. Science policy, research policy, technology policy, and economic policy increasingly focus on the promotion of clusters, and if they act unanimously, they have a good chance of succeeding. In some cases it is a distinctive policy goal to *create a cluster* with the help of special institutions (for example, science parks) and project managers, or, at least, to promote its formation.

Economists who believe strongly in the allocative power of markets may object to this apparently new and misplaced task of policy. In fact, it is neither new nor misplaced. More than 150 years ago, in 1841, Friedrich List published his famous book on *Das nationale System der politischen Ökonomie*, proposing networks to disseminate know-how, training of the workforce, and duties, to protect infant industries—all elements to create clusters within the underdeveloped German industry. The pioneer industries of England of course objected and voted for free trade, guaranteeing their lead, thereby revealing implicitly the potential power of List's arguments. Since that time governments have never ceased in helping to form clusters, and modern economics provides the scientific framework, as shown before. New growth theory demonstrates the importance of the acquisition and accumulation of specific knowledge, Krugman (1991) emphasises the likelihood of increasing divergence between core and periphery, if left to free market forces; Freeman (1997), Lundvall (1992), and Nelson (1993), among others, stress the importance of independent national systems of innovation. It is, therefore, not a strange thing for a government to participate in creating a cluster; nevertheless one should be aware that it is not an easy task.

If it is to be done, the government must build on existing strengths. A knowledge-intensive field has to be selected, in which the region is already competitive and in which *several firms* and *more than one research institution* are active; if only one research institution exists, it should be outstanding and, additionally, at least one firm should have a research department of its own. An already proven strength of firms and research activities in the planned cluster and a reasonable number of nodes are both preconditions to secure the necessary mix of information density and output diversity. A third important point is novelty, that is, innovative power. The product or the process at the centre of the cluster should be at an early stage of development, at a stage in which several different routes to success are still open, and the market is expected to grow. Fourth, the government has to refrain from following a sector-based industry approach: *the strengths should cut through different industries*, based on knowledge cutting through different branches and interacting among them (Jacobs and De Man, 1996). Therefore those clusters which rely on cross-section technologies or new process technology rather than on new products have a better chance. A last important point to obey refers to the relations among the members of the clusters. The optimal type is a *network cluster* (Tichy, 1997, page 251): the members cooperate in developing a range of different goods or services, based on the same stock of knowledge, but supplied to different customers in different branches. Definitely more risky is the *star cluster*, characterised by a dominant firm as the dynamic centre, surrounded by a circle of firms. Such a form is easier to create and it may even start very dynamically, but it is in serious danger of overspecialisation and inflexibility caused by one-sided knowledge. *Subsupplier relations* are sometimes called clusters as well, but this is definitely a misuse of the term; they are a poor starting point for a cluster.

Sometimes economic policy promotes the formation of clusters at a rather late stage of the product cycle. The Austrian province of Styria, for instance, straining to overcome a deep regional and structural crisis, tried hard to form a cluster developing and supplying auto parts. The promotion was quite successful. A network was formed between previously unconnected firms, partly situated in an old industrial area. The existing firms expanded, cooperation and specialisation increased, and foreign investment was attracted. To form a cluster of previously unconnected firms and to upgrade the products was surely good policy, particularly as a part of the cluster is based on top know-how in the construction of diesel engines and four-wheel-drive equipment. The traditional car is a mature product, however, and the

Styrian auto cluster is therefore risky and in danger of ageing over time. Policymakers will have to take care that it does not dominate the region, that the region's know-how is not overly concentrated in this special field, and they should try hard to create another cluster at an earlier stage of the product cycle.

4.2 Clusters in the growth phase

The growth phase of a cluster, in most cases identical with the growth phase of the product cycle of the cluster's supply, appears to be the best of all worlds to all participants. It is the phase, nevertheless, which may generate the first deviations which cause later troubles. Success is easy in this phase, so that little pressure exists to search for further development of the cluster's strengths, for other applications of its knowledge, etc. It is tempting to concentrate on the best-selling product and to produce it in ever-increasing quantity, utilising economies of scale. As a consequence economic policy must stand against concentration, try to avoid overspecialisation, and protect the region's information density. It must restrain the region from rushing into the *two-sided cluster trap*: the trap between full specialisation, with deep knowledge in a very small field and paucity of contacts and new ideas as a consequence on the one side, and nonspecialisation on the other side, with a consequent skill and knowledge deficit. In this phase it is most important to prevent the cluster concentrating on its core activities, and to ensure that it retains its diversity and flexibility.

Network components are complementary, but good dense networks allow for substitutes (alternatives). They provide substitutes, made from other complements, as the definition of clusters says: clusters are sets of production and service firms, of research institutions at various levels, and of education and training facilities, offering complementary components that are combined to produce various composite goods, which (normally) are substitutes of each other. At this point, however, a trade-off may arise: compatibility of network components makes complementarity actual, but it may impede flexibility, the adaptation of the network to new problems.

Several policy instruments are available towards these policy goals. To protect the cluster's information diversity government should stop promoting R&D directed towards the cluster's main specialities; rather it should investigate other uses of the dominant skills (and/or products), enabling the cluster to remain diversified and to shift its focus if necessary. The establishment of new research facilities or the promotion of spin-offs may prove appropriate. Competition policy should avoid an early concentration of enterprises, as concentration of enterprises gives rise to concentration of products and reduction of diversity. In addition, it may be feasible to build on the foundation of a new cluster to increase the region's portfolio, as it is extremely dangerous for a region's wealth to rely on one single cluster of activities.

4.3 The critical maturity phase

The likelihood of the petrification of a cluster depends on the nonadaptability of its underlying networks—all three of them. The best *indicator signalling ageing of clusters* is therefore the loss of the cluster's potential to react to market demand and to develop new offers. No easy way exists, however, to measure this potential; one must rely on specific detailed studies. Important aspects could be a shrinking number of firms and a reduction of their input-output relations, the tendency for outsourcing to decrease, a low number of newly founded firms within the cluster's competence, a reduction in the number and diversity of the goods and services produced by the cluster, a late product-cycle stage of the cluster's supply, decreasing price elasticities and income elasticities of the goods produced, the tendency of the relevant markets to shrink, and decreasing reliance on new processes.[2]

If a cluster has reached that stage, action is urgently needed. The cluster region is in serious danger of suffering under an overly restricted portfolio of activities. The instruments available to stem petrification do not differ from those available in the previous stage. It is most important to bring new information into the region. As Casson (1982) emphasises, an entrepreneur is a person who draws unconventional information from unconventional sources, processing it in an unconventional way, to take judgmental decisions about the coordination of scarce resources. As the cluster insiders have proven unable to utilise the cluster's still existing competence to combine complementary components to produce new composite goods, outside research institutions or project managers should be authorised to undertake this task. Most important, however, is the search for new clusters and the acquisition of new skills.

A good example of the problems of this stage is the pharmaceutical and biotechnical industry in Switzerland. It has been a most active cluster for a very long time; its firms have grown large and multinational. The scientific basis of the business has changed, however, in the last two decades: gene technology has gained in importance, and at the centre of gene technology are a few star scientists at US universities (Zucker et al, 1991). Swiss firms have therefore shifted their research centres to the United States and produce the most modern part of their portfolio in their foreign affiliates (Hotz-Hart and Küchler, 1996). It is interesting that these Swiss *firms* managed to keep up with the product cycle, but *not the Swiss regional cluster*. Only traditional pharmaproducts are produced in Switzerland—the *regional* cluster is probably on the brink of petrification.

4.4 What to do with petrified clusters?

Petrified clusters are well known as old industrial areas. They are characterised by large vertically integrated firms producing standardised goods at a late stage of their product cycle. During its growth phase the cluster has

---

[2] Clusters based on products are more likely to petrify than those based on process technology.

been successful, has attracted all the available workforce, and dominated R&D, education, and training. All the skills and facilities not matching the cluster have been driven out, and the matching ones have grown old, inflexible, and outdated with the cluster. The product-and-process portfolio has been reduced step by step, as only mass-produced goods can afford the heavy overhead costs of big bureaucratic firms, and the firms lack flexibility to serve market niches. If niches are served at all, the production of these articles is outsourced. As all information networks and skills different from those central to the core activities have been driven out, an indigenous potential for change is lacking. The region needs help from outside, but in most cases the willingness for a thorough change is also lacking, until the cluster is in a serious crisis.

Given this diagnosis, it is trivial to say that rejuvenation of a petrified cluster is a complicated task. Political, sociological, psychological, and economic aspects combine to form an impenetrable mesh. If a solution exists at all it must be tailor-made. As no general rules exist, a successful example will be given.

Upper Styria in Austria is a district which has relied on iron as long as mankind has used it. It was the largest European producer of iron in the 16th century (Tichy, 1994). Its European importance declined steadily as England learned to replace charcoal by coal; its local dominance remained; but the cost disadvantage (no nearby coal) and the mature product enforced concentration, as predicted by theory. The Europe-wide scarcity of raw materials during and after World War 2 caused a big boom which, however, ended in the late 1960s. The problems grew and they were reinforced by the fact that the industry was nationalised. In a last big concentration effort one big company with about 70 000 employees was formed, but could not prevent the demise of the industry (Tichy et al, 1982). After several experiments with holding-company solutions of various kinds, in the end, several small companies were formed. It was up to them to find a way to survive and to prepare themselves for privatisation within a few years. A large restructuring began in a decentralised way. Noncore activities were sold, foreign firms with matching activities acquired, matching firms merged. Today most of the firms are profitable, some of them dominant in market niches.[3]

Breaking up the large conglomerates is evidently *one aspect* of breathing new life into old clusters. But the new Austrian steel cluster employs only half of its previous staff (even with higher production). Restructuring the cluster's firms therefore resuscitates the cluster but not the region. The regeneration of the region must rely on *additional instruments*. The Styrian government choose two. First it made efforts to create an auto cluster (see section 4.1). This action was appropriate as some firms, a few of them even dominant in niches,

[3] One of these firms recently tried to buy a steel plant in one of the petrified steel areas of Germany; because of elections in this area the government prevented the deal and bought the firm, thus demonstrating the difficulties of dealing with the problems of old industrial areas.

already existed, but there was no network linking them and an auto cluster is a good customer of steel and steel products. The second action was to reorganise economic promotion as an agency independent in its operative actions, and to redirect promotion on high technology (Geldner, 1995). The rejuvenation was a hard business and it took three decades (Geldner, 1998). Today production and employment in the region are growing faster than in the rest of Austria.

References
Casson M 1982 *The Entrepreneur: An Economic Theory* (Martin Robertson, Oxford)
Coe D, Helpman E, 1993, "International R&D spillovers", WP-444, National Bureau of Economic Research, Washington, DC
Coombs R, Albert R, Saviotti P P (Eds), 1996 *Technological Collaboration: The Dynamics of Co-operation in Industrial Innovation* (Edward Elgar, Cheltenham, Glos)
Freeman C, 1997, "The 'national system of innovation' in historical perspective", in *Technology, Globalisation and Economic Performance* Eds D Archibugi, J Michie (Cambridge University Press, Cambridge) pp 24 – 49
Geldner N, 1995, "Evaluierung der steirischen Wirtschaftsförderung", Studie im Auftrag der Steirischen Wirtschaftsförderungsgesellschaft mbH, Wien
Geldner N, 1998, "Erfolgreicher Strukturwandel in der Steiermark" *Monatsberichte des Österreichischen Instituts für Wirtschaftsforschung* **71**(3) 167 – 172
Goto A, Suzuki K, 1989, "R&D capital, rate of return to R&D investment and spillover of R&D in Japanese manufacturing industries" *Review of Economics and Statistics* **81** 555 – 564
Griliches Z, 1992, "The search for R&D spillovers" *Scandinavian Journal of Economics* **94** supplement 29 – 47
Grossman G M, Helpman E, 1991 *Innovation and Growth in the Global Economy* (MIT Press, Cambridge, MA)
Hammer M, Champy J, 1990 *Reengineering the Corporation* (Harper and Row, London)
Heuβ E, 1979, "Wettbewerb", in *Handwörterbuch der Wirtschaftswissenschaften* (Gustav Fischer, Stuttgart) pp 679 – 697
Hotz-Hart B, Küchler C, 1996, "Das Technologieportfolio der Schweizer Industrie im In- und Ausland" *Schweizerische Zeitschrift für Volkswirtschaft und Statistik* **132** 317 – 334
Hutschenreiter G, 1994, "Cluster innovativer Aktivitäten in der osterreichischen Industrie, tip-Studie, Wien", Österreichisches Institut für Wirtschaftsforschung im Auftrag des Bundesministerium für Wissenschaft und Forschung und des Bundesministerium für Öffentliche Wirtschaft und Verkehr, Wien
Hutschenreiter G, 1995, "Intersektorale und internationale 'F&E-Spill-overs': Externe Effekte von Forschung und Entwicklung" *Monatsberichte des Österreichischen Instituts für Wirtschaftsforschung* **68** 419 – 427
Ichniowski C, Shaw K, 1995, "Old dogs and new tricks: determinants of the adoption of productivity-enhancing work practices", Brookings Papers on Economic Activity: Microeconomics, 1 – 65, The Brookings Institution, Washington, DC
Jacobs D, De Man A P, 1996, "Clusters, industrial policy and firm strategy: a menu approach" *Technical Analysis and Strategic Management* **4** 425 – 437
Klepper S, Simons K L, 1996, "Technological extinctions of industrial firms: an inquiry into their nature and causes", mimeo; copy from K L Simons, Royal Holloway, University of London, Egham, Surrey
Krugman P, 1991 *Geography and Trade* (MIT Press, Cambridge, MA)
List F, 1841 *Das nationale System der politischen Ökonomie* published in English in 1885 in *The National Systems of Political Economy* (Longman, London)

Lucas R E, 1988, "On the mechanics of economic development" *Journal of Monetary Economics* **22** 3–42

Lundvall B A (Ed.), 1992 *National Systems of Innovation: Towards a Theory of Innovation and Interactive Learning* (Frances Pinter, London)

Marshall A, 1920 *Principles of Economics* (Macmillan, London)

Myers M B, Rosenbloom R S, 1996, "Rethinking the role of industrial research", in *Engines of Innovation* Eds R S Rosenbloom, W J Spencer (Harvard Business School Press, Boston, MA) pp 209–228

National Association of Manufacturers, 1994, "Technology on the factory floor II, National Association of Manufacturers, 1331 Pennsylvania Avenue NW, Washington, DC 20004-1790)

Nelson R, 1993 *National Systems of Innovation: A Comparative Analysis* (Oxford University Press, Oxford)

Roever M, 1991, "Tödliche Gefahr Überkomplexität" *Manager Magazin* **10** 218–233

Romer P M, 1987, "Growth based on increasing returns due to specialization" *American Economic Review* **77**(2) 56–62

Romer P M, 1990, "Endogenous technological change" *Journal of Political Economy* **98**(5) S71–S102

Suárez F F, Utterback J M, 1995, "Dominant designs and the survival of firms" *Strategic Management Journal* **16** 415–430

Tichy G, 1987, "A sketch of a probabilistic modification of the product-cycle hypothesis to explain the problems of old industrial areas", in *International Economic Restructuring and the Regional Community* Ed. H Muegge, W Stöhr (Avebury, Aldershot, Hants), pp 64–78

Tichy G, 1991, "The product-cycle revisited: some extensions and clarifications" *Zeitschrift für Wirtschafts- und Sozialwissenschaften* **111**(1) 27–54

Tichy G, 1994, "Vom vergangenen, laufenden und künftigen Strukturwandel: zur Dynamik der Steirischen Wirtschaft", in *Steirische Statistiken 38* (Amt der Steiermärkischen Landesregierung, Graz) pp 13–36

Tichy G, 1997, "Cluster-Konzepte: ihre Bedeutung für die österreichische Wirtschafts- und Technologiepolitik" *Wirtschaftspolitische Blätter* **44**(3–4) 249–256

Tichy G, Österreichisches Institut für Raumplanung, Österreichisches Institut für Wirtschaftsforschung, 1982, "Regionalstudie Obersteiermark", study commissioned by the Chancellor and Government of Styria, Österreichisches Institut für Raumplanung, Vienna

Torstenson R M, 1997, "Growth, knowledge transfer and European integration", mimeo; copy from Department of Economics, Lund University, Box 7082, S-22007 Lund, Sweden

Zucker L, Darby M, Brewer M, 1991, "Intellectual capital and the birth of U.S. biotechnology enterprises", WP-4653, National Bureau of Economic Research, Washington, DC

Promoting Competitiveness for Small Business Clusters through Collaborative Learning: Policy Consequences from a European Perspective

J Cullen
Tavistock Institute, London

1 Introduction
Interest in organisational learning is nothing new. A recent development, however, has been the increasing interest in how small businesses engage in pooling know-how in order to promote collective learning and increase the competitiveness of the local economy. This focus on 'clusters' of small and medium-sized enterprises (SMEs) as learning organisations draws together several strands of thinking, incorporating approaches from the domains of regional development, pedagogic research, and vocational training. A common strand connecting these different disciplines has been the notion of 'sensemaking', or how the different world views of different actors within a bounded organisational or spatial setting merge together to form a common set of meanings, as the building blocks for 'organisational learning'.

Conceptualisations of the learning organisation have evolved significantly since the pioneering work of Follett in the 1940s, and that of the organisational development movement of the 1960s, reflecting a shift from conventional 'second wave' management thinking about organisational learning—the so-called 'MBA' approach, with its emphasis on linearity, quantification, and SWOT (strengths, weaknesses, opportunities, threats) analysis—towards a more constructivist approach that draws on a range of disciplines and perspectives, including the psychoanalytical (Starkey, 1998). Kets de Vries (1995), for example, using concepts such as the 'transitional object' developed by Winnicott to explain how children develop relationships between 'inner and outer realities', forcefully argues the case for fostering a 'creative' and 'playful' environment in organisations, in order to help them develop flexible options that are adaptive to sudden change. This widening of choices involves more than optimising rational decisionmaking, because it emphasises self-reflection and downplays pre-determination, or working to plan. Kets de Vries argues that organisational failure is seldom caused by lack of information about the latest techniques but by dysfunctional leadership. More recently, therefore, organisational learning has been seen as the key to making organisations more responsive to change, with an emphasis on management by participation rather than intimidation. According to Dixon (1994), organisational learning embodies a virtuous cycle with distinct stages. Stage 1 encompasses the collection of external data and the internal development of new ideas relating to both product and process. Stage 2 involves integrating this information into the organisational context on a 'total system' basis. Stage 3, the key stage, is about collectively interpreting shared information. Crucially, Dixon argues that organisational learning relies on the collective interpretation of meaning, rather than the insights of experts. This 'bottom-up' approach reflects the influence of theorists such as Vygostky, who saw learning as the product of sociocultural 'forms of mediation'. More recent approaches, reflected, for example, in the work of Grabher (1993), Senge (1990), and Weick (1995) depict the learning organisation as a dynamic human process, anchored in 'communities of interaction' that contribute to the amplification and development of learning and the creation of new knowledge.

This interest in the social construction of meaning within the organisational learning dynamic has also shaped understandings about regional and local development. There is an increasing recognition within the literature that the 'character' or distinctiveness of local economies reflects a process of merging together of 'local' and 'global' identities. In the influential work published by Amin and Thrift (1995), for example, a range of authors are quoted, each of whom have different interpretations of what globalisation is. They point out that globalisation has been alternatively described as a heterogeneous process of "intersecting universal and local narratives" (King, 1991), a 'space of flows' (Castells, 1989), as completely borderless and uniform (Ohmae, 1990), but also as a "necklace of localized production districts" (Storper, 1991). Furthermore, they argue, although there is evidence to support the view of a continuing process of global and cultural integration, shaped by powerful transnational companies, producing the simultaneous production and consumption of the same products and images around the world (Chesnaux, 1992), the global economy "continues to be constructed in and through territorially bounded communities".

The significance of local identity in globalisation does not necessarily mean that the 'local' itself constitutes 'sameness'. Thus Amin and Thrift give weight to Massey's (1993) contention that places do not have single pregiven identities, but are constructed out of emerging multiple social relations, and should be seen as "shared spaces ... riven with internal tensions and conflicts" (Amin and Thrift, 1995, page 5). Conflicts appear to be transformed into shared identities and competitiveness through the channeling into economic advantage of what Amin and Thrift refer to as the 'institutional thickness' of a locale. This 'institutional thickness' embodies social and cultural factors that live at the heart of economic success, and encompasses factors such as: institutional persistence (how local institutions are reproduced); the construction and deepening of an archive of commonly held knowledge of both the formal and the tacit kinds; institutional flexibility (the ability of organisations in a region both to learn and to change); the ability to extend trust and reciprocity; and the consolidation of a sense of inclusiveness (a widely held common project which serves to mobilise the region with speed and efficiency). Such attributes are very similar to those attributed to the 'learning organisation'.

How and why do small businesses that are frequently in situations of intensive competition against each other act collectively as learning organisations? One obvious reason is that collaborative pooling of resource by SMEs leads to significant economies of scale in learning terms. Although European SMEs account for 99% of the approximately 16 million European enterprises, and some 71% of European employment (DELOS, 1997; Eurostat, 1995), a number of studies indicate that they find it difficult to access, absorb, and apply information on the latest developments in their economic sector. Such studies also point to problems facing SMEs in formulating and applying relevant training strategies to suit their needs and the needs of European business generally, and in identifying and seeking to address 'skills gaps', such as lack of marketing and business management know-how, in their operations (DELOS, 1998). In addition, knowledge acquisition, skills development, and training tends to be idiosyncratic, rather than defined by a coherent human resource development strategy. This is mainly because of the role played by entrepreneurs in decisionmaking. As a recent study carried out by the Business School at Warwick University (Hendry et al, 1991), demonstrates "in positioning, building and sustaining the SME, it is personal networks that count" and, furthermore, "any outside help has to recognise that the entrepreneur is the central figure, both in the achievements and problems of the SME" (page ii).

Recent thinking around improving the competitiveness of SMEs [for example, the European Commission White Paper on growth, competitiveness and employment (EC, 1996)] has thus highlighted the need to develop cooperation between enterprises as a means of achieving economies of scale, in order to combine flexible production with continual innovation of products. In this regard, interest has focused on how certain clusters of SMEs concentrated in specific geographical locations (for example, the North Italian textile industry) achieve growth through strong integration between local institutions, service centres, education centres, and enterprise, fostering the development of a favourable innovation climate. Such clusters, it is argued, act as aggregated learning organisations, in which individual (and competing) SMEs collaborate in the pooling of knowledge, professional updating, and skills transfer in order to maximise competitive advantage for the cluster as a whole.

Cooperation between organisations within markets has long been identified as a factor in economic success. As Alter and Hage (1993) argue, networking between organisations contributes to stability and reduces uncertainty. Such networks can evolve over time—as 'natural' clusterings of firms, or can be 'induced' artificially as a result of interventions such as the development of business or science parks. However, there is little consensus on how industrial clusters can be defined and mapped. Some approaches [for example, those influenced by the work of Marshall (1920)], consider the structural relationships of networks, focusing on three main types of cluster: those generated through labour-market effects, those shaped by supply relationships, and those emerging as a result of the transfer of information between firms and research and development institutions (technological spillover). Others [for example, those influenced by the work of Porter (1990)] place more emphasis on the institutional nature of interorganisational arrangements, and particularly the extent to which collaborative 'learning' is facilitated through institutional frameworks imposed from above, or at a 'grass roots' level. As with SMEs individually, it is assumed that the nature of interrelationships within particular clusters will be shaped by features such as how long industrial districts have been established, their 'embeddedness' within the industrial milieu; existing institutional arrangements (for example, the prevalence of training infrastructure), and market sector and position.

Against this background, I explore the role of SME 'clusters' in promoting collaborative interaction between small businesses, and critically assess how far notions of the 'learning organisation' can be applied to such collaboration. Drawing on the results of recent research work in this field, for example the DELOS project[1], which was funded under the European Commission Targeted Socio-Economic Research Programme (TSER), I present a typology of organisational learning at the level of the individual SME organisation. Broadening the discussion to the level of the industrial district, or SME 'cluster', I then present a typology of cluster networks, and examine whether different types of cluster are associated with different types of organisational learning. Finally, I consider the policy implications of organisational learning within clusters, with particular reference to local and regional development.

2 Organisational learning within the SME

Across the sector as a whole, it needs to be said that small businesses are not the ideal setting in which to cultivate organisational learning. From a number of studies, including a 'baseline survey' of decisionmakers in over 320 SMEs in six different European countries, and intensive case studies of economic, sociocultural, and organisational learning processes and outcomes in different European 'SME clusters'[2], clear evidence emerges of a primarily 'crisis-management' culture dominated by the entrepreneur. For example, in the study of 320 SMEs carried out by the DELOS project, over 55% of those surveyed described the decisionmaking style of the firm as 'entrepreneurial', compared with the 40% of firms engaged in 'participative' decisionmaking, for example in terms of cross-job problem-solving. If we look at the evidence

[1] DELOS was a collaborative research project under the TSER Programme, financed by the European Commission, DGXII. Coordinating partner was IGT, Italy. Other partners included Tavistock Institute, United Kingdom; Joanneum Research, Austria; Formit, Italy; ECWS, The Netherlands; CCI, France; Infyde, Spain.
[2] These include old-established North Italian textile districts such as Biella, new industrial development in the Mirandola biomedical industry; established metals and engineering clusters such as automotive industrial districts in the UK West Midlands and in the Basque country; metals in Upper Styria, Austria; 'intelligent networks'; and development agency initiatives in Picardie, and East Brabant, Holland.

overall, three main 'levels' of organisational learning within the SME sector can be identified (table 1):
(1) information-gathering,
(2) knowledge acquisition,
(3) competence consolidation and development.

These levels broadly reflect the strategic-positioning, change-and-adaptation, and strategic-management forms of learning identified in the Warwick study described above. Information gathering can be conceived of as 'low-level' data monitoring, acquisition, and management, intended to ensure that the firm remains aware of changes and developments in the markets in which they operate. Knowledge acquisition can be defined as the process whereby firms define and acquire the skills, know-how, and strategic intelligence necessary to carry out day-to-day activities. Competence consolidation and development can be seen as the process whereby existing information and knowledge are converted into 'learning', through identifying skills deficits and acquiring new knowledge through training and collaboration.

Table 1. Classification framework for organisational learning in SMEs.

Element	Subelement	Key sources
Information gathering	Strategic	Technical magazines Exhibitions and fairs Data banks Seminars and training sessions Government sources R&D centres and universities Buying in research
	Informal	Local business associations Chambers of Commerce Other firms Informal learning Clients
Knowledge acquisition	In-firm: adopting predominantly on-the-job learning and mentoring inside the firm	Mentoring On-the-job experience Experts in the firm Incorporation of new workers
	External: adopting more conventional and formal learning processes	Sending staff to training courses Mobility initiatives from outside
	Community: predominantly drawing on expertise from within the local milieu	Mobility initiatives inside the industrial district Collaboration with other firms
Competence consolidation and development	Formal learning	Technical education Postgraduate specialisation Continuous training
	Informal learning	Family mentorship On-the-job learning Informal relations with businesses

In practice, the utilisation by SMEs of these forms of learning varies considerably. Table 2, which shows the results of the DELOS SME survey, illustrates the relative representation of the three levels of learning practised by SMEs. It suggests that SMEs are most actively involved in knowledge-acquisition activities, that is, acquiring and implementing the skills necessary to conduct day-to-day business activity. This type of organisational learning is primarily focused on using external sources (for example, sending staff on training courses), but there is a significant component of collaborative interaction with other SMEs within the surrounding industrial district. Participative knowledge acquisition (for example, cross-job problem-solving within the firm) is much less significant than external or community-based learning. In contrast, both basic information gathering (low-level monitoring of what is going on and opportunities within the industrial sector) and high-level competence consolidation (acquiring new skills, leading-edge R&D, and future-oriented innovation) are considerably less developed within SMEs. These findings would therefore appear to reinforce the picture painted by other research studies of SMEs as crisis-driven and reactive, rather than proactive learning organisations.

Table 2. Relative representation of forms of learning in SMEs.

Element	Low score (%)	Average score (%)	High score (%)	Top score (%)
Information gathering				
strategic	42	33	19	6
informal	28	38	25	9
Knowledge acquisition				
in-firm	0	3	59	38
external	1	2	29	68
community	1	1	13	86
Competence consolidation				
informal	32	35	14	19
formal	39	37	10	14

As a result, it is likely that SMEs on the whole lack expertise in a number of key competence areas, particularly 'core' skills and cross-job skills. Generic core skills among SMEs are concentrated in production activities. Skills in innovation and in marketing are less highly developed. Almost a quarter of the firms sampled in the DELOS survey did not carry out any form of auditing process to identify gaps in knowledge and competencies within the firm. Those that did tended to use informal methods of auditing, either in-house or by consulting with other firms in the area.

3 An organisational learning model at the SME level

Looking at the ways in which SMEs acquire and apply information and knowledge, and the ways in which information and knowledge are harnessed to strategic decisionmaking, it is hard to argue a case for the widespread utilisation of organisational learning within the European SME sector. What we find instead is a range of different organisational behaviours that may incorporate some of the elements attributed to the learning organisation but which are mediated by structural features, such as the size and age of the firm, and sociocultural features related to the characteristics of the 'locale' in which it operates. There is clear evidence, for example, that the larger the firm, the more likely it is to utilise strategic methods of information gathering, such as attendance at conferences, participation in trade fairs, and buying in research. Micro-enterprises, probably for reasons of cost, tend not to invest in formal training, such as technical education, whereas larger firms are significant users of training services. Organisational learning is also shaped by the decisionmaking or leadership 'style' of the SME, and by the degree to which it is embedded in local networks. Firms that are dominated by an entrepreneurial style of management tend to be significantly less likely to invest in external training for competence development. Similarly, the more 'embedded' the firm is within its local industrial district, and the more collaborative its relations and networks with other local firms, the less likely it is to invest in outside training. Market positioning is also a factor in shaping the nature of organisational learning within SMEs. The evidence suggests that SMEs which form part of a larger organisational network (for example, as a subsidiary of a larger company) are likely to draw on the resources of the parent or associated organisation to support their information-gathering activities.

On the basis of the evidence available, it is possible to suggest a typology of organisational learning within SMEs that reflects interactions between structural and sociocultural characteristics. The typology, shown in table 3 (over), identifies five main types of firm, and reflects how organisational learning within SMEs is shaped according to the interrelationship between information-gathering, knowledge-acquisition, and competence-development behaviours, and the relationship between the SME and its industrial milieu, and other structural characteristics, such as size, length of time established, and decisionmaking 'style'.

The first type, which might be described as *crisis-driven* in fact exhibits little evidence of organisational learning behaviour. Information-gathering practices, knowledge-acquisition strategies, and competence development appear to be either absent or rudimentary, and the firm typically responds to challenges and opportunities rather than pursues an active policy of human resource development and strategic management. This category shows a high representation of very small enterprises and new start-ups, whose decisionmaking strategies are typically shaped by a dominant personality—usually the entrepreneur. The evidence suggest that this type of firm constitutes the largest category of SME.

Table 3. Types of SME organisational learning.

Type	Information gathering	Knowledge acquisition	Competence development	Structural features
Crisis-driven	Unsystematic	Reactive	Not prioritised	Microenterprises New firms Entrepreneurial Disengaged from industrial milieu
Endogenous	Unsystematic	Mentoring On-the-job experience Buying in New workers	Not prioritised	Larger firms Disengaged from industrial milieu
Exogenous	Unsystematic	Externalised (mainly training courses)	High level of formal learning (continuous training)	Opportunistic use of local networks
Embedded—information centred	Strategic: exhibitions links with R&D centres	Unsystematic	High level of informal learning (family mentorship, links with local firms)	Close links with local networks Highly embedded in industrial milieu
Embedded—competence centred	Informal: Chambers of Commerce Other firms	Unsystematic	More formalised competence development	Highly embedded in industrial milieu Recently established

The second type of firm identified might be described as *endogenous* because learning within the firm is focused on knowledge-acquisition processes and behaviours, rather than information gathering or competence development, and these are derived from in-house practices rather than bought in from external sources, or supported by community networks. In this context, knowledge is acquired and utilised primarily through mentoring, on-the-job experience and head-hunting of appropriate qualified personnel.

In contrast, the third type of firm—the *exogenous* type—though operating outside the margins of its industrial milieu, is outward rather than inward looking and draws on external sources of expertise for developing its skills base. In this case, strategic management practices focus on systematic competence development on a continuing training basis and using specialised training providers.

The last two types of firm are highly *embedded* within the local industrial milieu, and use community-based networking for intelligence gathering, acquisition of new knowledge, and consolidation and enhancement of skills.

The distinguishing feature of the first type is the limited development of organisational learning. Strategic practices are largely confined to information gathering, whereas higher level organisational learning—competence consolidation and development—is rooted in the utilisation of community networks, family relationships, and informal networking with other businesses. In contrast, learning behaviour in the second type is primarily focused on competence development using formalised practices and processes (for example, technical education and specialised external training).

4 Organisational learning at the level of the SME 'cluster'

Can we extrapolate the types of organisational learning that appear to occur at the level of the firm to the level of the industrial district or SME 'cluster'? In order to look at learning at this aggregated level, it is necessary to focus on how organisational infrastructures within 'clusters' serve to encourage organisational learning at the aggregate level. Such structures can be best seen as boundary-spanning or bonding agents that serve to consolidate collaborative interactions between individual SMEs within particular clusters, and encompass many different forms, from regional authorities through to Chambers of Commerce. Ultimately, therefore, an assessment of aggregated organisational learning needs to consider interactions between structural, institutional, and organisational features of SME clusters and to assess whether such configurations—what might be termed the 'learning environment'—could be equated with particular patterns of organisational learning—what might be termed 'learning practices'—at the aggregated level of the cluster.

Organisational learning at the SME cluster level can thus be considered in terms of the three key dimensions shown in table 4. On this basis, five types of 'cluster configurations' can be identified.

The *porterian* type of cluster is situated in a clearly defined industrial milieu, which has well-grounded historical roots and a highly developed cultural identity [figure 1(a)]. The territorial cohesiveness of the cluster reinforces

Table 4. Key elements in organisational learning within SME clusters.

Dimension	Main elements
Industrial milieu	Social and cultural identity Degree of concentration or differentiation of activity base Historical and cultural links between firms and sociocultural context Degree of spatial 'boundedness'
Morphological dimension	Network structure (hierarchical, dispersed, central actor) Extent of regional or sectoral policies supporting networks Commonality of competence structure
Organisational infrastructure dimension	Cluster or network management structure Involvement of SME representation in policymaking and decisions

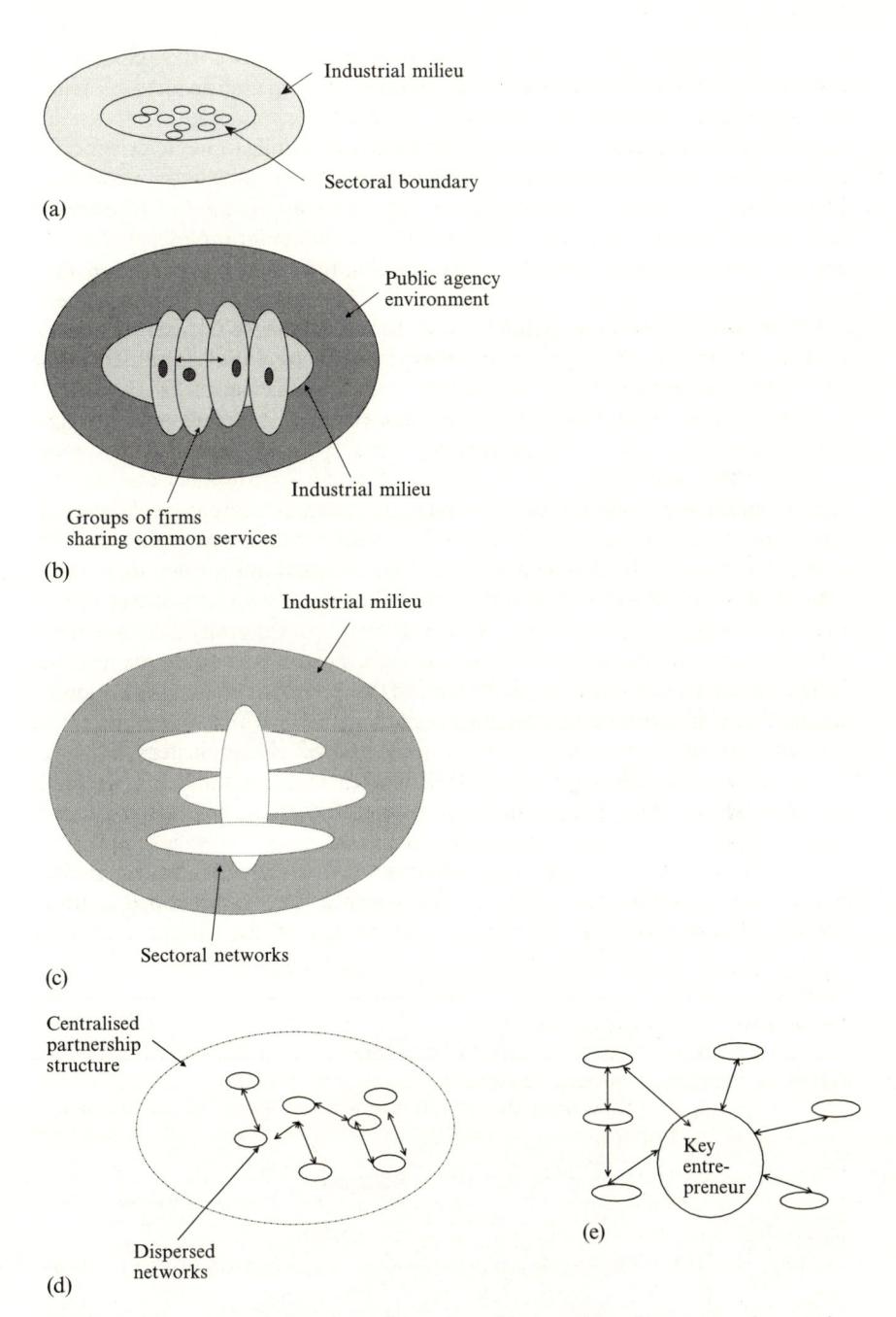

Figure 1. Five types of cluster configuration: (a) porterian, (b) segmented porterian, (c) interlocking, (d) induced partnerships, (e) virtual cluster.

and is reinforced by sectoral homogeneity that provides for collaborative networking between SMEs working in similar markets and production relations. Governance structures tend to be flexible and spontaneous.

The *segmented porterian* type of cluster shows similar characteristics to the first type, in that it occupies a well-defined sociocultural setting with a strong sense of local identity [figure 1(b)]. However, interactions between SMEs within the cluster are also shaped by differentiation in producer–supplier relations and in different market positions and niches within the market. Networking is therefore characterised by loose associations grouped around a central actor, professional associations, or a common service base. Governance structures and communication systems are more formalised than in the first type and will typically take the form of participatory management structures supported by local agencies, autonomous management through professional associations, or partnership structures administered by local authorities.

The *interlocking* type of cluster is spatially bounded, but its territoriality is not derived from a particularly anchored sociocultural identity [figure 1(c)]. Rather, constituent firms working within the cluster have forged links as a result of common interests related to their particular positions within a complex local economy. As a result, this type of cluster is sectorally differentiated rather than monosectoral, and its networking arrangements consequently are diverse, ranging from loose interest groups formed primarily for promotional purposes through to professional associations within a common project.

The main characteristic of the *induced* type of cluster is the key role played by external (non-community-based) agencies in formulating a common identity, and in coordinating organisational learning within the cluster [figure 1(d)]. Public service actors, such as development agencies, typically provide communications and decisionmaking structures that may also be reinforced and supported through central services (for example, R&D links with business associations).

The *virtual cluster* is primarily characterised by the absence (or low importance) of territoriality as a bonding agent or boundary spanner in the development and sustainability of collaborative networking [figure 1(e)]. In practice, this type of cluster tends to be represented by a national network of family enterprises bound together by a common history and common objectives, with a dominant role played by key entrepreneurial decisionmakers, and focused on a particular spatial 'nerve centre'—the historical origin of the network. Other examples, however, include virtual networks or associations with a common activity base, linked together through information and communication-technology infrastructure.

Research also suggests that the particular learning environments associated with these cluster types (in terms of factors such as the strength and nature of collaborative networks, provision of training infrastructure, etc) tend to promote a particular type of 'organisational learning'. For example, the interlocking cluster develops a shared set of competencies through informal collaborative learning, whereas competence development in the segmented cluster is primarily engineered through networks of training institutions embedded within the industrial district. Table 5 (see over) shows some of the associations that can be identified between the structural and sociocultural features that define a cluster or industrial district and which will shape its institutional thickness, and the type of organisational learning cluster that is likely to be associated with these features.

As table 5 shows, the 'learning capacity' of an industrial district—those factors that shape collaborative interactions between constituent small firms—are reflected in three main dimensions:

(a) territorial characteristics—how culturally homogenous (for example, the strength of collective identity) and economically homogenous (for example, the nature of production relationships between firms) the cluster is;

(b) morphological characteristics—the size base of the cluster, the volatility or stability of its economic base, collaborative structures and networks that promote engagement between firms, and the influence of external institutions such as development agencies;

(c) learning arrangements—the training infrastructure available, the strategies firms use to collaborate, and the key actors who promote and facilitate interaction.

However, although it is possible to show associations between the morphological and territorial features of SME clusters and particular forms of aggregated collaborative learning, it should be emphasised that there is no guarantee that promoting organisational learning in SME clusters will automatically help them to be competitive. A major conclusion from the work reported in this chapter is that there is only weak and in many respects contradictory evidence that the type of cluster in which an SME affects its competitiveness and economic performance. For example, the DELOS study indicated that some highly embedded clusters in North Italy have shown the lowest levels of performance, suggesting that membership of a cluster can in some cases be a barrier to success. The study did, however, indicate differences between levels of performance and the type of cluster. The interlocking type on average shows higher turnover growth than other types, and the segmented porterian type shows the lowest.

It should also be stressed that the concept of the industrial district as a context for the development of organisational learning amongst SMEs is not easily transferable. The evidence strongly suggests that there is no one universal type of 'industrial district' or SME cluster. I propose a typology of five cluster types (or 'organisational learning environments') that are broadly consistent with five types of organisational learning practices. These different types represent different configurations of sociocultural embeddedness, structural features, communications, and collaboration structures and learning arrangements.

Table 5. Factors associated with organisatonal learning in SME clusters.

Cluster type and structural and socio-cultural features	Porterian	Segmented Porterian	Interlocking	Induced	Virtual
Territorial					
Economic homogeneity	High	Differentiated Niche markets	Differentiated	Diversified	High
Cultural homogeneity	High	High	Low	Low	High
Morphological					
Size	Small New	Larger Well-established	Variable	Variable	Small Independent
Market Interaction	Stable Socially shared Entrepreneurial	Stable Loose associations	Dynamic Interest groups	Adaptive Induced	Stable Centralised
Agency effects	Self-governing	Participatory Professional associations	Variable Some local agency involvement	Top-down Agency promoted	Absent
Learning arrangements					
Process	Internal Work based	Acquired relevant external knowledge Strategic	Acquired relevant external knowledge Problem solving	Total quality management Problem solving	Internal Work-based
Mechanism	Informal exchange	Benchmarking Feedback	Benchmarking Feedback	Steering groups Agency based	Informal exchange
Actions	Oral Fora	Training systems	Variable, informal and formal	Databases Information technology systems	Oral Newsletters
Actors	Entrepreneurs Employers' groups	Partnerships Professional associations	Partnerships Professional associations	Agency– SME collaboration	Mentors

5 Conclusions and policy implications

The foregoing discussion poses a number of implications for training and employment policies in favour of SMEs.

First, there is no evidence that 'organisational learning', as reflected in collaboration between networks of SMEs sharing common geographical, cultural, or operational spaces, is a universal phenomenon amongst European SMEs. Nor is there evidence that such 'aggregated learning' will in itself necessarily provide 'added value' for SMEs, in terms of outcomes such as human resource development, strategic market positioning, and economic performance. Thus, training and logistical support policies and initiatives for SMEs need to be carefully targeted rather than generic, to take account of the varying structural features of SMEs, and the different types of learning behaviours they exhibit. Policy instruments developed by the European Commission and member states to facilitate support for SMEs could be more tightly targeted to reflect the different configurations of 'cluster' and learning organisation identified. In addition, training and labour-market observatories currently being developed through EU actions and initiatives, for example the LEONARDO Programme, could be used to capture, analyse, and disseminate rich data on the number, characteristics, and relative strengths and weaknesses of European 'industrial districts'. These data could contribute to further development of the 'cluster typologies' suggested above, and in the longer term to better targeting strategies for Structural Funds, for example.

In this context, three main constituent components of 'organisational learning' need to be targeted in relation to training and support policies: information gathering, knowledge acquisition, and competence consolidation and development. These components imply different training and logistical support capabilities, and should incorporate provision of 'formal' services, together with actions designed to enhance informal networking arrangements. Regional development agencies are in the best position to take a leading role in promoting formal information-gathering actions. This implies the development of distributed databases containing data on conferences, exhibitions, and developments in technology. Informal information-gathering support is naturally within the remit of SME institutions, such as local Chambers of Commerce. Such networks would benefit from assistance to act as 'communication hubs' within a locale or cluster in order to facilitate better communication between local firms and their clients.

By extension, SMEs need to be made aware of the need to balance these three different components in their human resource development planning and management. At present, SMEs are relatively active in knowledge-acquisition activities, but not in lower level market-intelligence gathering or higher level competence development. This underlines the need for awareness-raising campaigns, through policy instruments currently available to the European Commission and member states, aimed at encouraging small firms to consider these aspects of learning, together with the promotion of curriculum development and marketing policies of training support services by regional agencies and SME institutions to reflect the different components of learning.

The entrepreneur is a pivotal figure in decisionmaking, but the evidence suggests that a large proportion of SMEs are in crisis management rather than pro-active learning situations. Because entrepreneurial decisionmaking styles are closely associated with such crisis management, there is a need to encourage SMEs to adopt a more participative style of collective learning. This again highlights a need for awareness raising aimed at encouraging entrepreneurs and key decisionmakers in SMEs to consider 'alternative' forms of decisionmaking and human resources strategies. In addition, SME organisations and local training providers could profitably develop and run training support services aimed at providing key decisionmakers with the management skills necessary to support strategic management and human resource development.

Microenterprises and new start-ups are particularly prone to crisis management, and the lack of a coherent organisational learning strategy. Because this situation is almost certainly associated with lack of resources, it would suggest the need for support services that can provide pooled resources for SMEs. At the European level, such services might take the form of labour-market or sectoral monitoring observatories that could provide resource services to individual SMEs. These services could provide on-line information on local courses available, key contacts, intelligence reports, and links to on-line libraries. At the local level, SME organisation networks and regional agencies could provide the focal point for local resource centres providing libraries, databases of training courses and providers, and similar facilities to European-wide support centres.

There would appear to be significant gaps in the skills capabilities of European SMEs. Small firms tend to concentrate their efforts in developing and enhancing production-based skills, but there is a clear lack of competencies in marketing, and in cross-job skills that particularly needs to be addressed. SMEs appear to operate in general in highly localised rather than sectoral labour markets, which, as the DELOS fieldwork confirms, means they tend to buy in new staff rather than train people. In turn, training appears to be geared to short-term and firm-specific objectives. This highlights the need for awareness-raising campaigns at European and member-state level to encourage awareness of 'skills standards' issues, particularly in the area of cross-job competencies. In turn, regional agencies and SME organisations and training providers need to target and market courses in these skills areas. An integral part of the 'skills gap' for SMEs is the lack of expertise, and practices, in skills auditing amongst SMEs, their support organisations, and regional development agencies. This suggests a role for regional development in carrying out routine skills auditing and monitoring exercises within local industrial districts.

References
Alter C, Hage J, 1993 *Organisations Working Together* (Sage, London)
Amin A, Thrift N, 1995 *Globalisation, Institutions, and Regional Development in Europe* (Oxford University Press, Oxford)
DELOS, 1997 *Deliverable 1.1, 'Clusters of SMEs: A Synthesis'* European Centre for Work and Society, Maastricht
DELOS, 1998 *Deliverable 3.1, 'Modelling and Recommendations'* Tavistock Institute, London
Dixon N M, 1994 *The Organizational Learning Cycle: How We Can Learn Collectively* (McGraw-Hill, New York)
EC, 1996 *White Paper: Teaching and Learning: Towards the Learning Society* EC DGXXII, Brussels
Eurostat, 1995 *Eurostatistics* (Statistical Office of the European Commission, Luxembourg)
Grabher G (Ed.), 1993 *The Embedded Firm: On the Socioeconomics of Industrial Networks* (Routledge, London)
Hendry C, Jones A, Arthur M, Pettigrew A, 1991, "Human resource development in small to medium enterprises", RP 88, Warwick Business School, University of Warwick, Coventry
Kets de Vries M F R, 1995 *Life and Death in the Executive Fast Lane: Essays on Irrational Organisations and their Leaders* (Jossey-Bass, San Francisco, CA)
Marshall A, 1920 *The Principles of Economics* (Macmillan, London)
Porter M, 1990 *The Competitive Advantage of Nations* (Free Press, New York)
Senge P, 1990 *The Fifth Discipline—The Art and Practice of the Learning Organisation* (Doubleday, New York)
Starkey K, 1998, "What can we learn from the learning organisation?" *Human Relations* **51** 531 – 547
Weick K, 1995 *Sensemaking in Organisations* (Sage, London)

Clusters: New Developments in Austria and their Relevance in Economic Policy

W Clement
Industriewissenschaftliches Institut, Vienna

1 Introduction
Soon after the publication of Porter's *The Competitive Advantage of Nations* (1990) the basic ideas were introduced to Austria in a lecture by Porter's coauthor Enwright in Alpbach in 1991. Since that time, a large number of publications and empirical analyses have been written. Meanwhile clusters almost seem to be developing into an economic policy panacea. They can be found in regional initiatives as well as in the national priorities of technology or export initiatives. Quite obviously the cluster concept fills a gap in economic policy in Austria. Since World War 2 the Austrian economic system has been a mixture of relatively strong government intervention, free market system, and corporatism. A mesoeconomic formulation fits here, but not the French variety of the mesoeconomy of filières (de Bandt, 1989; Quelin, 1993), which since the 1970s has developed in the spirit of indicative national economic planning, being a variant of the mixed-economy market system. In this case, economic policy initiatives are being set both at the federal and at the regional level, in order to promote clusters, fields of competency, or strategic alliances. Since the earlier subsidisation policy has been considerably restricted, not least through EU competition law, reorientation is necessary. A 'cluster-oriented industrial policy' (CIP) comes at the right time. Although, admittedly, such cluster initiatives are operated quite pragmatically, a minimum theoretical and empirical foundation is necessary and useful. Stages of such applied cluster developments will be outlined below.

2 Theoretical arguments of cluster formation
In the background of the search for a new direction for indicative structural policy are the new forms of international competition. Regional economic blocs (first and foremost the European Union) are setting new rules of competition policy and negotiate in an increasingly closed manner (OECD, 1991). According to estimates of the UNCTAD (1995) up to two thirds of international trade or international production is dominated by 'intrafirm trade' amongst multinational companies. Clusters have one root in the 'development blocs' approach, which Dahmén (1950) termed an "anti-Keynesian and pro-Schumpeterian concept" and which above all analyses the transformation process and the dynamics (Amendola and Gaffard, 1988, page 142) between the cluster elements (that is, firms and other institutions). Dahmén refers to this as the mesolevel (similar to the French term), to distinguish it from neo-classical price theory and mainstream macroeconomic theory (Bellak, 1992;

Johansson and Mattson, 1987). The entrepreneurial and dynamic elements of competition are the main focus. Thus, along with the traditional justification for industrial policy, namely the correction of market failure, there are other justifications. In particular the following should be mentioned here.

2.1 Economies of scale and economies of scope
If the best solutions cannot be achieved through large enterprise units, one would like to achieve at least roughly similar results through the alliance of small and medium-sized enterprises (SMEs). The alliance effect of clusters should create economies of scale to a certain extent.

2.2 Spillovers
New growth theory introduced important arguments for spillovers, in particular from research and development, as well as in general for intangible investment (Clement, 1993). The aim of cluster formation is above all the creation of fields of competence and spillovers would be the expected result (Howells and Michie, 1997; Jaffe, 1989).

2.3 Rent shifting
New foreign trade theory states that protectionist ('strategic') measures can bring about welfare effects to the benefit of some nations and to the burden of others ('rent shifting'). Clusters are supposed to create such effects in international competition (Fagerberg et al, 1997; Feenstra et al, 1993; Grossman, 1992; Grossman and Helpman, 1993; Helpman, 1987; Krugman, 1986; Markusen and Rutherford, 1993).

2.4 Reciprocity
Because the international economy is characterised increasingly by strategic trade, more and more acts of reciprocity will be seen at the national level, regardless of whether a global welfare loss is to be expected. However, unilateral disarmament is a rare thing.

2.5 Structural change
Intensified international competition within the framework of regional integration and waves of mergers and acquisitions will cause an acceleration in structural change. Structural policy—to balance the resulting regional and industry-specific problems or to cope with future trends—will have an increased value beside competition policy (Laurencin, 1989). Clusters can be instruments for the elimination of structural and regional deficits or may generate positive external effects through their dynamics.

3 The IWI's 'bottom-up' approach to clustering
None of the empirical cluster studies closely follows Porter's study. Even if one wanted to do it, it would not be possible because the procedure is not precisely comprehensible. In the spirit of Porter, the object is, first, to detect competitive products, then to reveal input–output relationships, and, finally, to find factors to explain competitiveness (Fernau, 1997). One particular drawback

lies in the fact that statistically strict analyses cannot be accurately described. Furthermore, the most recent detailed input-output table for Austria dates back to 1983 and thus cannot be used for analyses of current competitiveness. Additional statistical problems are the limited disaggregation of data as well as the current change in classifications in the new EU statistics.

The main interest of the procedure used by the IWI (Institute for Industrial Research) lies in the description and evaluation of successful industrial clusters in Austria that cannot be illustrated with the traditional official statistics. In extremely simplified fashion, it is a matter of finding an answer to the question "Which are the industrial (economic) strengths of the country?", which refers not only to single businesses and branches, but also to dense fields (clusters) of international competitiveness. In the first comprehensive study for Austria (Weiss, 1992) a cluster-approach to international competitiveness was determined as sketched out in figure 1.

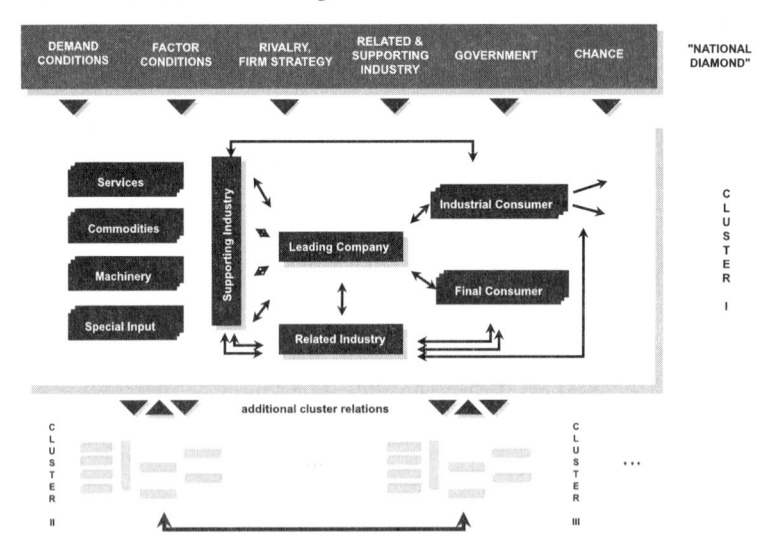

Figure 1. Cluster diagram (source: Clement and Fabris, 1997, page 258).

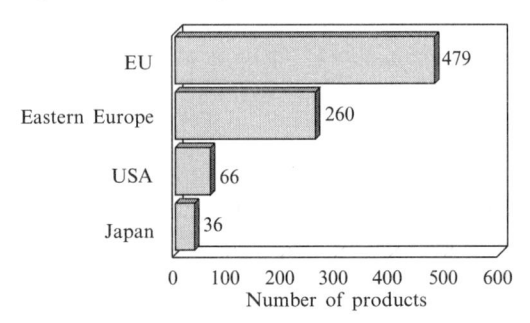

Figure 2. Geographical orientation of exports of competitive Austrian products in 1991 (source: WEISS, 1994).

Table 1. Competitive clusters in Austria in 1991 (source: Weiss, 1994).

Cluster	Exports (Austrian schillings)[a]			
	primary goods	machinery	special inputs	total
Metal	21 449 568	3 180 251	2 222 641	26 852 462
	(4.48)	(0.66)	(0.46)	(5.61)
Forest products	3 413 498	3 985 156	0	7 398 654
	(0.71)	(0.83)	(0.00)	(1.54)
Chemistry	3 336 876	1 834 802	0	5 171 678
	(0.70)	(0.38)	(0.00)	(1.08)
Computers	3 235 641	0	0	3 235 641
	(0.68)	(0.00)	(0.00)	(0.68)
Supplier branches	31 435 583	9 000 209	2 222 643	42 658 435
	(6.56)	(1.88)	(0.46)	(8.91)
Multiple branches	3 099 698	2 065 109	0	5 164 807
	(0.65)	(0.43)	(0.00)	(1.08)
Transport	7 200 113	1 269 948	4 582 096	13 052 157
	(1.50)	(0.27)	(0.96)	(2.72)
Energy sector	3 235 457	0	1 570 906	4 806 363
	(0.68)	(0.00)	(0.33)	(1.00)
Office services	93 697	0	0	93 697
	(0.02)	(0.00)	(0.00)	(0.002)
Telecommunications	0	0	0	0
	(0.00)	(0.00)	(0.00)	(0.00)
Defence	0	0	0	0
	(0.00)	(0.00)	(0.00)	(0.00)
Industrial consumption	13 628 965	3 335 057	6 153 002	23 117 024
	(2.85)	(0.70)	(1.28)	(4.83)
Food	1 449 734	3 348 346	757 601	5 555 681
	(0.30)	(0.70)	(0.16)	(1.16)
Textiles	12 655 375	624 627	48 462	13 328 464
	(2.64)	(0.13)	(0.01)	(2.78)
Building	6 001 737	2 985 931	3 268 329	12 255 997
	(1.25)	(0.62)	(0.68)	(2.56)
Medicine	4 339 858	469 508	0	4 809 366
	(0.91)	(0.10)	(0.00)	(1.00)
Private	5 219 840	0	0	5 219 840
	(1.09)	(0.00)	(0.00)	(1.09)
Leisure	4 614 950	0	0	4 614 950
	(0.96)	(0.00)	(0.00)	(0.96)
Final consumption	34 281 494	7 428 412	4 074 392	45 784 298
	(7.16)	(1.55)	(0.85)	(9.56)
Total exports	79 346 042	19 763 678	12 450 037	111 559 757

[a] Figures in parentheses are percentage of total Austrian exports in 1991.

On the basis of the very detailed and recent export statistics provided by ÖSTAT, the Central Statistical Office (rather than the UN SITC) 8608 export products were surveyed. A positive unit-value and a positive degree of coverage of imports were used as criteria. In addition, Austria's competitive strength in the countries of the triad (EU, USA, Japan) and Eastern Europe were given particular consideration. The RCA (revealed comparative advantages) values were also calculated on the basis of the SITC statistics. As a result, 604 internationally successful products were identified under this procedure for 1991. The regional distribution is shown in figure 2. An aggregation based on the analysis of export statistics results in the clusters shown in table 1.

After determining the clusters very mechanically, the main work of the IWI consisted (and consists) of enriching and completing this material. Because no more progress could be made by way of formal statistics, broad questionnaire surveys are being used, which are accompanied by numerous personal interviews in all the provinces (Clement and Fabris, 1997; Fabris and Terzer, 1997; Fabris et al, 1995; Weiss, 1994). Such interviews are especially important, because substantial additional information is obtained from the discussions and because actual interdependences can be traced. However, the large number of studies meant that a restricted number of clusters had to be selected for this chapter. The selected clusters are presented in the form of cluster charts and are described in detail.

The empirical–analytical work on clusters could be considered complete at this stage. However, in practice, this is where the actual work begins. To develop a functioning and successful integrated initiative on the basis of a statistical cluster, the difficult process of implementation has to follow (see below). Hence the IWI has recently attempted to include the first stages of cluster implementation (see IWI, 1998a; 1998b).

4 Regional examples

Cluster analyses and cluster policy have been especially successful in certain provinces of Austria. We will illustrate the IWI procedure with surveys performed in Styria, Vorarlberg, and Tyrol. In general, data about exports on a disaggregated level (as a proxy for the penetration of international markets) are calculated in relation to average values. This results in a portfolio that contains growing and shrinking, competitive and noncompetitive clusters.

The automobile cluster (part of the transport cluster in figure 3) proved to be the fastest growing cluster of Styrian industry. The essential core competency of this cluster lies in the production and research and development of engines, automobiles, and components. A consistent location policy and the promotion of cooperation with research and training institutions played a crucial role in the development of this cluster and economic policy measures had an important impact (Joanneum Research, 1997).

The analysis of the structure of the manufacturing sector revealed that the cluster 'technical know-how for building' is, along with the well-known textile industry, by far the most significant of the manufacturing industries in

Cluster developments in Austria 259

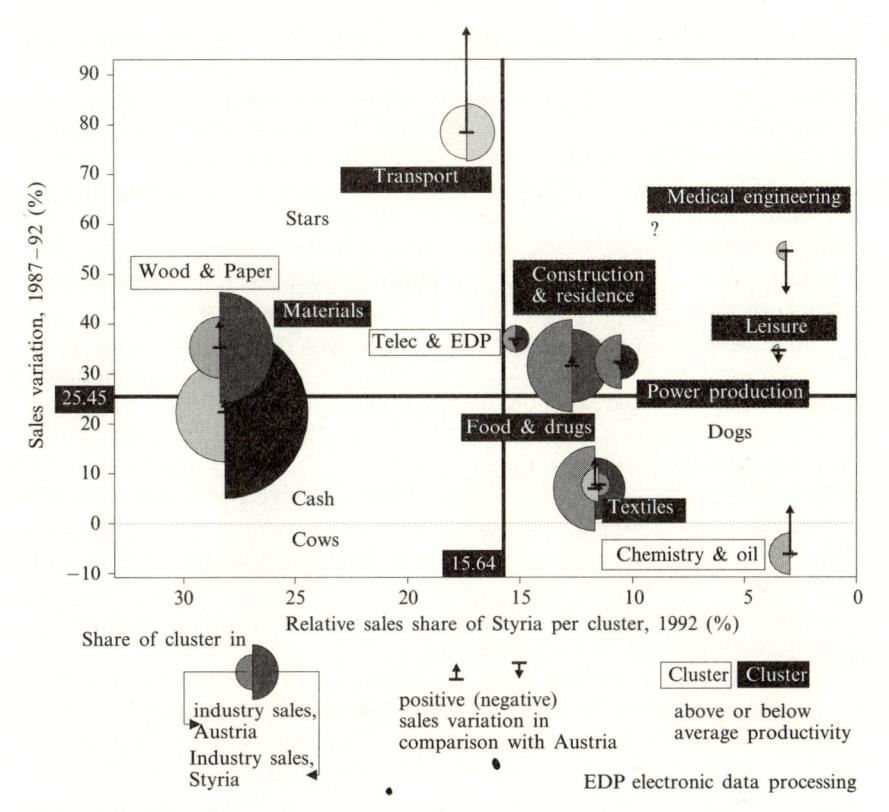

Figure 3. Portfolio of Styrian industry-clusters (source: Fabris et al, 1995, page 103).

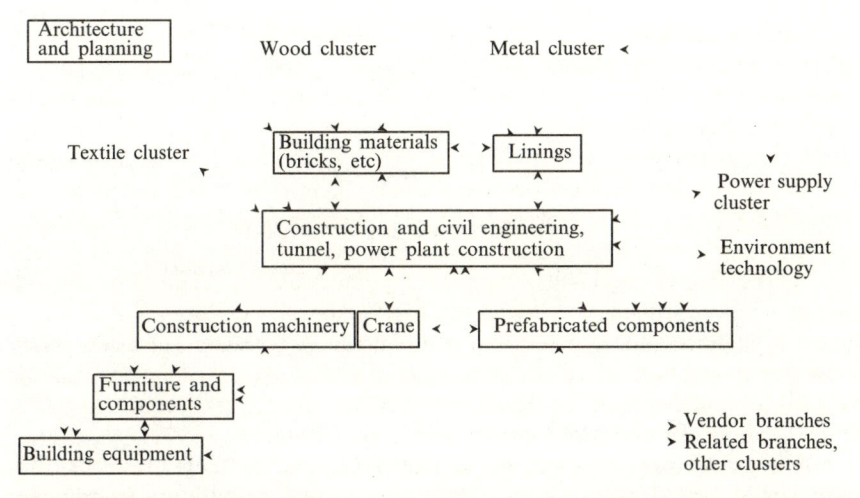

Figure 4. The 'building' cluster in Vorarlberg (source: Fabris and Terzer, 1997, page 181).

Vorarlberg (see figure 4). The cluster is shaped by a very broad spectrum of products and services. Along with planning and construction work in primary production (that is, construction and civil engineering with new developments in the branches of residential and industrial construction, tunnel and traffic infrastructure construction), main product lines are machine-tools and related supplies. Highly competitive positions also exist in related industries, such as in the field of construction engineering and building control systems.

The industrial structure of the Tyrol is characterised by, among other things, big internationally successful companies. Therefore, for the Tyrol it is especially worth considering a model which is illustrated by the term 'escort vessel', which is a smaller variant of clusters. This means building a strategic enterprise network with a leading industrial plant. The core firm is the stronghold in pursuing the objective of competitiveness, smaller firms being subcontractors to the 'flagship'.

The conventional statistical cluster analyses and questionnaires from the firms revealed that clusters such as 'Alpine wellness' (a recreation package including farming, outdoor activities, and traditional cultural events), 'Alpine-style housing', 'Tyrolean agricultural products', 'Tyrolean leisure and fashion', and 'Alpine technology' (see figure 5) are recognised as the most promising clusters.

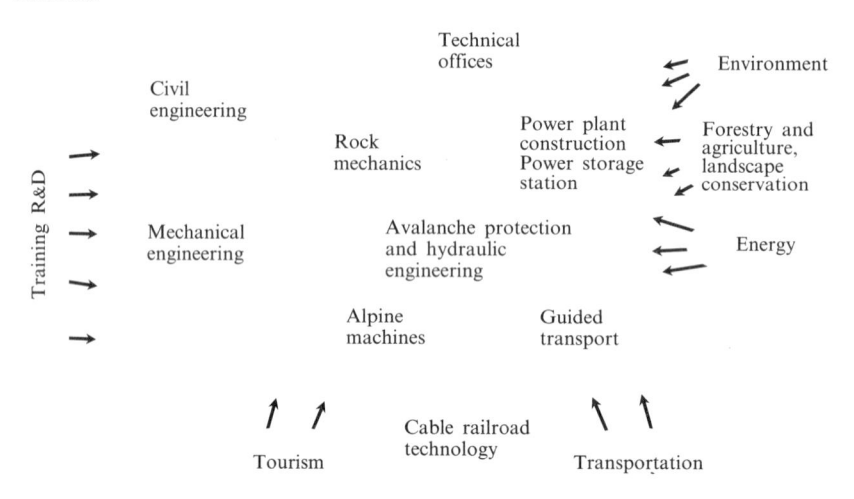

Figure 5. 'Alpine technology' cluster in the Tyrol (source: Clement and Fabris, 1997, page 274).

5 Difficulties of implementation

It would be a great illusion to expect that rational arguments in favour of a CIP would be sufficient on their own to generate enthusiasm among the individual enterprises and create a rush to build alliances, clusters, joint marketing operations, cooperative research ventures, or to set up general quality standards, educational associations, etc. On the contrary, the individual enterprise interest will [and must for reasons of efficiency and *Wirtschaftsrechnung* (von Hayek, 1931)] ultimately always follow economic principles. This priority should not be weakened by an ineffective cluster orientation.

Second, the cluster principle must not be interpreted by adherents of a planning approach to industrial policy as another name for similar measures, such as budget policies, curtailment of ownership functions and ownership rights, targeting, overmanning, preserving subsidies, etc. In spite of the present meaning used in theoretical discussions, neither the concept of 'strategic trade' nor that of clusters provides a carte blanche for government interventionism.

From these two caveats it follows that CIP involves a very delicate 'tightrope walk'. However, the advantages of a cluster-oriented policy strategy compared with a sectoral policy strategy should be a convincing principle for a reform of industrial policy (Clement, 1994). The following description, created in the framework of the cluster analysis of the Tyrol, serves as an example of the organised implementation of such a policy (figure 6).

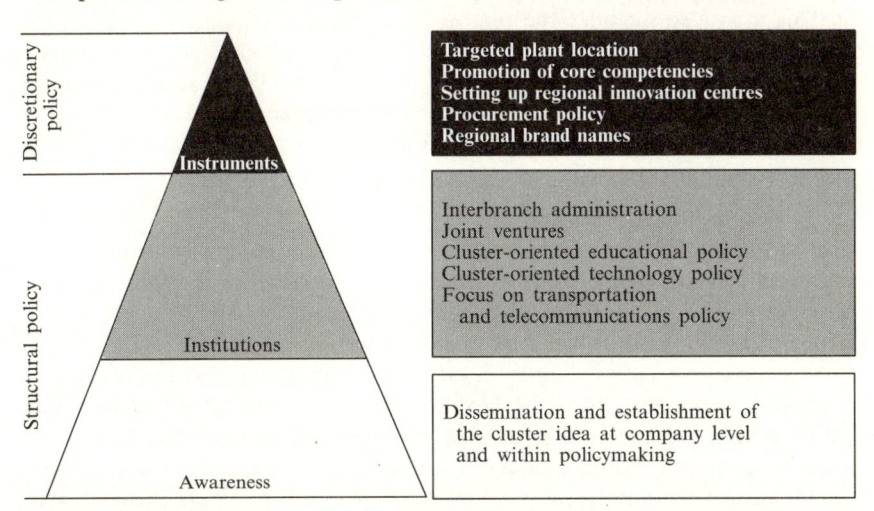

Figure 6. Stages of a cluster-oriented industrial policy (source: Clement and Fabris, 1997, page 280).

According to this principle, all forms and areas of industrial policy must be redesigned (Stadler, 1994; Tichy, 1994). We cannot accomplish this in this brief description. Instead, we will present a few hints (see also Weiss, 1994) and will consider the usual general typology of industrial policy (Bellak et al, 1997). Accordingly, as a structuring principle of industrial policy clusters can be viewed from the angle of a regulatory, competition, and location policy; a policy of institutional renewal; structural policy, and discretionary policy.

5.1 Regulatory, competition, and location policy

In this group of economic policy measures introducing a change along the lines of the cluster principle is very difficult. It is hardly possible to introduce cluster-oriented competition in tax, money, and environmental policies.

Regulations which take clusters into account are more likely in

Incomes policy: At a time when one may discuss and negotiate opening clauses in wage bargaining, it is probably not taboo to think about the influence of wage costs on clusters and possibly express this in a new organisation of trade unions and employers' associations.

Capital market and privatisation policy: For a long time Austrian industry has suffered from equity shortage, which was especially noticeable in international comparisons. Economic policy initiatives to create venture capital funds or the broader use of the stock market have only recently enabled Austrian industry to make up for this drawback. In spite of that, Austrian industry has made a significant effort to catch up both in equity-capital ratio and in cash-flow ratio.

Foreign trade and current account policy: Here (despite common EU foreign trade and current account policy) there is still room to promote national interests both in exports and in imports.

This area also includes the removal of inhibiting regulations. Above all, discriminatory tax regimes for intangible as compared with fixed investments should be removed (Clement, 1993; Hammerer et al, 1995), which would be very useful to the cluster concept based on competence; there are also many entry barriers in the so-called 'free' professions, where a stronger market position could also be achieved by larger networks of firms.

5.2 Policy of institutional renewal

This is an essential field of activity. As long as the economy is coordinated by traditional organisations (some of which already existed in the 19th century), the cluster idea will remain on paper only. Institutional reform requires a change of thinking. Thus, we need to enhance this process. If it really was the case, as some companies claimed in interviews, that the promotion of clusters is not the state's duty, then the next question to ask is according to which additional criteria should a policy including export support, financial guarantee systems, promotion of research, and regional brand names be designed. The usefulness of joint ventures, which to some extent resemble clusters, is no longer questioned. Examples which already exist in Austria may be seen in automotive suppliers (AOEM), rail vehicles and train operation and maintenance (ARE), wire and cable machine manufacturers (VÖDKM), and various research associations in the precompetitive domain or in joint ventures in sales, design, or export (Patsch, 1994). Over time such associations should also be able to give a stimulus to regrouping in the area of public administration and state-related representation of interests. Formulated more radically, all economic-related ministries, their subdivisions, chambers of commerce, and other associations should be checked to see whether they still reflect organisational principles that represent modern competitive fields of strength. Finally, it goes without saying, there needs to be a breakthrough to produce a cluster-oriented approach toward lobbying or international advertising of 'business location Austria' [compare the Austrian Business Agency, ABA (see Schröck et al, 1998)].

5.3 Structural policy
This is the area in which pattern recognition in the form of clusters should most obviously succeed. Thus, only a few examples are listed: priorities in transportation and telecommunications policy should be related to the strengthening of potential clusters. New educational institutions (specialised institutions of higher education) should be created according to cluster-specific fields of competency. The same applies to reforms of the curricula. The application of the cluster principle in setting the crucial goals in the area of applied research, including technology transfer and the diffusion of technology, in unconventional types of financial services, in regional and location policy, in the institutional promotion of business clusters (for example, of small and medium-sized plants), as well as in quality control policy all need to be emphasised. However, cluster orientation reaches a nationwide significance for foreign trade and capital flows. Because Austria has few domestic multinationals (Siegel, 1993), has few products with a brand name which is known worldwide, and does not have the image of a highly competitive industrial location, product and performance strengths need to be advertised to the world market with greater determination. To put in another way, one should fly the flag with clusters! However, this must be accompanied by definite actions, such as export initiatives, consulting coupled with product deliveries, a strategic presence on the spot, and even with political-diplomatic commitment.

5.4 Discretionary policy
There is no question that incremental measures will have the most direct effect on clusters. However, the following factors may harm their success: public policy failure in general and, in particular, discrimination against nonpreferred economic areas producing distortions of competition as well as possible welfare losses and incompatibility with EU regulations. Therefore we will not list possible discretionary policy measures for CIP. However, we should mention procurement policy, which is also subject to EU competition rules, but is ideal for cluster promotion. Another aspect of discretionary policy will clearly reveal the acid test of the CIP, that is, when and for how long can temporary government interventions and measures be justified, whose purpose is to support the survival of a competency network? On the one hand, this should prevent the loss of know-how, human capital, physical capital stock, and jobs through short-term turbulences, but on the other hand, must not turn into a permanent subsidy and lead to structural conservation.

6 Pragmatic aspects of the implementation
Economic policy institutions and entrepreneurs are in general too impatient to endure long discussions about system conformity of economic policy interventions. To allow at least a certain amount of rationality in the creation of regional cluster policies, the following points should be kept in mind:

To what extent are clusters embedded in an economic policy strategy at the national level?

How are duties or competency distributed between the territorial authorities? Is compatibility with EU law guaranteed?

How could a gradual process of cluster formation and development be organised?

How will the central problem of financing be approached?

Will a 'cluster manager' be nominated and how is his or her role vis-à-vis public authorities defined?

Although not complete, the following stages should guarantee successful policy measures for potential clusters: definition of core areas and peripheral areas; listing of companies; strength-weakness analysis of industries and companies; relationships to fundamental and applied research; existence of common generic technologies; relationship to the education system; networking with telecommunications; degree of existing quality standards; existing clusters; bidder groups; joint service packages of firms; joint organisation of subcontracting; joint brand names and logos; joint trade fairs; joint marketing operations; joint participation in international tenders; joint offer of packages and systems supply; demands for technology consulting; demands for broad education and curriculum development; demands for cluster management (external coordinator); demands for cluster subsidies ('cluster bonus'); cluster-related location policy; a global image campaign for fields of competence; development of cluster-related financial services.

7 Current projects and outlook
Clusters have developed into a central aspect of contemporary economic policy in Austria.

7.1 Export initiative
Within the framework of the federal government's export initiative the Federal Ministry of the Economy and other institutions have pledged to support the Austrian economy through consultation and services, to produce a short-term improvement in the size of firms, to reduce capital and sales weaknesses, and to develop higher quality products and know-how. Among other things, the participation of cluster firms at trade fairs to improve marketing should be promoted. On the basis of demands from firms, the organisation of joint marketing measures should create export impulses. The topics range from water supply, food, and the environment, to energy conservation, tunnel building, and software.

7.2 Technology initiative

The idea of clusters is also referred to in the technology initiative. The technology initiative begins from existing clusters, for example, the train cluster (Austrian Railway Equipment and the Federal Ministry of Science and Transport), drinking water (the Austrian Chamber of Commerce), the Styrian automobile cluster, AOEM (Austrian Original Equipment Manufacturers), and the wine marketing cluster. The development of high-technology clusters is favoured by the following factors: high-ranking universities and research institutions, highly qualified workers, venture capital funds, proximity to potential customers. The policy measures should concentrate on growth areas. The 'K+' initiative (see section 7.4) is an implementation step that should particularly benefit high-technology industries.

7.3 Financing

One central problem in exporting is financial services in difficult markets. Asia, Eastern Europe, and Latin America belong to this category and it is evident that investment in exports to these areas are only possible with structured financing. Figure 7 shows how complex a 'BOT model' (build – operate – transfer) can be. Clusters plus structured financing are termed 'export packages'. Cooperation in the technical and business areas should strengthen international competitiveness and is enhanced by an element of financial services. Measures in the export and technology initiative include elements of such an export package.

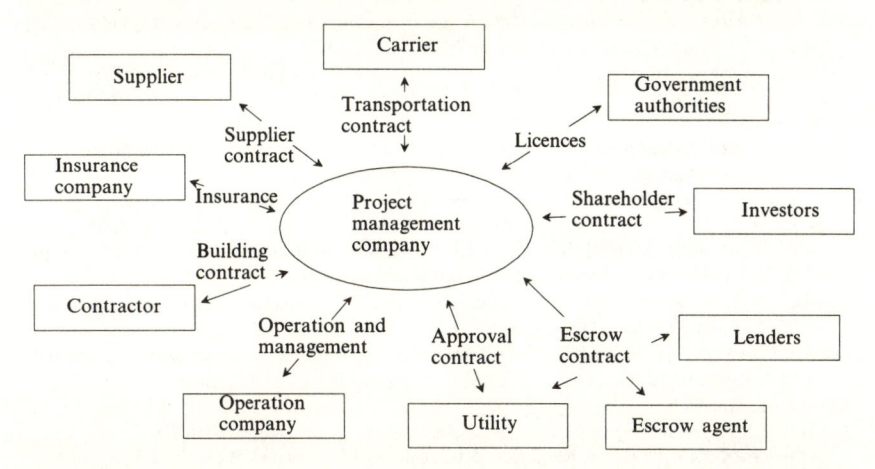

Figure 7. Actors in a build – operate – transfer model and their interactions (source: Kreid et al, 1996, page 10).

7.4 Setting up competency centres (the K+ initiative)

The Federal Ministry of Science and Transport is considering the creation "of research institutions planned and operated jointly by the academic and industrial community with the highest quality specifications and long-term orientation" (BMWV, 1997). This should create an investment incentive for foreign firms as well as offering SMEs the opportunity to cooperate. Furthermore, this should result in a better usage of existing knowledge from the universities and other research institutions through joint applied research. Clusters are named as one of the qualitative criteria for evaluating a centre of competence.

It might be too early to draw even preliminary conclusions about the success of CIP in Austria. However, there is no question that clusters are also a way to develop creativity and a disciplining instrument for fragmented policy responsibilities. In this respect they contribute to learning processes of modern economic policy.

References

Amendola M, Gaffard J-L, 1988 *La Dynamique Economique de l'Innovation* (Editions Economica, Paris)
Bellak C, 1992, "Der Netzwerkansatz—Alternative oder Egänzung traditioneller Ansätze im Beriech der Markt-Unternehmensdebatte?" *Der öffentliche Sektor—Forschungsmemoranden* **18**(1) 72 – 82
Bellak C, Clement W, Hofer R, 1997, "Wettbewerbs- und Strukturpolitik", in *Grundzüge der Wirtschaftspolitik Österreichs* 2nd edition, Eds E Nowotny, G Winckler (Manz, Wien) pp 127 – 165
BMWV, 1997 *K+ Forschungskompetenz und Wirtschaftskompetenz* (Bundesministerium für Wissenschaft und Verkehr, Wien)
Clement W, 1993 *Immaterielle Investitionen: Einbindung in die VGR, Steuerliche Behandlung, Förderungsmöglichkeiten* IWI studies, volume 11 (Industriewissenschaftliches Institut, Wien)
Clement W, 1994, "Cluster und ihre industriepolitischen Konsequenzen in Österreich", WP5, Industriewissenschaftliches Institut, Wien
Clement W, Fabris W, 1997 *Die produzierende Wirtschaft Tirols* IWI studies, volume 28 (Industriewissenschaftliches Institut, Wien)
de Bandt J, 1989, "A mesoeconomic approach to industrial economics", in *The Scope of Industrial Economics and Competition* (Industriewissenschaftliches Institut, Wien) pp 23 – 32
Dahmén E, 1950 *Entrepreneurial Activity and the Development of Swedish Industry 1919 – 1939* (Industriens Utredningsinstitut, Stockholm)
Fabris W, Terzer H, 1997 *Identifizierung zukunftsträchtiger Cluster in der Vorarlberger Wirtschaft* IWI studies, volume 27 (Industriewissenschaftliches Institut, Wien)
Fabris W, Hohl N, Mazdra M, Schick M, 1995 *Wirtschaftsleitbild Steiermark* IWI studies, volume 25 (Industriewissenschaftliches Institut, Wien)
Fagerberg J, Hansson P, Lundberg L, Melchior A, 1997 *Technology and International Trade* (Edward Elgar, Cheltenham, Glos)
Feenstra C, Yang T H, Hamilton G G, 1993, "Market structure and international trade: business groups in East Asia", WP4536, National Bureau of Economic Research, Cambridge, MA
Fernau A, 1997 *Werkzeuge zur Analyse und Beurteilung der internationalen Wettbewerbsfähigkeit von Regionen* dissertation, Universität Hochschule für Wirtschafts-, Rechts- und Sozialwissenschaften, St Gallen
Grossman G M (Ed.), 1992 *Imperfect Competition and International Trade* (MIT Press, Cambridge, MA)
Grossman G M, Helpman E, 1993, "Trade wars and trade talks", WP4280, National Bureau of Economic Research, Cambridge, MA
Hammerer G, Herzog A, Schwarz K, 1995, "Die Bedeutung Immaterieller Investitionen für Österreich", WP12, Industriewissenschaftliches Institut, Wien
Helpman E, 1987, "Imperfect competition and international trade: evidence from fourteen industrial countries" *Journal of the Japanese and International Economies* **1** (March) 28 – 43

Howells J, Michie J, 1997 *Technology, Innovation and Competitiveness* (Edward Elgar, Cheltenham, Glos)
IWI, 1998a, "Vienna-biotechnology-cluster", Industriewissenschaftliches Institut, Wien (forthcoming)
IWI, 1998b, "Bio-energy-cluster", Industriewissenschaftliches Institut, Wien (forthcoming)
Jaffe A B, 1989, "Characterizing the 'technological position' of firms, with application to quantifying technological opportunity and research spillovers" *Research Policy* **18**(2) 87 – 97
Joanneum Research, 1997, "Kooperationen in KMU-Netzwerken: Massnahmen zur Unterstützung verschiedener Funktionen von Technologietransfer bis Export", Institut für Technologie- und Regionalpolitik, Graz
Johansson J, Mattsson L G, 1987, "Interorganizational relations in industrial systems: a network approach compared with the transaction-cost approach" *International Studies of Management and Organization* **17**(1) 34 – 48
Kreid E, Küng K, Paparella C, Vlasits M, 1996 *Finanzierung und Absicherung von Anlagenexporten in asiatische Wachstumsmärkte* (Industriewissenschaftliches Institut, Wien)
Krugman P R (Ed.), 1986 *Strategic Trade Policy and the New International Economics* (MIT Press, Cambridge, MA)
Laurencin J-P, 1989, "L'impact sectoriel et régional du grand marché: le cas de l'industrie francaise" *Revue d'Economie Industrielle* **49** 67 – 95
Markusen J R, Rutherford T F, 1993, "Anti-competitive and rent-shifting aspects of domestic content provisions in regional trade blocks", WP4512, National Bureau of Economic Research, Cambridge, MA
McKelvey M, 1991, "How do national systems of innovation differ?: a critical analysis of Porter, Freemann, Lundvall and Nelson", in *Rethinking Economics* Eds G Hodgson, E Screpanti (Edward Elgar, Cheltenham, Glos) pp 117 – 137
OECD, 1991 *Strategic Industries in a Global Economy: Policy Issues for the 1990s* (OECD, Paris)
Patsch C, 1994 *Die österreichische Zulieferindustrie: Ihre volkswirtschaftliche Bedeutung und zukünftige Entwicklungsmölichkeiten* IWI studies, volume 21 (Industriewissenschaftliches Institut, Wien)
Porter M, 1990 *The Competitive Advantage of Nations* (Harvard University Press, Cambridge, MA)
Quelin B, 1993, "Les analyses de la filière: bilan et perspectives", WP5, Industriewissenschaftliches Institut, Wien
Schröck T, Cerny M, Farar D, Skrinner E, 1998 *Stärkenprofil der österreichischen Bundesländer* IWI studies (Industriewissenschaftliches Institut, Wien) forthcoming
Siegel D, 1993 *Multis Made in Austria* IWI studies, volume 9 (Industriewissenschaftliches Institut, Wien)
Stadler W, 1994, "Die Stunde der Industriepolitik" *Wirtschaftspolitische Blätter* **41**(1) 4 – 15
Tichy G, 1994, "Cluster-Konzepte—Ihre Bedeutung für die österreichische Wirtschafts- und Technologiepolitik" *Wirtschaftspolitische Blätter* number 3 – 4, 249 – 256
UNCTAD, 1995 *World Investment Report* (United Nations, Geneva)
von Hayek F, 1931 *Prices and Production* reprinted 1967 (Routledge and Kegan Paul, London)
Weiss A, 1992 *Konzeptuelle Grundlagen und empirische Messung der Wettbewerbsfähigkeit Österreichs* (Diplomarbeit, Wien)
Weiss A, 1994 *Österreich als Standort international kompetitiver Cluster* IWI studies, volume 13 (Industriewissenschaftliches Institut, Wien)
Wildner T, 1993 *Medizintechnik made in Austria—eine empirische Cluster-Studie* (Diplomarbiet, Wien)

Clusters in the Context of the European Union's Cohesion Policies†

R Shotton
Directorate General for Regional Policy and Cohesion, European Commission, Brussels

1 Introduction
In this chapter I set out some considerations arising from practical experience in the management of the European Union's Structural Funds. I discuss the significance of the concept of clusters for the European Union's cohesion policies and the implications for the management of the Structural Funds in particular.

It is hard to define with precision what is meant by the concept of clusters of competence. But in its most simple expression many see the significance for regional development as being the idea that high-cost, high-welfare societies can compete in an enlarged Union and in global markets by seeking competitive advantage through advanced know-how in specific fields, sometimes quite narrowly defined. A second and equally important message in the dawning of the information society is that physical proximity is important for the successful building of clusters. Therefore an integrated regional policy for competitiveness is relevant and naturally complements and reinforces national sector policies.

Regional competitive advantage requires high standards of infrastructure, excellent education and training facilities, and a dynamic business sector often involving, though not always, major international companies as benchmarkers for global standards and gateways to global markets, and last but not least an aware and involved political class. But the essential binding element is cooperative behaviour and networking both locally and in the wider economy. Without this extra ingredient, the critical mass for effectiveness in terms of international competitiveness will not be achieved, despite the excellence that individual components may have. This is the bonus from cluster policies.

The foregoing might lead to the conclusion that clusters are a policy which is only relevant for those who are already successful. Only they have the enabling conditions in place in terms of infrastructure, human resources and business activity. Only they have the sophistication in political and social organisation that can transcend the traditional barriers between business and the public sector, between business and academia, and which fosters intelligent cooperation amongst businesses that are competing in other respects.

Can the European Union's structural policies in support of less favoured areas promote social organisation skills in regions where they are presently weak? Can they establish self-sustaining islands of modernity in a sea of

---

† The views expressed in this chapter are those of the author, and do not necessarily represent those of the European Commission.

tradition and reach out from that gradually to transform the wider society? In short, are clusters a relevant regional development concept for the less favoured areas of the Union also?

2 Cluster-building policies as a challenge to traditional methods

Cluster-building policies challenge many of the preconceptions of traditional approaches to structural policies. Cluster policies are relatively cheap, but a large amount of effort has to be put into the management of relatively small amounts of development funding. For this reason alone, cluster-building policies may not be attractive to administrations which are rich in financial resources but relatively poor in administrative resources.

Companies have to be weaned off conventional funding support for individual projects, and encouraged to develop collaborative behaviour as a condition of access to public money. There is no doubt that this message is not always easy to pass, especially amongst small and medium-sized enterprises, many of whom identify real or potential competitors more readily than confident partners. How can companies compete and yet share resources in certain fields? Large companies have been developing this approach for some years now; smaller companies may still have a simpler view of economic life.

Within administrations there is a reallocation of funds and responsibility from those traditionally responsible for relations with industry, trade, and commerce, to others who in the past had only relatively limited links to the business world (for example, education, research). The emphasis in public policy is shifted to institution building, to establish platforms that bring together business, academic, and public sector interests in diverse ways to support the design and implementation of cooperation and networking. Policy priorities are no longer autodetermined within the public sector, but in genuine dialogue with the business sector through the various cooperation platforms established.

More fundamentally even, a culture of cooperation has to be developed between the private and public sectors—which is no easy task. The private sector wants quick decisionmaking, flexible funding, and quick results—but also the freedom to change direction or even abandon a project at short notice if it is no longer judged worthwhile. The private sector has to learn to participate in long-term policies based on a shared understanding of the common interest. Building a competence cluster from basic beginnings may well take ten to fifteen years of sustained effort.

In contrast, the public sector offers a relatively constant strategic direction (subject of course to political events), sometimes indeed to an excessive degree in the face of changing circumstances. The public sector imposes strict requirements for the definition in advance of expenditure purposes, and burdensome controls during project implementation directed towards expenditure inputs rather than results obtained, because this is the way public expenditures are traditionally controlled. Modifications in the course of implementation are treated as deviations that need to be fully justified and approved case by case, rather than as the normal evolution of a project. The public sector has yet to learn to focus on results not on inputs, to accept uncertainty and its implications, and to be ready to abandon projects that are not working, without any implication of mismanagement of public funds necessarily being attached.

Cooperation between business and the world of education is a special challenge. Introducing the world of business into higher education, especially in technical fields, is not such a new idea nowadays—even if much remains to be done to translate the slogans into reality. Still more remains to be done at earlier stages in the education system. Adapting education systems to supply interdisciplinary skills is still relatively unusual (for example combining engineering, business administration, economics, languages, and computing skills). Opportunities for short-term detachments from firms to research institutes or universities, which combine a contribution to lifelong education with specific project work of direct business relevance, are still more unusual. In general, the mechanisms for easy movement between the worlds of education and business need to be multiplied. Many of the cooperation platforms referred to above will have this as one of their specific aims.

But the greatest challenge inherent in cooperation and networking is to manage complexity and multiplicity of partners in the decisionmaking process. Rather than controlling access to information, information is diffused. Rather than operating by the hallowed principles of divide and rule, networks and partnerships imply overlapping relationships that can only be managed by consensus building. In short, making a reality of clusters requires a different mentality from that traditionally prevailing in our societies, both in the private and in the public sectors.

3 The relevance of the European Union's structural policies
Behind every successful cluster, there are individuals with vision and drive who are dedicated to its development and success In the beginning, there is often a spark—a single, strong, and capable personality that starts the operation, drawing in others progressively until the networking takes on a life of its own. Clearly the European Union's structural policies cannot provide that spark if it is not present. In practice, all they can do is to encourage those on the ground and occasionally offer useful advice—for example, about contacts elsewhere in the Union, or within the European institutions.

The general context is of great importance for the success of cluster policies. In practice, although it is possible to create a small island of modernity in a generally hostile social context, experience suggests that it is very difficult in practice to reach out from that small island and change the wider reality. There are examples in the least favoured areas of the Union (called Objective 1 areas in the Fund regulations), where large amounts of public money have backed highly competent researchers or top-class academic institutions. They have built up business innovation centres in the same region. Efforts have been made to reform the vocational training system. In short many of the building blocks of a potential cluster are in place. But the essential extra ingredient—a widespread understanding both of the value and of the rules of the game for networking and cooperative behaviour—is not there, or is only shared by too narrow a group in society. The spark exists, but the lift-off does not follow.

The European Union's structural policies do try to promote cooperative behaviour. They encourage cooperation between different parts of the administration dealing with business-support policy and vocational training and education, for example. They encourage cooperation between the public administration and the social partners, and in particular with the private sector. They encourage the involvement of nongovernmental organisations, particularly those dealing with environmental and social issues. The difficulty, however, is that failure to develop cooperative models of behaviour in practice does not imperil the flow of funding from the funds, provided formal satisfaction is given, for example, regarding the composition of monitoring committees. Better partnership and more coordination are recommendations which will only take root if the groundwork has already been done. Again the European Union is not well placed to create this method of work where there is resistance or incomprehension. In practice, such barriers can only be overcome by local initiative and local enthusiasm.

In this respect, a word of warning—people must be free to choose whether they wish to work together and with whom they wish to work. That is why it is counterproductive to use public policy to push one single structure or model of cooperation rather than allowing for a plurality of approaches, even if this is untidy and results in some overlapping. The European Union should not fall into the trap of offering one standard formula for interregional and transnational cooperation on a take-it or leave-it basis. In the same way, it should discourage any tendency at the regional level to privilege one type of platform to the exclusion of others. Efficiency in the use of public resources will not be served by such a simplistic approach.

The European Union's structural policies have developed mechanisms for the exchange of best practice through the financing of innovatory actions. Cooperation and networking are a fundamental condition to be met for funding to be given—in other words, a project has to have partners in several regions, typically in different countries, networked with the project leader. These actions therefore combine innovation with the requirement for cooperation and networking in an attractive way. They offer an additional opportunity for regions to benchmark their proposals and management methods by comparison with standards of excellence elsewhere in the Union.

The exchange of best practice and networking supported by the European Union is a useful opportunity and of real value, but some of the best performing regions are not involved, or not sufficiently involved, in networks sponsored by Structural Funds. To some extent, top-class networks lead a separate existence drawing on funding elsewhere in the range of European Union policies (for example, the fifth Research and Development Framework Programme). Rather than creating separate layers of cooperation (first and second class), the Union's structural policies should do more to mix the two layers together, stretching the capacity and imagination of actors in the less favoured regions. And because human talent is evenly distributed, there is no doubt that such networks will prove valuable for the most advanced regions as well, even if there may be some initial prejudice to be overcome.

However successful, such policies cannot substitute for the stimulus of a dynamic outward-looking business sector in the regional economy. How then to acquire a dynamic business sector if that is not the present reality? Can the European Union's structural policies help? One approach is to encourage inward investment from the right type of companies and then to work hard with them to link them into a local network of subcontracting companies (in short the Irish model). This is the best approach if it is successful. However, it requires a rigorously realistic assessment of the relative attractiveness of the region—and a commitment to remedying the negative features perceived by potential inward investors. This will raise issues which go beyond conventional understanding of what regional policy is about—the quality of the local schooling system, for example, or the high cost of air transport, or telecommunication tariff structures. Marketing based on simple regional patriotism is out of place when competing for inward investors.

An additional approach (but experience suggests this is not an alternative) is to encourage companies to spin off from local centres of excellence in education or research in business incubators. But a few isolated successes are not likely to suffice as a catalyst for wider change. They may survive, even develop in their protected environment, but that is all.

Will the Commission's proposals for revisions to the regulations governing the Structural Funds change the situation from the year 2000 onwards? There are large transfers of funds through the European Union's Structural Funds mainstream programmes in support of assisted areas. In principle, these programmes could be designed to support cluster-building policies in a more explicit way and, undoubtedly, in the next generation of programmes beginning from the year 2000 there will be proposals to do just that.

Three issues can be raised in this respect. The first is that the changes proposed for the European Social Fund (ESF), contained in the Commission's recent proposals for revised Structural Fund regulations, open the door to a wider range of eligible expenditures relevant to cluster-building. That is why the ESF could be well placed to play a major role as a source of Union funding for cluster-building operations (building cooperation platforms, for example) in the next generation of programmes.

The second concerns the European Regional Development Fund (ERDF). Outside Objective 1, the ERDF will be constrained by a geographically more concentrated map of areas eligible for assistance. Many urban centres outside Objective 1 will be excluded from ERDF-assisted areas in order to be able to respect the more restrictive ceilings on overall population coverage. Many of the core activities of a cluster are located in or near the main urban centres of a region. Unless there can be some flexibility for the ERDF to finance networking and cooperation within adjacent areas outside the designated assisted areas, the region concerned will have to rely on its own resources and on the ESF to finance the linking of assisted areas to economic motors elsewhere in the region. Within the assisted areas the ERDF can be used to support inward investment or to encourage indigenous high-tech companies to locate there in the traditional way.

A third issue for the next programming period will be whether it is possible to build a cluster of clusters—that is, an interrelated set of specialisations, building out from the original one or two core areas. It seems that some regions in Northern Europe are ready to try this very ambitious level of strategy: for example, in Austria and in Northern Scandinavia. Where future Objective 1 areas are adjacent to well-established clusters such as Oulu in Northern Finland (electronics and medical equipment), the European Union's Structural Funds are particularly well placed to support that ambition.

4 Towards a conclusion

Returning to the questions asked in the introduction to this chapter, we can begin by underlining again the point that the European Union's structural policies cannot substitute for the dynamism and social organisational skills that must exist locally for cluster-building policies to succeed. All it can do is offer support to those who have decided for their own reasons to cooperate amongst themselves and with others according to a cluster method of working. If the spark is not there, the lift-off cannot be artificially generated.

In other words, regions and economic and social actors have to be strongly motivated themselves to cooperate transnationally. They have to be ready to be measured against high standards. If, on the contrary, they are basically content with their present reality, if it is enough for them to do a little better each year, but if the goal is not really to join the first league, then Union-wide cooperation platforms, however well designed, will not be successful.

Second, attempts to build up clusters in less favoured regions have often been frustrated by the absence of a sufficiently dynamic outward-looking business sector. A few companies with this mentality are not enough. That is why such regions have to build their capabilities in successive phases.

The classic starting point is to aim for excellence in research and education through public funding in targeted areas. This has to take account of the human resources available—in other words, to start it is a question of backing individuals with global standards of competence. This is a difficult path to follow if the culture is one of 'fair shares'.

Once these points of excellence are on the map, it is possible to try to attract companies who want to make use of this resource and to try to spin off companies from the education and research centres. This implies pursuing a business-friendly policy. A problem for some regions will be that their own attitudes and initiatives are partly nullified by difficulties in the broader national context (for example, as regards labour-market regulations). These wider obstacles may, in extreme cases, make it impossible to create a population of modern firms sufficient to be able to go further, despite the best efforts of the region concerned.

At a third stage, once a sufficient population of firms has been established, regions can try to develop formal platforms for cooperation and networking between business, education, and research, drawing in the public administration and local politicians, as well as other important actors in the local society. These platforms have to be demand driven, with the flexibility to respond to the interests and working methods of those wishing to cooperate. Potential partners cannot be forced into some preconceived straitjacket laid down by the public policy acting as 'tutor'. All the participants in the cooperation networks have to be ready to accept the 'rules of the game'—seeking win–win situations not zero-sum games where a winner implies a loser. Indeed, if these rules of the game have to be the subject of formal agreements, then this is perhaps a signal that the participants are not yet mature enough to move to the third stage.

Fourth, we can conclude that the relevance of cluster-building policy is not restricted to those who are already successful, and who are not beneficiaries of the European Union's Structural Funds or only to a marginal degree. Amongst the assisted regions, however, it is most relevant to those who already have most of the building blocks in place for global competitiveness, but who lack still the extra ingredient of well-organised cooperation and networking. Those who are further away still from this situation will need to concentrate on placing the necessary building blocks, but with a strong emphasis on human competencies as well as economic infrastructures. They need to identify in particular the exceptional talents and institutional capacities in place, and to back them fully (but without drowning them in easy money!) This can be the starting point for the development of high levels of specialised know-how in the local economy.

The whole process from the first to the third stage might take fifteen to twenty years of sustained effort. Experience suggests that it would be extremely difficult during the first phase to make the jump straight to a cluster approach in a small island of modernity within unreformed traditional structures. Attempts to do this do not appear to have succeeded so far. This step-by-step approach to cooperation could be built into the European Union's support mechanisms in a more conscious fashion.

In summary, if one essential ingredient of success should be proposed it could be based on the well-known military watchword: invest in success whenever and wherever it occurs. Abandon failure resolutely before significant resources are sucked into it. In that way the capabilities for global excellence will gradually emerge. How can we reconcile that with fair shares and political balance? It is difficult indeed if not impossible. It is better therefore that the cluster-building part of a regional development programme should be free to manage its resources with its own criteria and methods of work, leaving other actions managed in other ways to provide complementary actions as necessary.

Index

Adopters 116–118
Adoption technology 123
Advantages 12, 92, 226–227, 231
Agency for Biomedical Technologies 52
Agglomeration 2, 24, 33, 93, 111, 184, 229
Agglomeration economies 3, 8, 30–31, 35, 94, 111–112, 183, 212–213, 223
Air transport 75–76
Area plans 76
Arrow's impossibility theorem 84
Asymmetric interests 89
Auction system 147
Autonomy 81
Awareness-raising campaigns 14, 251–252

Backing losers/winners 5, 13
Backwash effects 33-34
Barriers 64, 94, 112, 117, 124, 204–207, 249, 262
Behaviourist – reductionist perspective 212
Best-practice 11, 196, 271
Big bang approach 129
Biological and medical poles 51
Bottom-up methods 11, 20, 239
Branch plants 194, 204
Bureaucrats 81–89
Business cycles 12
Business incubators 272

Capital market 262
Chain analysis 20
Chambers of Commerce 57, 78, 112, 245, 251
Cluster
  distribution 131
  financial 229
  formation 8, 10, 129–131, 144–147, 154, 159, 211, 231, 254–255
  growth 18
  identification of 1, 5, 10, 216
  induced 246–250
  interlocking 246–250
  leading firm 131
  manager 264
  network 131, 136, 232
  policy 12–5, 18–19, 23
  porterian 246–250
  process 131, 229
  product 132, 229
  R&D 132
  raw-material 131, 229
  requirements 131
  segmented porterian 246–250

Cluster (continued)
  skill 131, 229
  socialist 130–137
  strategies 5
  supplier 131
  theory 8, 23
  typology of 4, 8, 11, 19
  virtual 246–250
Clustering of innovations 2, 29
Coase 4
Cognitive processes 46
Collaborative arrangements 95
Collective action problem 9, 82–85
Collective good 82, 88
Comanufacture/co-makership 7, 62, 78
Combines 132–144
Communication costs 60
Communication hubs 251
Competence consolidation/development 10, 14, 242–249
Competency centres 265
Competition policy 254–255
Competitiveness 3–4, 11–13, 19, 24–28, 32–33, 60–78, 141, 190, 202, 205, 217, 226, 231–233, 238–239, 249, 255-256, 268
Cooperation 3–4, 8–12, 59–60, 69, 82–90, 95, 189, 197, 211, 216, 221–223, 232, 271, 274
Cooperative behaviour 65, 271
Core activities 226
Corporate venturing 206
Corporatism 254
Cost-benefit calculations 83
Crisis-management 14, 241
Cross-job skills 243, 252
Cross-section technologies 232
Cumulative causation theories 34

Data mining 119
Decentralisation 15, 63, 67, 81–90
Decisive player 88
DELOS project 241–252
Dependency model 205
Distance calculations 113, 126
Distribution services 13
Dysfunctional leadership 238

Economic performance 6
Economic space 27, 33
Economic success 64, 78, 226
Economies of scope 255

Economies of specialisation 227
Education 14, 78, 105, 184, 203, 234, 244, 263–264, 268–274
Efficiency 84
Efficiency frontier 186
Embeddedness 3, 32, 241–249
Employment 57–63, 195, 207, 250
Endogenous development model 58
Endogenous growth model 64
Environment 21, 129–130, 140–141, 148, 200, 261, 271
Environmental regulation 21
Environmental technologies 21–22
EU cohesion policies 268
EU programmes/policies 130, 196–199, 251, 268–274
EU Structural Funds 199, 250, 268–274
European Monetary Union 59
European Recovery Programme 130
European Regional Development Fund 272–273
European Social Fund 272
Evaluation 6, 144
Export base 74
Export initiative 264
External economies 2, 31–32
External effects 6, 211
External scale economies 18, 31
Externalities 30–35
Extractive industry 25

Face-to-face relations 8, 41–48, 124, 227
Fairness 84
Federalist reform 15, 79
Feedback mechanisms 4
Filière approach 2, 4, 62, 94
Financial portfolio 185–186
First-tier suppliers 115
Flexibility 18, 64–65, 78, 231–233, 239, 274
Flexible specialisation 32
Folk theorem 85
Framework conditions 20
Free-riding 15, 47, 88–90
Fuzziness 182

Game
    assurance 83–87
    chicken 83–87
    noncooperative 83
    rules of the 61, 83, 270–274
    theory 9, 14, 82–83
    zero-sum 274

Gateways 75
Geographic space 29
Globalisation 2, 7, 11, 70, 78, 181–184, 201, 228, 239
Globalisation trap 12
Goodwill 15, 208
Governance structure 58, 92
Greenfield investments 228
GREMI 3
Group size 86, 90
Growth centres 33
Growth phase 233–234
Growth poles 2, 27–33
Growth theory 5, 34–35, 226, 231

Hierarchical firms 9, 228
Hierarchy of power 82
High-tech 81, 203–204, 236, 265
Homogeneity 6
Hub-and-spoke 182
Hubs 76
Human resources 6, 13, 34–36, 42, 51, 68, 70, 78, 184, 244, 251–252, 268, 273

Incentives 4, 14, 47, 82
Incomplete information 85
Increasing returns 23, 31, 94
    to scale 182
Index of diversity 185
Individual behaviour 9, 82
Industrial atmosphere 196
Industrial cores 136–139
Industrial districts 3–6, 18, 30–37, 57–59, 77, 98, 192, 243, 249, 252
Industrial location theory 30
Industrial milieu 10, 241–248
Industrial mix 105, 185
Industrial organisation 3, 59, 67
Industrial parks 81, 134
Industry–research collaborations 18
Infant industries 231
Information
    and communication technologies (ICTs) 8, 42–48, 55
    flows 112, 184
    gathering 10, 14, 242–250
    provision 19
    set 83
    processing model 212
Innovation diffusion 9, 42
Innovative (post-Fordist) industries 81
Innovative milieu 41

Index

Innovative networks 8
Input – output 3 – 5, 26, 94 – 97, 109, 182, 187, 215, 226 – 227, 256
Institution building 269
Institutional frameworks 193
Institutional persistence 239
Institutional renewal 261
Institutional thickness 239
Interfirm cooperation 3, 7
Intermediate cities 74
Intermunicipal planning 76
Internal scale economies 31
Intrafirm trade 254

Job creation 61, 77, 192

Knowledge 214 – 221, 232, 238
   acquisition 10, 14, 242 – 244
   clusters 1
   codified 42 – 44, 53
   industry-specific 32
   infrastructure 1
   intensification 5
   spillovers 32 – 36, 182, 211 – 220
   related externalities 36
   tacit 3, 42-44, 53

Laager mentality 230
Labour
   division of 2 – 3, 12, 31 – 32, 46, 60, 226
   market effects 182, 226, 240
   mobility 58
   turnover 65
   unions 231
Leader firm 62
Leading industries 4
Learning 2 – 6, 14 – 15, 34, 211 – 223, 238 – 251, 266
   aggregated 251
   capacity 249
   environment 246
   organisational 14, 244
   practices 246
   processes 29
Less favoured areas 269 – 273
Linear causalities 4
Linkages 9
Links 214 – 223
Local development policies 42

Local labour market 57
Local networks 49
Local production systems 6, 7, 57 – 80
Localisation economies 30 – 31, 93
Lock-in effect 36
Logistic structures 70
Logistics 13

Macrolevel 10, 20, 28, 92 – 96
Market globalisation 70
Market structure 31
Market-oriented development strategy 71
Mass production 60, 230 – 235
Maturity phase 234
Mesolevel 10 – 11, 20, 95, 130
Metropolitan regions 34, 74
Microlevel 10 – 11, 20, 92 – 106, 130
Milieu 3, 205
Mobility of workers 65
Multinationals 66 – 69, 201 – 208, 263
Multinational trading blocs 94
Multiperson decision problems 82

Nash equilibrium 83 – 84
National innovation systems 29, 228
Needs assessment 20
Network
   brokering 19
   creation 15
   model of organisation 67, 73
   of innovation 42
   of technological diffusion (NTDs) 50 – 55
   snowflake 218
   web 218
Nonadaptability 234
Non-profit-making third sector 129

Old industrial areas 12, 183, 211, 226, 234 – 235
Organisational principles 9, 228
Outsourcing 60, 69, 78, 228
Overachieving 4
Overspecialisation 232 – 233

Patent diffusion 94
Payoff matrix 84 – 86
Periphery 13, 33, 194 – 197, 205, 232
Petrification 12, 230 – 234

Policy
  cluster 15
  discretionary 261
  incomes 262
  interventions 5
  location 261
  privatisation 262
  procurement 263
  quality control 263
  regulatory 261
  structural 261
Political autonomy 15
Poor media 47
Portfolio quality 106
Principal agent relations 144
Prisoner's dilemma 10, 83 – 89
Private sector 129, 269 – 271
Privatisation 81, 130, 135, 147, 234 – 235
Probabilistic artifacts 95
Production costs 63
Production function approach 3
Production services 58
Productivity differences 3
Profit, expected 114
Profit maximisation 154
Property transfers 10
Propulsive industries 27
Proximity 3, 5, 29 – 31, 41, 47, 268
  geographical 8 – 9, 13, 35, 41 – 55, 111 – 112, 182 – 183, 227
  organisational 8, 42 – 55
Public choice 9, 82 – 83
Public sector 129, 268
Public – private partnerships 61, 77, 143 – 144

Quality circles 220

Rational players 82
Regional development agencies 197, 200, 251 – 252
Regional development theory 23 – 24
Regional divergence 196
Regional economic blocs 54
Regional growth 6, 33
Regional inequality 13
Regional portfolio 11, 181 – 189, 233
Regional production system 5, 60
Regional specialisation 1 – 2, 13 – 15, 68
Regional welfare function 189
Regionalisation 81 – 90
Regulations 65, 89, 262, 273

Rejuvenation 226
Rent shifting 255
Rent seeking 85
Research 14, 46, 62, 144, 184, 194, 264, 273 – 274
  and development 1, 62, 78, 118, 121, 205, 215 – 222, 227 – 235, 241 – 247
  foreign 228
  domestic 228
  institutions 3, 63, 228, 241, 270
  programmes 228
Resource leveraging 23
Resource targeting 23
Restructuring 57
Return – variance space 186
Risk 1, 7, 12, 181 – 190, 205, 226
  return trade-off 12, 187 – 189
Rural areas 7, 77

Sales contracts 10
Sales model 11, 130
Scale economies 32
Schumpeter 182, 230, 254
Science parks 41, 228
Science-based industry 226
Sclerotic club 231
Scope economies 32
Scottish Enterprise 15, 192 – 207
Search for excellence 4
Sectorial concentrations 1
Self-interest 81 – 88
Sensemaking 238
Similarity space 9, 113, 122
Skilled labour force 61
Skills 240, 252
Social distance 9
Social embeddedness 32, 249
Social networks 3
Social presence 45
Social pressure 86
Soft location factors 73
Solutions
  external 85
  indirect external 86 – 89
  internal 85
  Nash 84, 149 – 180
  Pareto-optimal 149 – 180
Spatial complexes 2
Spatial concentration 24, 73, 105
Spillovers 3, 183, 21 – 23, 226 – 227, 241, 255
Spin-offs 63, 78, 133 – 143, 203 – 5, 217, 233
Spread effects 34

Stackelberg 146, 154
Star cluster 232
Stochastic dependence 186
Strategic alliances 254
Strategic trade 261
Strategy
   birth-rate 192 – 193, 208
   cluster-informed 21
   cluster-specific 21
   holistic development 15, 21
   tit-for-tat strategy 85 – 86
Supporting industries 24 – 26
Supporting institutions 19
Sustainable regional cooperation 89
Synergies 1 – 6, 19 – 26, 51 – 55, 183
System of national accounts 5

Technology
   adoption 20
   districts 41
   initiative 265
   parks 81
   policies 20
   transfer 8 – 10, 19, 41 – 42, 69
Technopoles 18, 41
Telecommunications 20, 112, 264
Teleworking 76
Templates 95 – 98
Territorial concentration 7
Territorial organisation 72
Territorial quality 73 – 74
Thick markets 226
Top-down approach 11
Total quality management 220 – 222

Total system basis 238
Trade barriers 94
Trade theory 24
Traded interdependencies 93
Trade-offs 11, 60, 82, 233
Traditional (Fordist) industries 32, 81
Traditional regional policy 4
Transaction costs 58
Transactional approach 61
Transfer model 130
Transformation 7, 10, 129
Transport costs 10, 60, 148, 151
Treuhand 129 – 180
Trust 3, 49, 197, 208
Two-sided cluster trap 13

Uncertainty 85
Underachieving 4
Unlearning 224
Untraded interdependencies 3
Urban centres 7, 13, 73 – 78
Urban policies 74
Urban size 30
Urban system 13
Urbanisation economies 30 – 31
Utility 82, 148

Value chains 2, 11, 19-26, 62, 94, 215 – 221
Vertical disintegration 2
Vertical integration 60 – 62, 132, 234
Vertical production chains 1, 4
Videoconferencing 46 – 48
Virtuous cycle 238